Queueing Theory 2

SCIENCES

Mathematics, Field Director – Nikolaos Limnios

Queuing Theory and Applications, Subject Head – Vladimir Anisimov

Queueing Theory 2

Advanced Trends

Coordinated by
Vladimir Anisimov
Nikolaos Limnios

iSTE

WILEY

First published 2020 in Great Britain and the United States by ISTE Ltd and John Wiley & Sons, Inc.

ISTE Ltd
27-37 St George's Road
London SW19 4EU
UK

www.iste.co.uk

John Wiley & Sons, Inc.
111 River Street
Hoboken, NJ 07030
USA

www.wiley.com

Library of Congress Control Number: 2019957522

British Library Cataloguing-in-Publication Data
A CIP record for this book is available from the British Library
ISBN 978-1-78945-004-0

ERC code:
PE1 Mathematics
PE1_21 Application of mathematics in industry and society

MIX
Paper from responsible sources
FSC® C013604

Contents

Preface

Vladimir ANISIMOV[1] and Nikolaos LIMNIOS[2]
[1] *Amgen Inc., London, United Kingdom*
[2] *University of Technology of Compiègne, France*

Queueing theory is a huge and very rapidly developing branch of science belonging to probability theory and stochastic modelling that originated a long time ago from the pioneering works by Erlang (1909) on the analysis of the models for telephone communication using Poisson processes. Later on, these results were extended further in different directions in the works of such famous mathematicians as Pollaczek, Khinchin, Kendall, Kleinrock and many others.

Nowadays, queueing theory is rapidly growing in various areas including a theoretical analysis of queueing models and networks of rather complicated structure using rather sophisticated mathematical models and various types of stochastic processes. It also includes very wide areas of modern applications: computing and telecommunication networks, traffic engineering, mobile telecommunications, etc.

The aim of this second volume, together with Volume 1, is to reflect the current cutting-edge thinking and established practices in the analysis and applications of queueing models.

This volume includes 8 chapters written by experts well-known in their areas.

Two chapters, Chapters 1 and 7, are devoted to investigating a stability analysis of some types of multiserver regenerative queueing systems with heterogeneous servers and a regenerative input flow using synchronization of the input and majorizing output flows; and a stability analysis of regenerative queueing systems based on a renewal analysis technique, which is illustrated on classical $GI/G/1$ and $GI/G/m$ queueing systems.

Chapter 2 considers a few selected queueing models that are useful in service sectors using both analytical and simulation approaches; and highlights the significant role played by the correlated arrivals, which may occur due to the fact that customers/jobs arrive from different sources. Some interesting observations based on the analytical models of some well-known queueing systems are also reported.

Chapter 3 is devoted to the discussion of similarity between probability distributions and random processes related to queueing and reliability models and their use in economics, industry, demography, environmental studies. Some open and challenging new problems for further research directions are marked, as well as various possible areas of practical applications.

Chapter 4 is devoted to an important problem of how a social planner or a monopolist should act to incite customers to adopt a desirable behaviour, that is, to increase the social welfare or the monopolist's revenue/profit, respectively, using the concepts of free market. Several techniques for the control of information in queueing systems and their impact on strategic customer behaviour using different mechanisms, e.g. pricing structures, priority systems and non-standard queueing disciplines, are considered and illustrated using various examples.

Chapter 5 is devoted to applications of Renyi's and Tsallis's non-extensive maximum entropy methods of inductive inference for the analysis of state probabilities of the stable $M/G/1$ queue with long-range interactions (heavy tails) of order q $(0.5 < q < 1)$. Consequently, novel q-dependent state probabilities of the $M/G/1$ queue with heavy tails are derived by maximising the respective non-extensive maximum entropy functionals. Some numerical examples are considered.

In Chapter 6 a survey of the inventory models with positive service time including perishable items, stock supplied using various control policies such as $(s; S)$; $(r; Q)$, random reorder quantity, etc. is presented. Stochastic decomposition results are also discussed. A brief description of the work done in queues with the requirement of additional items for service, is also provided.

Chapter 8 examines a transient analysis of Markovian queueing systems focussing on the derivation of closed-form solutions for the main transient state distributions and the development of numerical techniques with the aim to position and underline the role of the uniformization technique.

The two volumes of *Queueing Theory* will be useful for graduate and PhD students, lecturers, and also the researchers and developers working in mathematical and stochastic modelling and various applications in computer and communication networks, science and engineering in the departments of Mathematics & Applied Mathematics, Statistics, or Operations Research at universities and in various research and applied centres.

1

Stability Analysis of Queueing Systems based on Synchronization of the Input and Majorizing Output Flows

Larisa AFANASEVA
Department of Probability, Faculty of Mathematics and Mechanics, Lomonosov Moscow State University, Russia

This chapter is focused on the stability conditions for a multiserver queueing system with heterogeneous servers and a regenerative input flow X. The main idea is constructing an auxiliary service process Y, which is also a regenerative flow and determination of the common points of regeneration for both processes X and Y. Then the traffic rate is defined in terms of the mean of the increments of these processes on a common regeneration period. It allows us to use well-known results from the renewal theory to find the instability and stability conditions. The possibilities of the proposed approach are demonstrated by examples. We also present the applications to transport system capacity analysis.

1.1. Introduction

This chapter deals with the study of stability conditions for a heterogeneous multiserver queueing system with regenerative input flow (Afanaseva 2019).

We consider queueing systems with regenerative input flow for three reasons. First, a process describing the performance of the system under some natural conditions

Queueing Theory 2,
coordinated by Vladimir ANISIMOV, Nikolaos LIMNIOS. © ISTE Editions 2020.

turns out to be a classical regenerative process (Asmussen 2003; Thorisson 2000) and the renewal theory gives very effective tools for asymptotic analysis of the system. Second, the class of regenerative flows is rather wide and includes fundamental flows from queueing theory. Finally, regenerative flow has some useful properties that allow us to investigate various applied models (Afanasyeva and Bashtova 2014).

The main objective of our study is to determine conditions under which the process describing the performance of the system is stochastically bounded and therefore, under some additional assumptions, a stable process.

Let us note that stability results for the classical homogeneous multiserver queue are very well known. We refer to the works of (Kiefer and Wolfowitz 1955) for the $GI|GI|m$ system and the general ergodicity results of (Loynes 1962). As it was shown in Sadowsky (1995), the proposed approach could not be applied for heterogeneous systems. Instead, in the work of (Sadowsky 1995), the stability is examined from the point of view of Harris recurrent Markov chain theory. Based on the works of (Malyshev and Menshikov 1982; Meyn and Tweedie 2009; Georgiadis and Szpankowski 1992; Szpankowski 1994), the author obtained two results concerning the irreducibility and positive Harris recurrence of the corresponding process.

The heterogeneous multiserver queueing system with regenerative flow was investigated in the papers of (Afanasyeva and Tkachenko 2014, 2016, 2018). Under some assumptions, the necessary and sufficient stability condition was obtained there.

We consider m-server queueing system S with regenerative input flow X with intensity λ_X. The main assumption is that the process X does not depend on the stochastic sequences and processes determining the work of the servers, such as service time, the moments of breakdowns and recovery of the servers, and the random numbers of the servers required for the service of the customers.

We construct an auxiliary system S_0 with input flow X_0. The latter does not depend on X and is determined by the stochastic sequence describing the work of the servers. The flow X_0 is constructed in such a way that there are always customers for service. We introduce an auxiliary flow Y, which is the number of customers served in the system S_0 up to time t. Then we establish the conditions of Y being the regenerative flow with the intensity λ_Y.

The basic idea is constructing the common points of regeneration for both processes X and Y. Then we define the traffic rate of the system in terms of the first moments of increments of these processes on the common regeneration period. Under some conditions, we estimate the increments of the service process $\tilde{Y}(t)$ in the original system S (that is the number of customers served in the system up to time t)

with the help of increments of an auxiliary process $Y(t)$. It allows us to prove instability results and to find conditions of stochastic boundedness of the number of customers $Q(t)$ at the system at time t as $t \to \infty$.

One more traditional approach to the analysis of stability of the queueing systems is based on the regenerative structure of the processes describing the states of the system, such as the number of customers in the system or workload process. We make assumptions providing the presence of the points of the regeneration for the processes. Then, it is required to establish the finiteness of the mean of the times between the regenerations in accordance with the theorem of (Smith 1955). It is quite a complex problem for non-Markovian processes. Exactly this approach is employed in several works of Morozov and colleagues (Morozov *et al.* 2011; Morozov 2004, 2007; Morozov and Dimitriou 2017; Morozov and Rumyantsev 2016) for establishing the conditions for stability of a wide circle of models. Although our approach is also based on the notions of regeneration and synchronization, it significantly differs from the approach employed in the works mentioned earlier. Our main objective is to establish necessary and sufficient conditions of the stochastic boundedness of the process, describing the system. Thus, this process does not have to be regenerative. Let us point out that a similar situation emerges in the classic models (Borovkov 1976). The analysis of the stochastic boundedness of the process is important for the applications, since precisely the existence of the stochastic boundaries manifests that the system successfully deals with the task. Our conditions may seem too restrictive to be useful in applications. Therefore, we consider four models with service interruptions as examples. In particular, in section 1.7, a discrete-time heterogeneous multiserver queueing system with a regenerative input flow and service interruptions, which are described by independent renewal processes for various servers, is discussed. It is shown that the traffic rate ρ is not expressed in terms of the first moments of the random variables defining the model for the preemptive repeat different service discipline (Gaver 1962). We also prove that $Q(t) \to \infty$ as $t \to \infty$ when $\rho \geq 1$ and $Q(t)$ is a stochastically bounded process when $\rho < 1$.

We should consider that models with service interruptions occur in numerous applications including manufacturing processes, multiprocessor computer networks, telecommunicating networks and various types of service counters.

White and Christie (1958) were the first to study queueing systems with interruptions. Those authors investigated the $M|M|1$ model with preemptive resume priority discipline. Their results were later extended by (Avi-Itzhak and Naor 1963) and (Thiruvengadam 1963) who study queues with general service times. Gaver (1962) studied single-server queues with batch Poisson arrivals and generally distributed service times. So far there is extensive literature concerning queueing systems with interruptions. There are some review papers that cover most of the literature in this sphere. Some of the important papers on the single-server case are presented in Fiems and Bruneel (2013).

The most extensive literature survey on systems with interruptions both for single-server and multiserver cases is given by (Krishnamoorthy *et al.* 2012). This paper also covers some non-Markovian multichannel systems with homogeneous servers. There are some other articles with extensive literature survey as well (Pechinkin *et al.* 2009; Morozov *et al.* 2011). Nevertheless, to the best of our knowledge, there are no papers that study the stability problem for multichannel queueing systems with heterogeneous servers in non-Markovian case with general input flow and service times. Synchronization method combined with the regenerative theory is one of the powerful approaches to obtain stability conditions for such systems.

Let us also mention the fluid approximation approach as an alternative to the synchronization approach followed here. Such an approach has lent to significant progress in stability analysis of multiclass queueing networks (Dai 1995; Chen 1995; Chen and Yao 2001). See also Foss and Konstantopoulos (2004) for a survey of various approaches to stability of queueing systems with a focus on the fluid approach. Nevertheless, our analysis does not rely on a fluid approach because to the best of our knowledge, the synchronization method with regard to regenerative structure of the processes turns out to be suitable for obtaining complete and transparent proofs as well as natural stability conditions for the model at hand.

We also note that stability analysis is an essential and challenging stage of investigation of a stochastic model, however, stability conditions may be of independent interest. In particular, the stability criterion of the multiserver model can be used for the capacity planning of the model at the design stage to obtain the lower bound of the capacity that keeps system stable. As an example of applications of our results, we estimate the carrying capacity of the automobile road, intersected by a crosswalk. Under capacity we mean the upper bound of the intensity of the flow of cars when the queue of cars does not tend to infinity. This means that the analysis can be based on the results obtained in this chapter.

1.2. Model description

We begin from the definition of the regenerative flow (Afanasyeva and Bashtova 2014)). Assume a stochastic process $\{X(t),\, t \geq 0\}\, (X(0) = 0)$ taking values $0, 1, 2\ldots$ to be defined in a probability space $(\Omega, \mathscr{F}, P)$. The process has non-decreasing and right-continuous sample paths. There exists filtration $\{\mathscr{F}^X_{\leq t}\}_{t\geq 0}$, $(\mathscr{F}^X_{\leq t} \in \mathscr{F}$ for all $t \geq 0)$ exists such that X is measurable with respect to $\{\mathscr{F}^X_{\leq t}\}_{t\geq 0}$.

DEFINITION 1.1.– *The stochastic flow X is called regenerative if there is an increasing sequence of Markov moments $\{\theta_j^{(1)}, j \geq 0\}(\theta_0^{(1)} = 0)$ with respect to $\{\mathscr{F}_{\leq t}^X, t \geq 0\}$ such that the sequence*

$$\{\varkappa_j\}_{j=1}^{\infty} = \{X(\theta_{j-1}^{(1)} + t) - X(\theta_{j-1}^{(1)}), \theta_j^{(1)} - \theta_{j-1}^{(1)}, t \in (0, \theta_j^{(1)}) - \theta_{j-1}^{(1)}]\}_{j=1}^{\infty}$$

consists of independent identically distributed (iid) random elements on $(\Omega, \mathscr{F}, P)$.

The random variable $\theta_i^{(1)}$ is said to be the ith regeneration point of X and $\tau_i = \theta_i^{(1)} - \theta_{i-1}^{(1)}$ is the ith regeneration period $(i = 1, 2, \ldots)$. Let $\xi_j^{(1)} = X(\theta_j^{(1)}) - X(\theta_{j-1}^{(1)})$ be the number of customers arrived during the jth regeneration period. Assume that $\mathbf{E}\tau_1^{(1)} < \infty, \mathbf{E}\xi_1^{(1)} < \infty$. The limit $\lambda_X = \lim\limits_{t\to\infty} \frac{X(t)}{t}$ with probability 1 (w.p.1) is called the rate of X. It is easy to prove that $\lambda_X = \frac{\mathbf{E}\xi_1}{\mathbf{E}\tau_1^{(1)}}$ (Smith 1955; Thorisson 2000). The class of regenerative flows contains most of fundamental flows that are exploited in the queueing theory. First of all, the doubly stochastic Poisson process (Grandell 1976) with a regenerative process as a stochastic intensity is a regenerative flow. There are many other examples of regenerative flows, for instance, semi-Markovian, Markov-arrival, Markov-modulated and other processes (Afanasyeva and Tkachenko 2012).

We consider discrete-time queueing systems as well as continuous-time queueing systems (Avi-Itzhak and Naor 1963). In the first case, time is divided into fixed length intervals or slots and all arrivals and departures are synchronized with respect to slot boundaries. Moreover, in the case of some events being synchronized at one slot these events are ordered as follows: arrival and departure. The system is observed at the end of the slot.

We consider a queueing system S with m servers, FCFS discipline and regenerative input flow X. We assume that a server may simultaneously serve only one customer so that at any time the number of customers on the servers is not more than m. A customer leaves a system only after completion the service. The system is defined by the sequence $\{\overrightarrow{\eta}_n\}_{n=1}^{\infty}$ consisting of iid random vectors and multidimensional stochastic process $\{V(t), t \geq 0\}$, which are independent of the input flow X. The vector $\overrightarrow{\eta}_n$ determines the characteristics of the nth customer, that is service times by various servers or necessary number of servers for service. The process V describes the states of the servers. For example, for the systems with unreliable servers this process defines the moments of breakdowns and restorations of the servers. The state of the system at time t is described by the stochastic process $\overrightarrow{Z}(t) = (Z_1(t), \ldots, Z_k(t))$ where one of the coordinates is the number of customers in the system. We assume that the relation

$$\overrightarrow{Z}(t) = \Phi(\overrightarrow{Z}(0), t, \{X(s), s \leq t\}, \{\overrightarrow{\eta}_n\}_{n=1}^{X(t)}, \{V(s), s \leq t\}) \qquad [1.1]$$

takes place for some function $\Phi(\cdot)$ on the corresponding space. For example, for the system $Reg|G|1|\infty$ we consider the process $\overrightarrow{Z}(t) = (W(t), Q(t))$ where $W(t)$ is the virtual waiting time and $Q(t)$ is the number of customers in the system at time t. Let $\{\eta_j\}_{j=1}^{\infty}$ and $\{t_j\}_{j=1}^{\infty}$ be the sequences of service and arrival times of customers, respectively, and $U(t) = \sum_{j=1}^{X(t)} \eta_j - t$. Assuming that $Q(0) = 0$, we have $W(t) = \sup_{0 \le s \le t} (U(t) - U(s))$, $Q(t) = \sum_{j=1}^{X(t)} \mathbb{I}(t_j + W(t_j) + \eta_j > t)$ where $\mathbb{I}(\cdot)$ is an indicator function (Borovkov 1976).

The main goal of this chapter is the determination of the conditions of the stochastic boundedness of the number of customers $Q(t)$ in the system as $t \to \infty$. Our analysis is based on the construction of the auxiliary service process.

1.3. Auxiliary service process

For the system S, we define an auxiliary system S_0 with input flow X_0 such that when the number of customers in the system becomes less than m a new customer immediately arrives in the system. Therefore, there are always customers for service in S_0. Other characteristics such as the initial state, the sequence $\{\overrightarrow{\eta}_n\}_{n=1}^{\infty}$, stochastic process $\{\overrightarrow{V}(t), t \ge 0\}$ and a functional Φ are the same as for the system S. If in the system S the initial number of customers $Q(0) < m$, then the process X_0 has the jump $m - Q(0)$ at zero instant. We determine an auxiliary service process $Y(t)$ as the number of customers served in S_0 during $(0, t)$. Since the flow Y is defined by the processes $\{\overrightarrow{\eta}_n\}_{n=1}^{\infty}$ and V and these processes do not depend on the input flow X at the system S, we conclude that X and Y are independent flows.

We also need additional assumptions.

CONDITION 1.1.– For the continuous-time case, Y is a strongly regenerative flow with the sequence $\{\theta_n^{(2)}\}_{n=0}^{\infty}(\theta_0^{(2)} = 0)$ as points of regeneration.

We call the regenerative flow Y strongly regenerative if the regeneration period $\tau_n^{(2)} = \theta_n^{(2)} - \theta_{n-1}^{(2)}$ has the form

$$\tau_n^{(2)} = v_n^{(1)} + v_n^{(2)} \qquad [1.2]$$

where $P(v_n^{(1)} > x) = e^{-\delta x}(\delta \in (0, \infty))$, $v_n^{(1)}$ and $v_n^{(2)}$ are independent random variables and $Y(\theta_{n-1}^{(2)} + v_n^{(1)}) = Y(\theta_{n-1}^{(2)})$.

CONDITION 1.2.– For the discrete-time case, processes X and Y are regenerative aperiodic flows. As usually, aperiodicity means that the greatest common divisor (GCD)

$$GCD\{k : P(\theta_1^{(i)} = k) > 0\} = 1, \quad i = 1, 2.$$

Then we may determine common points of regeneration $\{T_n\}_{n=1}^{\infty}$ for both processes X and Y letting in the discrete-time case

$$T_n = \min\left\{\theta_j^{(1)} > T_{n-1} : \bigcup_{l=1}^{\infty}\{\theta_j^{(1)} = \theta_l^{(2)}\}\right\}, \; T_0 = 0 \qquad [1.3]$$

and in the continuous-time case

$$T_n = \min\left\{\theta_j^{(1)} > T_{n-1} : \bigcup_{l=1}^{\infty}\{\theta_{l-1}^{(2)} < \theta_j^{(1)} \le \theta_{l-1}^{(2)} + v_l^{(1)}\}\right\}, \; T_0 = 0. \quad [1.4]$$

LEMMA 1.1.– Let for the continuous-time (discrete-time) condition 1.1 (condition 1.2) be fulfilled. Then the sequence $\{T_n\}_{n=0}^{\infty}$ consists of common regeneration points for X and Y and

$$\mathbf{E}T_1 = \delta\mathbf{E}\tau_1^{(1)}\mathbf{E}\tau_1^{(2)} < \infty \qquad [1.5]$$

for the continuous-time case,

$$\mathbf{E}T_1 = \mathbf{E}\theta_1^{(1)}\mathbf{E}\theta_1^{(2)} < \infty \qquad [1.6]$$

for the discrete-time case.

PROOF.– Since the proof of [1.5] is almost the same as the proof of [1.6], we consider the discrete-time case only. Let

$$\nu_k = \min\left\{j > \nu_{k-1} : \bigcup_{l=1}^{\infty}\{\theta_j^{(1)} = \theta_l^{(2)}\}\right\}, \nu_0 = 0,$$

so that $T_k = \theta_{\nu_k}^{(1)}$. Then $\{\nu_k - \nu_{k-1}\}_{k=1}^{\infty}$ is a sequence of iid random variables and in accordance with Wald's identity $\mathbf{E}T_1 = \mathbf{E}\theta_1^{(1)}\mathbf{E}\nu_1$ (Feller 1971). Therefore, we need to prove the finiteness of $\mathbf{E}\nu_1$. Denote by $h_2(t)$ $(h(t))$ the mean number of renewals at time t for the renewal process $\{\theta_n^{(2)}\}_{n=0}^{\infty}$ $(\{\nu_k\}_{k=0}^{\infty})$ so that

$$h_2(t) = \sum_{l=0}^{\infty} P(\theta_l^{(2)} = t)$$

and

$$h(t) = \sum_{k=0}^{\infty} P(\nu_k = t).$$

Taking into account condition 1.2, we derive from Blackwell's theorem (Thorisson 2000)

$$h_2(t) \xrightarrow[t\to\infty]{} \frac{1}{\mathbf{E}\theta_1^{(2)}}, \quad h(t) \xrightarrow[t\to\infty]{} \frac{1}{\mathbf{E}\nu_1}$$

if $\mathbf{E}\nu_1 < \infty$ and $h(t) \to 0$ as $t \to \infty$ if $\mathbf{E}\nu_1 = \infty$.

Because of X and Y independence

$$h(j) = P(\bigcup_{l=0}^{\infty}\{\theta_j^{(1)} = \theta_l^{(2)}\}) = \mathbf{E}\left(\sum_{l=0}^{\infty} P(\theta_j^{(1)} = \theta_l^{(2)} | \theta_j^{(1)})\right) = \mathbf{E}h_2(\theta_j^{(1)}).$$

Since $\theta_j^{(1)} \xrightarrow[j\to\infty]{} \infty$ w.p.1, then $h_2(\theta_j^{(1)}) \xrightarrow[j\to\infty]{} \frac{1}{\mathbf{E}\theta_1^{(2)}}$ w.p.1. Therefore, from Lebesgue's dominated convergence theorem, we obtain $\mathbf{E}\nu_1 = \mathbf{E}\theta_1^{(2)} < \infty$. ■

Later we consider both cases (discrete-time and continuous-time) together. We only have to take condition 1.2 instead of condition 1.1.

Let

$$\Delta_Y(n) = Y(T_n) - Y(T_{n-1}),$$
$$\Delta_X(n) = X(T_n) - X(T_{n-1}).$$

Then

$$\lambda_X = \frac{E\Delta_X(n)}{E(T_n - T_{n-1})}, \lambda_Y = \frac{E\Delta_Y(n)}{E(T_n - T_{n-1})}.$$

We define the traffic rate as follows:

$$\rho = \frac{\lambda_X}{\lambda_Y} = \frac{E\Delta_X(1)}{E\Delta_Y(1)}. \qquad [1.7]$$

We think of λ_X and λ_Y as the arrival and service rate, respectively. Intuitively, it is clear that the number of customers in the system S is a stochastically bounded process if $\rho < 1$ and it is not the case if $\rho \geq 1$. The main stability result of this chapter consists of the formal proof of this fact.

We define the stochastic flow $\tilde{Y}(t)$ as the number of customers served at the system S during time interval $[0, t)$.

CONDITION 1.3.– The following stochastic inequalities take place:

$$\Delta_{\tilde{Y}}(n) = \tilde{Y}(T_n) - \tilde{Y}(T_{n-1}) \leq \Delta_Y(n), \ (n = 1, 2, \ldots).$$

Let $Q(t)$ be the number of customers in the system S including the customers on the servers at time t so that

$$Q(t) = Q(0) + X(t) - \tilde{Y}(t).$$

CONDITION 1.4.– There are two possible cases:

i) $Q(t)$ is a stochastically bounded process, i.e.

$$\lim_{x\to\infty} \liminf_{t\to\infty} P(Q(t) \leq x) = 1;$$

ii) $Q(t) \xrightarrow[t\to\infty]{P} \infty.$

CONDITION 1.5.– If $Q(t) \xrightarrow[t\to\infty]{P} \infty$, then for any $\epsilon > 0$ there exists n_ϵ such that for $n > n_\epsilon$

$$E\Delta_{\tilde{Y}}(n) \geq E\Delta_Y(1) - \epsilon. \qquad [1.8]$$

1.4. Instability result for the case $\rho \geq 1$

THEOREM 1.1.– Let conditions 1.3 and 1.1 (1.2) for the continuous-time (for the discrete-time) case be fulfilled. If $\rho \geq 1$, then

$$Q(t) \xrightarrow[t\to\infty]{P} \infty. \qquad [1.9]$$

PROOF.– Consider the embedded process $Q_n = Q(T_n)$ and denote $Z_k = \sum_{j=1}^{k} (\Delta_X(j) - \Delta_Y(j))$ $(Z_0 = 0)$.

Define the auxiliary sequence $\{\hat{Q}_k\}_{k=0}^{\infty}$ by the recursion

$$\hat{Q}_k = \max[0, \hat{Q}_{k-1} + \Delta_X(k) - \Delta_Y(k)],\ \hat{Q}_0 = Q(0).$$

Because of the equality $Q_k = Q_{k-1} + \Delta_X(k) - \Delta_{\tilde{Y}}(k)$ and condition 1.3 we get the stochastic inequality $Q_k \geq \hat{Q}_k$. It is well known (Feller 1971) that

$$\hat{Q}_k = \max_{0 \leq j \leq k} (Q(0) + Z_k, Z_k - Z_j)$$

and in distribution

$$\hat{Q}_k \geq \max_{0 \leq j \leq k} Z_j.$$

For $\rho \geq 1$, the sequence $\{Z_j\}_{j \geq 0}$ is a random walk with a non-negative drift. Hence, except when $\Delta_X(1) = \Delta_Y(1) = c$ (c is a constant), $\max_{0 \leq j \leq k} Z_j \xrightarrow[k \to \infty]{} \infty$ (w.p.1) (Feller 1971) that completes the proof. ■

1.5. Stochastic boundedness for the case $\rho < 1$

Our objective here is to establish stochastic boundedness of the process Q when the traffic rate $\rho < 1$. Under some additional assumptions providing the regenerative structure of the process Q, this property has its stability as a consequence.

THEOREM 1.2.– Let conditions 1.4 and 1.5 and condition 1.1 (1.2) for the continuous-time (for the discrete-time) case be fulfilled. If $\rho < 1$, then Q is a stochastically bounded process.

PROOF.– Because of condition 1.4 there are two possible cases: $Q_n = Q(T_n)$ is either stochastically bounded or $Q_n \xrightarrow[n \to \infty]{P} \infty$. Assume that the second case takes place and $\rho < 1$. Because of condition 1.5 for $0 < \epsilon < E(\Delta_Y(1) - \Delta_X(1))$, there is n_ϵ such that for $n > n_\epsilon$

$$\begin{aligned} EQ_{n+1} &= EQ_n + E\Delta_X(1) - E\Delta_{\tilde{Y}}(n) \leq \\ &\leq EQ_n + E\Delta_X(1) - E\Delta_Y(1) + \epsilon \leq EQ_n \end{aligned}$$

that contradicts our assumption that $Q_n \xrightarrow[n \to \infty]{P} \infty$. ■

In the next sections, we discuss some examples to verify our results and to compare them with previous works. First, we consider two queueing models with service interruptions. These models occur in numerous applications and there is extensive literature concerning queueing system with interruptions. Let us mention some papers in which detail description of the literature in this sphere is given (Krishnamoorthy *et al.* 2012; Fiems and Bruneel 2013; Pechinkin *et al.* 2009; Morozov *et al.* 2011). However, to the best of our knowledge there are no papers that study the stability problem for multichannel systems with heterogeneous servers for the non-Markovian case: with general input flow and general distribution of blocked and available periods.

1.6. Queueing system with unreliable servers and preemptive resume service discipline

We consider a continuous-time queueing system with regenerative input flow X and m heterogeneous servers that may be not available for operation from time to time. We also propose that the velocity of the service may be dependent on the state of the server. Assume that for the ith server a stochastic process $n_i(t)$ with state space $(0; r_1^{(i)}; \ldots; r_{k_i}^{(i)})$, $r_j^{(i)} > 0$, $j = \overline{1, k_i}$ is defined. If $n_i(t) = 0$, then the ith server is in unavailable state, for instance it is broken; if $n_i(t) = r_j^{(i)}$, then the ith server is working with the velocity $r_j^{(i)}$ $(j = \overline{1, k_i}, i = \overline{1, m})$. Service times of customers by the ith server in the case when the velocity of the service is equal to one constitute a sequence $\{\eta_{in}\}_{n=1}^{\infty}$ of iid random variables, which does not depend on the input flow and service times by other servers, $b_i = E\eta_{in} < \infty$, $B_i(x) = P(\eta_{in} \leq x)$, $i = \overline{1, m}$.

It is possible that an unavailable period starts while a customer is receiving service. Then service of the customer is immediately interrupted. There are various disciplines for continuation of the service after restoration (Gaver 1962). Here, we consider the preemptive resume service discipline assuming that interrupted service continues when the server returns from a blocked period and the service velocity is the next state of the process $n_i(t)$.

CONDITION 1.6.– The stochastic process $\vec{n}(t) = (n_1(t), \ldots, n_m(t))$ is strongly regenerative with regeneration points $\{\theta_n^{(2)}\}_{n=1}^{\infty}$ $(\theta_0^{(2)} = 0)$, $\tau_n^{(2)} = \theta_n^{(2)} - \theta_{n-1}^{(2)}$, $E\tau_n^{(2)} < \infty$ with an exponential phase $v_n^{(1)}$ so that $\tau_n^{(2)} = v_n^{(1)} + v_n^{(2)}$. We also assume that $n_i(\theta_{n-1}^{(2)} + t) = 0$ for $t \in [0, v_n^{(1)}]$, $i = \overline{1, m}$.

It follows from condition 1.6 and Smith's (1955) theorem that there exist the limits

$$\lim_{t \to \infty} P(\vec{n}(t) = (j_1, \ldots, j_m)) = \pi_{j_1, \ldots, j_m}$$

and

$$\lim_{t\to\infty} P(n_i(t) = j_i) = \pi_{j_i}^{(i)},$$

where j_i takes values $0, r_1^{(i)}, \ldots, r_{k_i}^{(i)}, i = \overline{1,m}$.

To define an auxiliary process $Y_i(t)$ for the ith server, we introduce a counting process

$$K_i(t) = \max\{j : \sum_{l=1}^{j} \eta_{il} \leq t\}.$$

Then

$$Y_i(t) = K_i(\int_0^t n_i(y)dy) \qquad [1.10]$$

and

$$Y(t) = \sum_{i=1}^{m} Y_i(t).$$

CONDITION 1.7.– Service times have the first exponential phase, i.e.

$$\eta_{in} = \eta_{in}^{(1)} + \eta_{in}^{(2)}$$

where $\eta_{in}^{(1)}$ and $\eta_{in}^{(2)}$ are independent random variables and $P(\eta_{in}^{(1)} \leq x) = e^{-\alpha_i x}$ $(\alpha_i \in (0,\infty))$.

As regeneration points for Y we take subsequence $\{\theta_{n_k}^{(2)}\}_{k=1}^{\infty}$ of the sequence $\{\theta_n^{(2)}\}_{n=1}^{\infty}$ such that at time $\theta_{n_k}^{(2)}$ interrupted services for processes Y_i $(i = \overline{1,m})$ are in the exponential phase. Because of conditions 1.6 and 1.7, Y is a strongly regenerative flow and we may define the common sequence $\{T_n\}_{n=1}^{\infty}$ of regeneration points for both processes X and Y with the help of formula [1.4]. We need only to take $\{\theta_{n_k}^{(2)}\}_{k=1}^{\infty}$ instead of $\{\theta_n^{(2)}\}_{n=1}^{\infty}$.

Because of [1.10] we can easily obtain from the renewal theory the formula for the rate of the auxiliary process

$$\lambda_Y = \sum_{i=1}^{m} b_i^{-1} \sum_{j=1}^{k_i} r_j^{(i)} \pi_j^{(i)}. \qquad [1.11]$$

Now we may calculate the traffic rate ρ and under some assumptions we get the necessary and sufficient stability condition for the system based on theorems 1.1 and 1.2. As an example, we consider the famous case (Morozov *et al.* 2011) when $r_i^{(j)} = 1$, $j = \overline{1, k_i}$, $i = \overline{1, m}$, i.e. a server may be in an available or unavailable state. Let $\{s_{i,n}^{(2)}\}_{n=1}^{\infty}$ be moments of breakdowns and $\{s_{i,n}^{(1)}\}_{n=1}^{\infty}$ moments of restorations for the ith server. Here

$$0 = s_{i,0}^{(2)} < s_{i,1}^{(1)} < s_{i,1}^{(2)} < \dots . \qquad [1.12]$$

Then $u_{i,n}^{(1)} = s_{i,n}^{(1)} - s_{i,n-1}^{(2)}$ and $u_{i,n}^{(2)} = s_{i,n}^{(2)} - s_{i,n}^{(1)}$ denote the length of the nth blocked and the nth available period for the ith server, respectively, $(i = \overline{1, m})$. The sequence $\{u_{i,n}^{(1)}, u_{i,n}^{(2)}\}_{n=1}^{\infty}$ consists of iid random vectors (for all $(i = \overline{1, m})$) and these sequences do not depend on the input flow X and service times. Let $u_{i,n} = u_{i,n}^{(1)} + u_{i,n}^{(2)}$ be the length of the nth cycle for the server i. A cycle consists of a blocked period followed by an available period. We assume that

$$Eu_{i,n}^{(1)} = a_i^{(1)} < \infty,\ Eu_{i,n}^{(2)} = a_i^{(2)} < \infty,\ a_i = a_i^{(1)} + a_i^{(2)}\ (i = \overline{1, m}).$$

We put $n_i(t) = 0$ if the ith server is in an unavailable state at time t and $n_i(t) = 1$, otherwise $(i = \overline{1, m})$. If a blocked period $u_{i,n}^{(1)}$ has an exponential phase, i.e. $u_{i,n}^{(1)} = v_{i,n}^{(1)} + v_{i,n}^{(2)}$ where $v_{i,n}^{(1)}$ and $v_{i,n}^{(2)}$ are independent random variables and $v_{i,n}^{(1)}$ has an exponential distribution with a parameter α_i, then we may define the sequence $\{\theta_n^{(2)}\}_{n=1}^{\infty}$ of regeneration points for the regenerative process $\overrightarrow{n}(t) = (n_1(t), \dots, n_m(t))$ as above. Therefore, condition 1.6 holds. Under condition 1.7, the auxiliary process Y is strongly regenerative and we can construct the common points of regeneration $\{T_n\}_{n=1}^{\infty}$ for X and Y and apply theorems 1.1 and 1.2 for this model. Since

$$\pi_i = \lim_{t \to \infty} P(n_i(t) = 1) = \frac{a_i^{(2)}}{a_i}$$

we have from [1.11]

$$\rho = \frac{\lambda_X}{\sum\limits_{i=1}^{\infty} b_i^{-1} \frac{a_i^{(2)}}{a_i}}.$$

If $b_i = b$, then we get the same stability condition as obtained in Morozov *et al.* (2011) for a queueing system $GI|G|m$ with a common distribution function of service times for all servers.

COROLLARY 1.1.– For a queueing system with $r_j^{(i)} = 1,\ j = \overline{1, k_i},\ i = \overline{1, m}$

$$Q(t) \xrightarrow[t\to\infty]{P} \infty$$

if $\rho > 1$.

Under condition 1.4, the process is stochastically bounded if $\rho < 1$.

PROOF.– Let, as before, $\tilde{Y}_i(t)$ be the number of customers actually served on the ith server up to time t. It is evident that stochastic inequality

$$Y_i(t) \geq \tilde{Y}_i(t),\ i = \overline{1, m}$$

for $t \geq 0$ takes place and hence

$$Q(t) = Q(0) + X(t) - \tilde{Y}(t) \geq Q(0) + X(t) - Y(t).$$

Since $\rho > 1$, then $Q(t) \xrightarrow[t\to\infty]{P} \infty$.

To prove the second statement, we first assume that conditions 1.6 and 1.7 hold. Then condition 1.1 for the process Y takes place. We also may organize the performance of the systems S and S_0 in such a way that inequality [1.8] is realized when $Q(t) \xrightarrow[t\to\infty]{P} \infty$. Thus, conditions 1.1, 1.4 and 1.5 are satisfied and because of theorem 1.2 the process Q is stochastically bounded.

If conditions 1.6 and 1.7 (or one of them) are not valid, we construct a system S_δ satisfying conditions 1.6 and 1.7 and majorising our system S, so that in distribution

$$Q(t) \leq Q_\delta(t) + m. \qquad [1.13]$$

Here, $Q_\delta(t)$ is the number of customers in the system S_δ at instant t. Let us introduce independent sequences $\{\{v_{i,n}\}_{n\geq 1}, \{\gamma_{i,n}\}_{n\geq 1}\}_{i=1}^m$ of iid random variables with exponential distribution with a rate δ. Assume that repair time $\tilde{u}_{i,n}^{(1)}$ in the system S_δ has the form $\tilde{u}_{i,n}^{(1)} = v_{i,n}^{(1)} + u_{i,n}^{(1)}$ and service time $\tilde{\eta}_{i,n}$ by the ith server has the form $\tilde{\eta}_{i,n} = \eta_{i,n} + \gamma_{i,n}$.

Then S_δ satisfies conditions 1.6 and 1.7. Since $\rho_\delta = \lambda x(\sum_{i=1}^m \frac{\delta}{1+\delta b_i} \frac{a_i^{(2)}\delta}{1+\delta a_i})^{-1}$ and $\rho = \rho_\delta < 1$ we may choose δ so that $\rho_\delta < 1$.

The proof of [1.13] is based on the “so-called” probability space method (Belorusov 2012).

Let us note that condition 1.4 may be provided in various ways. For instance, assume that blocked (or available) period has an exponential phase and

$$B_1(x) > 0 \text{ for all } x > 0. \tag{1.14}$$

Then Q is a regenerative process with points of regeneration $\{\theta_{n_k}^{(1)}\}_{k=1}^{\infty}$ that is a subsequence of the sequence $\{\theta_n^{(1)}\}_{n=1}^{\infty}$ such that $Q(\theta_{n_k}^{(1)}) = 0$ and all servers are in the exponential phase of their blocked (or available) periods. Now condition 1.4 follows directly from theorem 1 in Afanasyeva and Tkachenko (2014). We also note that in this case Q is a stable process if $\rho < 1$. If only assumption [1.14] takes place with the help of the majorising system S_δ, we obtain the stochastic boundedness Q when $\rho < 1$. ■

1.7. Discrete-time queueing system with interruptions and preemptive repeat different service discipline

Here, we consider the system with interruptions described in the previous section for the discrete-time case. The moments of breakdowns $\{s_{i,n}^{(2)}\}_{n=1}^{\infty}$ and moments of restorations $\{s_{i,n}^{(1)}\}_{n=1}^{\infty}$ for the ith server satisfy [1.12]. The input flow X is an aperiodic discrete-time regenerative flow with rate λ_X.

We consider the preemptive repeat different service discipline that means that the service is repeated from the start after restoration of the server and the new service time is independent of the original service time (Gaver 1962).

To define the process Y_i for the ith server in the auxiliary system S_0, we introduce the collection $\left\{\{\eta_{i,n}^{(j)}\}_{n=1}^{\infty}\right\}_{j=1}^{\infty}$ of independent sequences $\{\eta_{i,n}^{(j)}\}_{n=1}^{\infty}$ consisting of iid random variables with distribution function B_i. Of course, we assume that $P(\eta_{i,n}^{(1)} > u_{i,n}^{(2)}) > 0$, $(i = \overline{1,m})$. Let $\mathscr{K}_{i,j}(t)$ be the counting process associated with the sequence $\{\eta_{i,n}^{(j)}\}_{n=1}^{\infty}$, i.e. $\mathscr{K}_{i,j}(t) = \max\{k : \sum_{n=1}^{k} \eta_{i,n}^{(j)} \leq t\}$, $(\mathscr{K}_{i,j}(0) = 0)$ and $\mu_i(t)$ be the number of cycles for the ith server during $[0, t]$, i.e. $\mu_i(t) = \max\{j : \sum_{n=1}^{j} u_{i,n} \leq t\}$, $(\mu_i(0) = 0)$. Then the process Y_i is defined by the relation

$$Y_i(t) = \sum_{j=1}^{\mu_i(t)} \mathscr{K}_{i,j}(u_{i,j}^{(2)}) + \mathscr{K}_{i,\mu_i(t)+1}(\max[0, t - s_{i,\mu_i(t)+1}^{(1)}]) \tag{1.15}$$

and $Y(t) = \sum\limits_{i=1}^{m} Y_i(t)$. We denote by $H_i(t)$ the renewal function for $\mathscr{K}_{i,j}$, i.e. $H_i(t) = \mathbf{E}\mathscr{K}_{i,j}(t)$.

LEMMA 1.2.– There exists the limit

$$\lambda_{Y_i} = \lim_{t\to\infty} \frac{Y_i(t)}{t} = \frac{\mathbf{E}H_i(u_{i,n}^{(2)})}{a_i} \quad \text{w.p.1.}$$

The proof easily follows from the evident inequalities

$$g_i(\mu_i(t)) \le Y_i(t) \le g_i(\mu_i(t) + 1)$$

where $g_i(n) = \sum\limits_{j=1}^{n} \mathscr{K}_{i,j}(u_{i,j}^{(2)})$, the strong law of large numbers and convergence $t^{-1}\mu_i(t) \xrightarrow[t\to\infty]{} a_i^{-1}$ w.p.1.

From lemma 1.2, we have

$$\lambda_Y = \lim_{t\to\infty} \frac{Y(t)}{t} = \sum_{i=1}^{m} \frac{\mathbf{E}H_i(u_{i,1}^{(2)})}{a_i}. \qquad [1.16]$$

We introduce the counting processes

$$N_0(t) = \max\{k : \theta_k^{(1)} \le t\},$$

$$N_i(t) = \max\{k : s_{i,k}^{(2)} \le t\}, \quad i = \overline{1,m}.$$

CONDITION 1.8.– The counting processes $N_0(t)$ and $N_i(t)$ ($i = \overline{1,m}$) are aperiodic.

Then Y is a regenerative aperiodic flow with points of regeneration

$$\theta_j^{(2)} = \min\{t > \theta_{j-1}^{(2)} : \bigcap_{i=1}^{m} [N_i(t) - N_i(t-1) > 0]\}, \quad \theta_0^{(2)} = 0.$$

In other words, $\theta_j^{(2)}$ is a point of regeneration of Y if all the servers get out of the order simultaneously at this moment. Taking into account condition 1.8, we conclude from lemma 1.1 that $\mathbf{E}(\theta_j^{(2)} - \theta_{j-1}^{(2)}) < \infty$. Now we construct the sequence $\{T_n\}_{n=1}^{\infty}$ of common points of regeneration for processes X and Y with the help of [1.3]. Because of lemma 1.1 $\mathbf{E}(T_n - T_{n-1}) < \infty$ and the traffic rate ρ of the system is defined by [1.7].

COROLLARY 1.2.– Let condition 1.8 be fulfilled. Then

i) $Q(t) \xrightarrow[t\to\infty]{P} \infty$ if $\rho \geq 1$;

ii) $Q(t)$ is a stochastically bounded process if $\rho < 1$.

PROOF.– The first statement follows from theorem 1.1 since conditions 1.2 and 1.3 are realized.

Let $\rho < 1$. For the system S, we introduce the embedded process $x_n = (Q_n, \zeta_1(n), \ldots, \zeta_m(n))$, $n \geq 0$, where Q_n is the number of customers in the system on time T_n and $\zeta_i(n) = 1$ if there is a customer on the ith server and $\zeta_i(n) = 0$ otherwise. In a view of the service discipline after service restoration and properties of the synchronization epochs $\{T_n\}_{n\geq 0}$, the process $\{x_n\}_{n\geq 1}$ is a Markov chain with countable set of states $\mathscr{R} = \{\{0\}, (j, e_1, \ldots, e_m), j = \overline{1, m-1}; \{j\}, j \geq m\}$. Let R_0 be the set of unessential states and R_j $(j = \overline{1, r})$ irreducible classes of communicating states. It follows from the condition $\rho < 1$ that the number of classes $r < \infty$.

For any aperiodic class $\mathscr{K}_l$ of states based on Foster's criterion (Meyn and Tweedie 2009), we may easily prove that this class is ergodic (Afanasyeva and Tkachenko 2016, 2018). Therefore, the process Q_n is stochastically bounded if $Q_0 \in \mathscr{K}_l$. It is also true if $\mathscr{K}_l$ is a periodic class. Since the number of classes $r < \infty$, we obtain the stochastic boundedness of the process Q_n and therefore $Q(t)$.

We may obtain the upper bound of the traffic rate ρ providing the stochastic boundedness of the process Q. It is known from (Borovkov 1976) that

$$H_i(t) \geq \frac{t}{b_i} - 1.$$

Therefore

$$\sum_{i=1}^{m} H_i(u_{i,n}^{(2)}) \geq \sum_{i=1}^{m} \frac{a_i^{(2)}}{b_i a_i} - \sum_{i=1}^{m} \frac{1}{a_i}$$

and sufficient condition of the stochastic boundedness of Q has the following form

$$\lambda + \sum_{i=1}^{m} a_i^{-1} \leq \sum_{i=1}^{m} \frac{a_i^{(2)}}{b_i a_i}.$$

If $b_i = b$, then we have the same condition as obtained in Morozov *et al.* (2011). ∎

1.8. Queueing system with a preemptive priority discipline

In this section we study a continuous-time queueing system with two independent regenerative input flows X_1 and X_2 with intensities λ_1 and λ_2 and m servers. The customers of the second type (which belong to X_2) have an absolute priority with respect to customers of the first type. Service interruption for the low priority customer occurs when a high priority customer arrives during a low priority customer's service time. If at an arrival time of the second type customer there are m_1 free servers, m_2 servers occupied by customers of the first type and $m - m_1 - m_2$ servers occupied by customers of the second type, then an arriving customer randomly chooses any server from $m_1 + m_2$ servers, which are not busy by customers of the second type. Service times by the ith server for high(low) priority customers have distribution function B_0 $(B_i,\ i = \overline{1,m})$ with mean b_0 $(b_i,\ i = \overline{1,m})$. Therefore, for high priority customers we have a system $Reg|G|m$ with homogeneous servers and for low priority customers a system with interruptions and preemptive resume service discipline considered in section 1.6.

Denote by $Q_i(t)$ the number of customers of the ith type at the system including the customers on the servers at time t $(i = 1,2)$. Let $\{\theta_j^{(1)}\}_{j=1}^{\infty}$ $(\theta_0^{(1)} = 0)$ and $\{\theta_j^{(2)}\}_{j=1}^{\infty}$ $(\theta_0^{(2)} = 0)$ be the sequences of regeneration points for X_1 and X_2, respectively. Under some additional conditions, for example, when the inequality [1.14] is valid for the function B_0 (other sufficient assumptions are given in Afanasyeva and Tkachenko (2014)), the process Q_2 is regenerative with points of regeneration

$$S_n^{(2)} = \min\{\theta_j^{(2)} > S_{n-1}^{(2)} : Q_2(\theta_j^{(2)}) = 0\},\ n > 0;$$

$$S_0^{(2)} = 0.$$

The stability condition for the process Q_2 has the form (Afanasyeva and Tkachenko 2014)

$$\rho_2 = \frac{\lambda_2 b_0}{m} < 1 \qquad [1.17]$$

that is supposed to be fulfilled. We now want to get the stability condition for the process Q_1.

We start with the definition of the process of interruptions. Let $n_i(t) = 0$ if at instant t the ith server is occupied by a high priority customer and $n_i(t) = 1$ otherwise, $i = \overline{1,m}$. As regeneration points for $\vec{n}(t) = (n_1(t), \ldots, n_m(t))$, we take subsequence $\{\theta_{j_k}^{(2)}\}_{k=1}^{\infty}$ of the regeneration points sequence $\{\theta_j^{(2)}\}_{j=1}^{\infty}$ for the input flow X_2 such that $Q_2(\theta_{j_k}^{(2)}) = 0$. As before, we assume that [1.14] holds for B_0. Since X_2 is a strongly regenerative flow, condition 1.6 is fulfilled.

To obtain the traffic rate for low priority customers, we need to find $\pi_i = \lim_{t\to\infty} P(n_i(t) = 1)$. Because of the rule of the server choose by an arriving high priority customer, we have $\pi_i = \pi_1 = \pi$ for all $i = \overline{1,m}$. To calculate π, we define for high priority customers the following processes. Let $w_i(t)$ be the residual service time (virtual waiting time) on the ith server at instant t and $Z_i(t)$ the total service time of customers which arrived up to time t and have to be served on the ith server. Thus

$$\sum_{i=1}^{m} Z_i(t) = \sum_{j=1}^{X_2(t)} \eta_j$$

where η_j is the service time of the jth arrived customer. We note that w.p.1

$$\lim_{t\to\infty} \frac{Z_i(t)}{t} = \frac{1}{m} \lim_{t\to\infty} \sum_{j=1}^{X_2(t)} \eta_j = \frac{\lambda_2 b_0}{m}$$

and

$$\frac{w_i(t)}{t} \xrightarrow[t\to\infty]{} 0$$

because of the stability condition [1.17].

Since

$$w_i(t) = Z_i(t) - \int_0^t (1 - n_i(y))dy$$

and w.p.1

$$\lim_{t\to\infty} t^{-1} \int_0^t (1 - n_i(y))dy = 1 - \pi$$

then $\pi = 1 - \frac{\lambda_2 b_0}{m}$. Therefore, the traffic rate for low priority customers has the form

$$\rho_1 = \lambda_1((1 - \frac{\lambda_2 b_0}{m})\sum_{i=1}^{m} b_i^{-1})^{-1}.$$

From corollary 1.1 we obtain corollary 1.3.

COROLLARY 1.3.– Assume that X_2 is a strongly regenerative flow, B_0 satisfies [1.14] and $\rho_2 = \frac{\lambda_2 b_0}{m} < 1$. Then $Q_1(t) \xrightarrow[t\to\infty]{P} \infty$ if $\rho_1 > 1$. If additionally B_i satisfies [1.14] and $\rho_1 < 1$, then $Q_1(t)$ is a stable process.

The first statement follows from corollary 1.1 since $\overrightarrow{n}$ is a strongly regenerative process. To prove the second statement, we note that Q_2 is a regenerative process and its stochastic boundedness means stability.

1.9. Queueing system with simultaneous service of a customer by a random number of servers

Here, we consider a system S assuming that the nth customer requires service from ζ_n server simultaneously ($\zeta_n = \overline{1,m}$). The customer arrived in an empty queue begins service immediately if the number of available servers is more or equal ζ_n. Otherwise the customer becomes the first in a queue and the service begins when the required number of servers becomes available. A customer who arrives in an nonempty queue takes the last place in the queue. When service begins, each server's completion time is independent of all other servers and has an exponential distribution with rate μ. The sequence $\{\zeta_n\}_{n=1}^{\infty}$ consists of iid random variables and $\alpha_j = P(\zeta = j)$, $j = \overline{1,m}$, $\sum_{j=1}^{m} \alpha_j = 1$.

Queueing systems with simultaneous service have been studied in a number of works (Rumyantsev and Morozov 2017; and references therein). The stability conditions in an explicit form have been obtained for systems with a Poisson input flow and independent exponentially distributed service times by (Gillent and Latouche 1983). The main goal of this section is an extension of the stability condition to the model with a regenerative input flow X based on theorems 1.1 and 1.2. Thus, we consider the system S described in section 1.2 with $\overrightarrow{\eta}_n = (\eta_{n1}, \ldots, \eta_{n\zeta_n}, \zeta_n)$ where $\{\eta_{nj}\}_{j=1}^{m}$ is a sequence of independent exponentially distributed random variables not depending on ζ_n. Let S_0 be an auxiliary system defined in section 1.3. Instead of the process Y we consider the auxiliary flow Z that is the number of service completions by all m servers up to time t in the auxiliary system S_0. Denote by $U(t)$ the number of occupied servers at time t in the system S_0. Then U is a Markov chain and Z is a doubly stochastic Poisson process (Grandell 1976) with a random intensity $\lambda_Z(t) = \mu U(t)$. We note that the process U hits the state $\{m\}$ from any state $j = 1, 2, \ldots, m$ with a positive probability. It means that all states attainable from the state $\{m\}$ constitute the finite class $\mathcal{K}$ of communicating states. Therefore, there are limits

$$\lim_{t\to\infty} P(U(t) = j) = P_j > 0 \text{ for } j \in \mathcal{K}.$$

Let $s_0 = \max\{s = \overline{1,m} : \alpha_s > 0\}$ and $j_0 = m - s_0 + 1$, $\mathcal{K} = (j_0, j_0+1, \ldots, m)$.

We have the system of equations for $P_j = \lim_{t\to\infty} P(U(t) = j)$, $j = \overline{j_0, m}$.

$$\begin{cases} jP_j = (j+1)\frac{\beta_j}{\beta_{j+1}} P_{j+1} \quad \text{for } j_0 \le j < m \\ m(1-\alpha_1)P_m = \sum\limits_{j=j_0}^{m-1} \frac{\alpha_{m-j+1}}{\beta_j} P_j \\ \sum\limits_{j=j_0}^{m} P_j = 1 \\ P_j = 0 \quad \text{for } j < j_0 \end{cases} \qquad [1.18]$$

where $\beta_j = \alpha_m + \alpha_{m-1} + \ldots + \alpha_{m-j+1}$, $j = \overline{1,m}$. We may easily verify that the solution of [1.18] has the form

$$P_j = \frac{C}{j} P(\zeta > m - j),\ C = (\sum_{j=1}^{m} \frac{1}{j}\beta_j)^{-1},\ j = \overline{1,m}.$$

Since

$$\lambda_Z = \lim_{t\to\infty} \frac{Z(t)}{t} = \mu \sum_{j=1}^{m} jP_j$$

we get

$$\lambda_Z = \frac{\mu E\zeta}{\sum\limits_{j=1}^{m} \frac{1}{j} P(\zeta > m - j)}$$

and the traffic rate for the system S

$$\rho = \frac{\lambda_X E\zeta}{\lambda_Z} = \frac{\lambda_X}{\mu} \sum_{j=1}^{m} \frac{1}{j} P(\zeta > m - j). \qquad [1.19]$$

Let us note that this formula is the same as obtained by (Gillent and Latouche 1983) for queueing systems with a Poisson input flow. To employ theorems 1.1 and 1.2, we have to verify conditions 1.1, 1.3, 1.4 and 1.5. First of all we note that Z is a strongly regenerative flow. As points of regeneration, we may take the sequential hitting times $\{t_n^{(j)}\}_{n=1}^{\infty}$ of $U(t)$ into the fixed state $j \in \mathcal{K}$. (We take $t_1^{(j)} = 0$ if

$U(0) = j$.) Then $t_{n+1}^{(j)} - t_n^{(j)} = v_n^{(j)} + \gamma_n^{(j)}$, where $v_n^{(j)}$ is the sojourn time in the state j and $\gamma_n^{(j)}$ is the return time to this state after exit from it for $U(t)$. The random variables $v_n^{(j)}$ and $\gamma_n^{(j)}$ are independent and $v_n^{(j)}$ has an exponential distribution. Moreover, $Z(t_{n+1}^{(j)} + v_n^{(j)}) = Z(t_{n+1}^{(j)})$, therefore, condition 1.1 holds. Let $q(t)$ be the total number of servers that are already busy or will be busy by service of the $Q(t)$ customers, which are present at the system S at time t. Then $q(t)$ as well as $Q(t)$ is a regenerative process with a sequence $\{\theta_{n_k}^{(1)}\}_{k=1}^{\infty}$ of points of regeneration that is a subsequence of $\{\theta_n^{(1)}\}_{n=1}^{\infty}$ such that $q(\theta_{n_k}^{(1)}) = Q(\theta_{n_k}^{(1)}) = 0$. Let us recall that $\{\theta_n^{(1)}\}_{n=1}^{\infty}$ is a sequence of points of regeneration for X. Therefore, because of theorem 1 from (Afanasyeva and Tkachenko 2014), condition 1.4 is fulfilled. Now for any fix $j \in \mathcal{K}$ we define the common points of regeneration $\{T_n^{(j)}\}_{n=1}^{\infty}$ for the input flow X and auxiliary flow Z by the relation

$$T_n^{(j)} = \min\{\theta_l^{(1)} > T_{n-1}^{(j)} : \bigcup_{s=1}^{\infty}\{t_s^{(j)} \le \theta_l^{(1)} < t_s^{(j)} + v_s^{(j)}\}\},\ T_0^{(j)} = 0.$$

Let

$$\Delta_Z^{(j)}(n) = Z(T_{n+1}^{(j)}) - Z(T_n^{(j)}),$$

$$\Delta_X^{(j)}(n) = X(T_{n+1}^{(j)}) - X(T_n^{(j)}), \qquad [1.20]$$

$$\Delta_{\tilde{Z}}^{(j)}(n) = \tilde{Z}(T_{n+1}^{(j)}) - \tilde{Z}(T_n^{(j)}),$$

where $\tilde{Z}$ is the process Z for the system S that is the number of service completions by all m servers up to time t. Now we formulate the main result of this section.

COROLLARY 1.4.– For the system S, the process Q is a stable process if and only if $\rho < 1$.

PROOF.– Consider the case $\rho \ge 1$ and take m instead of j in [1.20]. It is evident that the stochastic inequality

$$\Delta_{\tilde{Z}}^{(m)}(n) \le \Delta_Z^{(m)}(n) \quad (n = 1, 2, \ldots)$$

takes place. Therefore condition 1.3 is fulfilled. Based on theorem 1.1, we obtain the convergence

$$Q(t) \xrightarrow[t\to\infty]{P} \infty. \qquad [1.21]$$

For the case $\rho < 1$, we take j_0 instead of j in [1.20]. Then if convergence [1.21] takes place for any $\epsilon > 0$, there is n_ϵ such that

$$E\Delta_{\tilde{Z}}^{(j_0)}(n) \geq E\Delta_{Z}^{(j_0)}(1) - \epsilon \quad \text{for } n \geq n_\epsilon.$$

The proof is based on the approach described in Afanasyeva and Tkachenko (2014). Thus, conditions 1.1, 1.4 and 1.5 of theorem 1.2 are fulfilled and Q is a stochastically bounded process when $\rho < 1$. Since Q is a regenerative process, this means that it is stable.

We see that for the model under consideration, the stability condition does not depend on the structure of the input flow. ■

1.10. Applications to transport systems analysis

The study of traffic flows has a long history (Gideon and Pyke 1999; Grinbeerg 1959; Greenshields 1935; Inose and Hamada 1975; and references therein). Various methods such as cellular automate (Maerivoet and de Moor 2005), statistical mechanics and mathematical physics (Blank 2003; Chowdhury 1999; Fuks and Boccara 2001; Helbing 2001; Schadschneider 2000) or queueing theory (Afanasyeva and Bulinskaya 2009, 2010, 2011, 2013; Afanasyeva and Mihaylova 2015; Afanasyeva and Rudenko 2012; Baycal-Gursoy and Xiao 2004; Baycal-Gursoy *et al.* 2009; Caceres and Ferrari 2007) were used.

The purpose of the proposed study is an estimation of the carrying capacity of the automobile road, intersected by a crosswalk. Under the capacity, we mean the upper bound of the intensity of the flow of cars, when the queue of cars does not tend to infinity. This means that the stability condition for the process determining the number of these cars is satisfied, so our analysis will be based on the results obtained in section 1.6.

Let us move to the description of the models.

We consider an automobile road with two directions of traffic and m traffic lanes in each. The flow of cars X_i in the ith direction is a regenerative flow with intensity $\lambda_X^{(i)}$ $(i = 1, 2)$. The road is intersected by a two-directional pedestrian crossing (Figure 1.1). We denote that pedestrians following from A to B have the first type, and from B to A have the second type. The flow of pedestrians of the ith type is a Poisson flow with intensity λ_i $(i = 1, 2)$. Pedestrians cross the road independently of each other with a random (but constant during the entire time of being at the crosswalk) speed.

First, we assume that there is no traffic light at the crossing and pedestrians have an absolute priority over cars. In this case the number of pedestrians at the crosswalk is the number of customers in the infinite-channel service system of $M|G|\infty$ type.

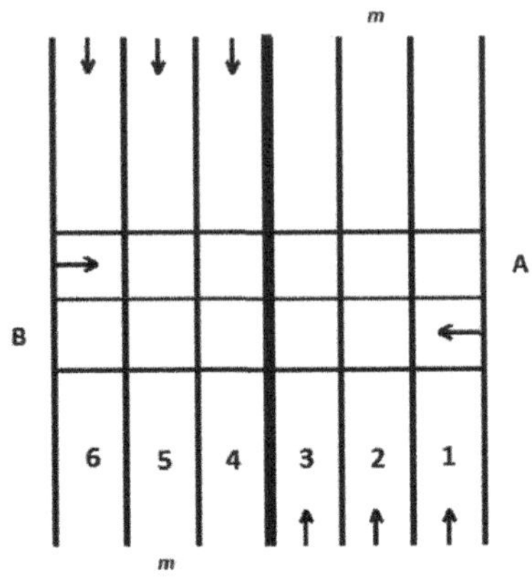

Figure 1.1. *A road intersected by a pedestrian crossing*

Let us assume that $2b$ is an average time of crossing the road by a pedestrian. Then the probability P_0 that there is no pedestrian at the crosswalk in a stationary regime is defined by the expression

$$P_0 = e^{-2(\lambda_1+\lambda_2)b} \tag{1.22}$$

Saaty (1961).

First assume that a car can cross a pedestrian crossing only if there are no pedestrians on it. Let us assume that the cars in the lanes $1, 2, \ldots, m$ are going in one direction and the cars in the lanes $m + 1, \ldots, 2m$ – in another. We consider the process $Q_1(t)$ – the number of cars in the lanes $1, 2, \ldots, m$ at time t (the consideration of lanes $m + 1, m + 2, \ldots, 2m$ is analogous).

Denote $H_j(t)$ the mathematical expectation of the number of cars that pass through the crosswalk at the lane j during time t under the condition that there are always cars at this lane and the crosswalk is free. Also denote $H(t) = \sum_{j=1}^{m} H_j(t)$. In relation to the process $Q_1(t)$, we have a single-channel service system with an unreliable server. The operating time u_1 has an exponential distribution with the parameter $\lambda_1 + \lambda_2$, and the unavailable time u_2 is the period of the system $M|G|\infty$ being busy.

Since

$$P_0 = \frac{Eu_1}{Eu_1 + Eu_2},$$

then

$$a = Eu_1 + Eu_2 = (\lambda_1 + \lambda_2)^{-1} e^{2(\lambda_1+\lambda_2)b}.$$

It is not difficult to show that under the assumptions made, the results of section 1.6 are correct and the traffic rate ρ_1 is determined by the expression

$$\rho_1 = \frac{\lambda_X^{(1)} e^{2(\lambda_1+\lambda_2)b}}{(\lambda_1+\lambda_2)h(\lambda_1+\lambda_2)}, \qquad [1.23]$$

where $h(\lambda) = \lambda \int_0^\infty e^{-\lambda y} H(y) dy$.

The necessary and sufficient condition for the stability of the process Q_1 is the fulfillment of the inequality $\rho_1 < 1$, and the capacity $\bar{\lambda}_X^{(1)}$ is defined as

$$\bar{\lambda}_X^{(1)} = (\lambda_1+\lambda_2)h(\lambda_1+\lambda_2)e^{-2(\lambda_1+\lambda_2)b}.$$

If, for example $H(t) = m\nu t$, which corresponds to the assumption that each car crosses the pedestrian crossing during an exponentially distributed time with a parameter ν, then

$$\bar{\lambda}_X^{(1)} = m\nu e^{-2(\lambda_1+\lambda_2)b}.$$

When the real intensity $\lambda_X^{(1)}$ is less than, but close to $\bar{\lambda}_X^{(1)}$, large queues accumulate before the crosswalk.

Their asymptotic analysis, as well as some results concerning characteristics of the process Q_1 in a stationary regime, when $\rho^{(1)} < 1$ can be found in the papers (Afanasyeva and Rudenko 2012; Afanasyeva and Mihaylova 2015).

Now we will consider model 2, in which the rules for crossing the crosswalk by a car are weakened. We assume that the car can move along the jth lane ($j = \overline{1,m}$) if there are no pedestrians of the first type (going from A to B) on the lanes $1, 2, \dots, j$, and there are no pedestrians of the second type on the lanes $j, j+1, \dots, 2m$. Denote $P_0(j)$ as the probability of this event in a steady-state.

Since the number of pedestrians of the first type on the lanes $(1, 2, \dots, j)$ is the number of customers in the system $M|G|\infty$ with the intensity λ_2 and with an average service time $(2m - j + 1)\frac{b}{m}$, then

$$P_0(j) = e^{-j\lambda_1 \frac{b}{m} - (2m-j+1)\lambda_2 \frac{b}{m}}. \qquad [1.24]$$

So we have a queueing system with m unreliable servers. All servers break when a pedestrian of the first (second) type appears on lane 1 (the $2m$th).This means that the available time $\tau_j^{(1)}$ of the jth server is exponentially distributed with the parameter $\lambda_1 + \lambda_2$. Let $\tau_j^{(2)}$ be a block time of the jth server and $a_j = E\tau_j^{(1)} + E\tau_j^{(2)}$. Then $P_0(j) = \frac{E\tau_j^{(1)}}{a_j}$, so

$$a_j = (\lambda_1 + \lambda_2)^{-1} e^{j\lambda_1 \frac{b}{m} + (2m-j+1)\lambda_2 \frac{b}{m}}.$$

Assuming that $h_j(\lambda) = \lambda \int_0^\infty e^{-\lambda t} H_j(t)dt$ and using the results of section 1.6, we can find the traffic rate ρ_2 for model 2.

$$\rho_2 = \lambda_X^{(1)} [(\lambda_1 + \lambda_2) \sum_{j=1}^{m} P_0(j) h_j(\lambda_1 + \lambda_2)]^{-1}. \qquad [1.25]$$

If $h_j(\lambda) = \frac{1}{m} h(\lambda), j = 1, \bar{m}$, then [1.25] can be written as

$$\rho_2(m) = \begin{cases} \frac{m\lambda_X^{(1)}}{(\lambda_1+\lambda_2)h(\lambda_1+\lambda_2)} \frac{e^{\lambda_2 \frac{b}{m}} - e^{\lambda_1 \frac{b}{m}}}{e^{-(\lambda_1+\lambda_2)b} - e^{-2\lambda_2 b}}, & \text{when } \lambda_1 \neq \lambda_2 \\ \frac{\lambda_X^{(1)} e^{2\lambda b + \frac{\lambda b}{m}}}{2\lambda h(2\lambda)}, & \text{when } \lambda_1 = \lambda_2. \end{cases}$$

When $m = 1$, we get

$$\rho_2(1) = \frac{\lambda_X^{(1)} e^{2\lambda_2 b + \lambda_1 b}}{(\lambda_1 + \lambda_2) h(\lambda_1 + \lambda_2)} < \frac{\lambda_X^{(1)} e^{2(\lambda_2 + \lambda_1) b}}{(\lambda_1 + \lambda_2) h(\lambda_1 + \lambda_2)} = \rho_1.$$

It is easy to show that for all $m \geq 1$, the inequality $\rho_2(m) < \rho_1$ holds. Weakening the rules of crossing the crosswalk increases the capacity of the route. To estimate this effect, we consider the ratio

$$\frac{\rho_2(m)}{\rho_1} = \begin{cases} \frac{m(e^{\lambda_2 \frac{b}{m}} - e^{\lambda_1 \frac{b}{m}})}{e^{(\lambda_1+\lambda_2)b} - e^{2\lambda_1 b}}, & \text{when } \lambda_1 \neq \lambda_2 \\ e^{-2\lambda b + \frac{\lambda b}{m}}, & \text{when } \lambda_1 = \lambda_2 = \lambda \end{cases}.$$

Putting $x = e^{\lambda_1 b} \geq 1$, $\lambda_2 = \alpha\lambda_1$, we have

$$\phi(x) = \frac{\rho_2(m)}{\rho_1} = \begin{cases} \frac{m(x^{\frac{\alpha}{m}} - x^{\frac{1}{m}})}{x^{1+\alpha} - x^2}, & \textit{when } \alpha \neq 1 \\ x^{-2+\frac{1}{m}}, & \textit{when } \alpha = 1 \end{cases}.$$

After drawing the graphs for $\phi(x)$ ($\phi(1) = 1$) for $\alpha = 0.5; 1.5; 2$ (see Figure 1.2), we can see that the effect of the weakened rule (model 2) in comparison with the standard rule (model 1) is stronger, as the number of the lanes increases and the intensity of the number of pedestrians increases.

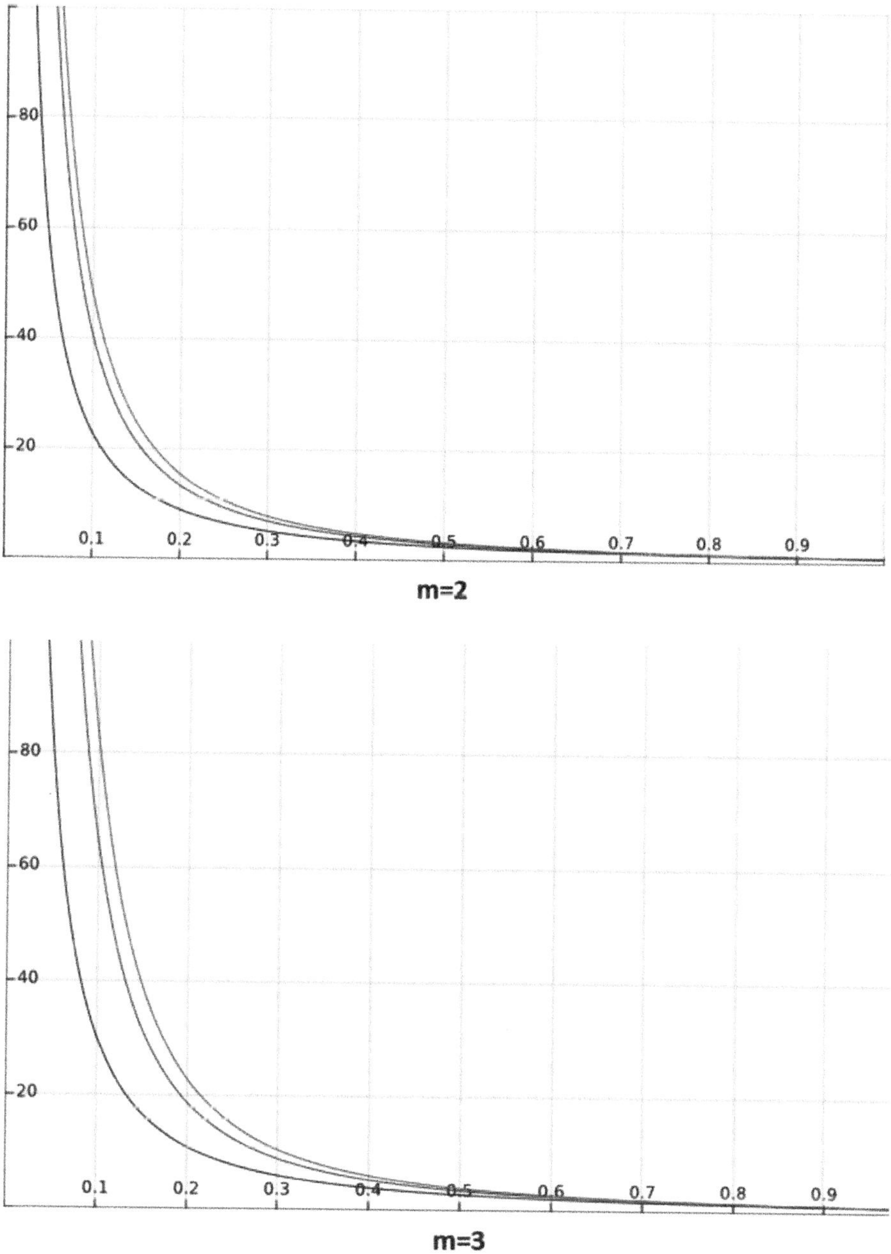

Figure 1.2. *Plots for* $\alpha = 0.5, 1.5, 2$

Currently there is no algorithm that estimates the number of cars before the crosswalk in model 2, however we can obtain asymptotic expressions for the average number of expected cars when $\rho_2 \uparrow 1$. It turns out that $EQ \sim \frac{c}{1-\rho_2}$, where c is a constant.

If the length of the queue of cars is unacceptably high, it is necessary to make organizational decisions. One of these decisions is to install a traffic light. Then, in relation to the cars, we again get a one-channel service system with an unreliable server, but now the server will not work if the red light is on (for cars) and will work if the green light is on. This model has been studied in papers (Afanasyeva and Bulinskaya 2013, 2010), in which the algorithms for estimating the queue length were proposed and the number of the asymptotic results were received. It can happen that, with the available traffic intensities of the cars and the pedestrians, the installation of a traffic light, even at the optimum interrelationship between switching intervals, does not provide an acceptable level of queues of pedestrians and cars. This may be used as the basis for the construction of an underground (or overground) pedestrian crossing, its elimination or transfer to another location.

1.11. Conclusion

In this chapter, we considered the stability problem for multiserver queues with a regenerative input flow. Let us note that stability analysis is an essential and challenging stage of the investigation of stochastic models. However, stability conditions may be of independent interest. In particular, the stability criterion of the multiserver model can be used for the capacity planning of a model and estimation of the upper bound of potential energy saving. The main contribution of our research is an extension of the stability criterion to the model with a regenerative input flow. The method we use has the following steps. First, we define an auxiliary process Y that is the number of customers that are served up to time t at the auxiliary system S_0 in which there are always customers for service. Second, assuming that this process is a regenerative flow not depending on the input flow X under some additional conditions, we construct the common points of regeneration of Y and X. We call this step synchronization of the processes. This approach allows us to use results from the renewal theory for the stability analysis of the systems satisfying additional conditions. These conditions may seem too restrictive to be useful for the analysis of the real models. Therefore, we apply our approach to the stability analysis of two classical systems with service interruptions (sections 1.6 and 1.7) of the queueing system with preemptive priority discipline (section 1.8) and of the system with simultaneous service of a customer by a random number of servers (section 1.9). It follows from our results that the structure of the input flow does not affect the stability condition. We only need to know the intensity of this flow to estimate the traffic rate. But for the preemptive repeat different service discipline, the distribution of the service time plays an essential role, since the traffic rate is expressed with the help of the renewal function corresponding to this distribution. We obtain the upper bound for the traffic rate providing the sufficient stability condition. Note that this condition is the same as the condition obtained in Morozov *et al.* (2011) for a more simple model. Finally, in section 1.9 we give applications of our results for the estimation of the capacity of the automobile road, intersected by a crosswalk.

1.12. Acknowledgment

This work is partially supported by RFBR grant 17-01-00468.

1.13. References

Afanaseva, L. (2019). Asymptotic analysis of queueing models based on synchronization method. *Methodology and Computing in Applied Probability*, 1–22, https://doi.org/10.1007/s11009-019-09694-9.

Afanasyeva, L.G., Bashtova E.E. (2014). Coupling method for asymptotic analysis of queues with regenerative input and unreliable server. *Queueing Systems*, 76(2), 125–147.

Afanasyeva, L., Bashtova, E., Bulinskaya, E. (2012). Limit theorems for semi-Markov queues and their applications. *Communications in Statistics Part B: Simulation and Computation*, 41(6), 688–709.

Afanasyeva, L.G., Bulinskaya, E.V. (2009). Some problems for the flow of interacting particles. *Modern Problems of Mathematics and Mechanics*, 2, 55–68.

Afanasyeva, L.G., Bulinskaya, E.V. (2010). Mathematical models of transport systems based on queueing systems methods. *Proceedings of Moscow Institute of Physics and Technology*, 2(4), 6–21.

Afanasyeva, L.G., Bulinskaya, E.V. (2011). Stochastic models of transport flows. *Commun. Stat. Theory Methods*, 40(16), 2830–2846.

Afanasyeva, L.G., Bulinskaya, E.V. (2013). Asymptotic analysis of traffic lights performance under heavy-traffic assumption. *Methodology and Computing in Applied Probability*, 15(4), 935–950.

Afanasyeva, L.G., Mihaylova, I.V. (2015). Two models of the highway intersected by a crosswalk. *Survey of Applied and Industrial Mathematics*, 22(5), 520–532.

Afanasyeva, L.G., Rudenko, I.V. (2012). $GI|G|\infty$ queueing systems and their applications to the analysis of traffic models. *Theory of Probability. Applications*, 57(3), 427–452.

Afanasyeva, L., Tkachenko, A. (2014). Multichannel queueing systems with regenerative input flow. *Theory of Probability and Its Applications*, 58(2), 174–192.

Afanasyeva, L., Tkachenko, A. (2016). Stability analysis of multi-server discrete-time queueing systems with interruptions and regenerative input flow. In *New Trends in Stochastic Modeling and Data Analysis*, Manca, R., McClean, S., Skiadas, C.H. (eds). ISAST: International Society for the Advancement of Science and Technology, Athens, 13–26.

Afanasyeva, L., Tkachenko, A. (2018). Stability of discrete multi-server queueing systems with heterogeneous servers, interruptions and regenerative input flow. *Reliability: Theory and Applications*, 13(1), 63–75.

Asmussen, S. (2003). *Applied Probability and Queues*. Springer-Verlag, New York.

Avi-Itzhak, B., Naor, P. (1963). Some queueing problems with the service station subject to breakdown. *Operations Research*, 11(3), 303–320.

Baycal-Gursoy, M., Xiao, W. (2004). Stochastic decomposition in $M|M|\infty$ queues with Markov-modulated service rates. *Queueing Systems*, 48, 75–88.

Baycal-Gursoy, M., Xiao, W., Ozbay, K. (2009). Modeling traffic flow interrupted by incidents. *Eur. J. Oper. Res.*, 195, 127–138.

Belorusov, T. (2012). Ergodicity of a multichannel queueing system with balking. *Theory of Probability and Its Applications*, 56(1), 120–126.

Blank, M. (2003). Ergodic properties of a simple deterministic traffic flow model. *J. Stat. Phys.*, 111, 903–930.

Borovkov, A.A. (1976). *Stochastic Processes in Queueing Theory*. Springer-Verlag, New York.

Caceres, F.C., Ferrari, P.A., Pechersky, E. (2007). A slow to start traffic model related to a $M|M|1$ queue. *J. Stat. Mech.* arXiv:cond-mat/0703709 v2 [cond-mat.stat-mech].

Chen, H. (1995). Fluid approximation and stability of multiclass queueing networks: Work-conserving disciplines. *Annals of Applied Probability*, 5, 637–665.

Chen, H., Yao, D. (2001). *Fundamentals of Queueing Networks*. Springer, New York.

Chowdhury, D. (1999). Vehicular traffic: A system of interacting particles driven far from equilibrium. arXiv:arXiv:cond-mat/9910173 v1 [cond-mat.stat-mech].

Dai, J. (1995). On positive Harris recurrence of multiclass queueing networks: A unified approach via fluid limit models. *Annals of Applied Probability,* 5, 49–77.

Feller, W. (1971). *An Introduction to Probability Theory and Its Applications*, 2nd ed. John Wiley & Sons, New York.

Fiems, D., Bruneel, H. (2013). Discrete-time queueing systems with Markovian preemptive vacations. *Mathematical and Computer Modeling*, 57(3-4), 782–792.

Foss, S., Konstantopoulos, T. (2004). An overview on some stochastic stability methods. *Journal of the Operations Research Society of Japan*, 47(4), 275–303.

Fuks, H., Boccara, N. (2001). Convergence to equilibrium in a class of interacting particle system evolving in discrete time. *Phys. Rev. E.*, 64, 016117.

Gaver Jr., D. (1962). A waiting line with interrupted service, including priorities. *Journal of the Royal Statistical Society. Series B (Methodological)*, 24, 73–90.

Georgiadis, L., Szpankowski, W. (1992). Stability of token passing rings. *Queueing Systems*, 11, 7–33.

Gideon, R., Pyke, R. (1999). Markov renewal modeling of Poisson traffic at intersections having separate turn lanes. In *Semi-Markov Models and Applications*, Janssen, J., Limneos, N. (eds). Springer, New York, NY, 285–310.

Gillent, F., Latouche, G. (1983). Semi-explicit solution for $M|PH|1$ – like queueing systems. *European Journal of Operational Research*, 13(2), 151–160.

Grandell, J. (1976). *Double Stochastic Poisson Process*, Lecture Notes in Mathematics, 529, Springer, Berlin.

Greenshields, B.D. (1935). A study of highway capacity. *Proc. Highway Res.*, 14, 448–477.

Grinbeerg, H. (1959). An analysis of traffic flows. *Oper. Res.*, 7, 79 85.

Helbing, D. (2001). Traffic and related self-driven many-particle systems. *Rev. Mod. Phys.*, 73, 1067–1141.

Inose, H., Hamada, T. (1975). *Road Traffic Control*. University of Tokyo Press, Tokyo.

Kiefer, J., Wolfowitz, J. (1955). On the theory of queues with many servers. *Trans. Amer. Math. Soc.*, 78, 1–18.

Krishnamoorthy, A., Pramod, P., Chakravarthy, S. (2012). Queues with interruptions: A survey. TOP, 1-31 doi:10.1007/s11750-012-0256-6.

Loynes, R.M. (1962). The stability of a queue with non-independent inter-arrival and service times. *Proc. Cambr. Phil. Soc.*, 58(3), 497–520.

Maerivoet, S., de Moor, B. (2005). Cellular automata models of road traffic. *Phys. Rep.*, 419, 1–64.

Malyshev, V.A., Menshikov, M.V. (1982). Ergodicity continuity and analyticity of countable Markov chains. *Trans Moscow Math*, 1, 1–48.

Meyn, S.P., Tweedie, R.L. (2009). *Markov Chains and Stochastic Stability*. Cambridge University Press, New York.

Morozov, E. (2004). Weak regeneration in modeling of queueing processes. *Queueing Systems*, 46, 295–315.

Morozov, E. (2007). A multiserver retrial queue: Regenerative stability analysis. *Queueing Systems*, 56(3-4), 157–168.

Morozov, E., Dimitriou, I. (2017). Stability analysis of a multiclass retrial system with coupled orbit queues. In *Computer Performance Engineering. EPEW 2017*, Reinecke, P., Di Marco, A. (eds). Lecture Notes in Computer Science. Springer, Cham, 85–98.

Morozov, E., Fiems, D., Bruneel, H. (2011). Stability analysis of multiserver discrete-time queueing systems with renewal-type server interruptions. *Performance Evaluation*, 68(12), 1261–1275.

Morozov, E., Rumyantsev, A. (2016). Stability analysis of a $MAP|M|s$ cluster model by matrix-analytic method. *European Workshop on Computer Performance Engineering*, 63–76.

Neuts, M. (1989). *Structured Stochastic Matrices of M/G/1 Type and Their Applications*. Marcel Dekker, New York.

Pechinkin, A., Socolov, I., Chaplygin, V. (2009). Multichannel queueing system with refusals of servers groups. *Informatics and Its Applications*, 3(3), 4–15.

Rumyantsev, A., Morozov, E. (2017). Stability criterion of a multi-server model with simultaneous service. *Annals of Operations Research*, 252(1), 29–39.

Saaty, T.L. (1961). *Elements of Queueing Theory with Applications*. McGraw-Hill, Inc, New York, NY.

Sadowsky, J.S. (1995). The probability of large queue lengths and waiting times in a heterogeneous multiserver queue: positive recurrence and logarithmic limits. *Adv. Appl. Prob.*, 27, 567–583.

Schadschneider A. (2000). Statistical physics of traffic flow. arXiv:arXiv:cond-mat/0007418 v1 [cond-mat.stat-mech].

Smith, W. (1955). Regenerative stochastic processes. *Proceedings of the Royal Society of London. Series A. Mathematical and Physical Sciences*, 232(1188), 6–31.

Szpankowski, W. (1994). Stability conditions for some distributed systems. Buffered random access systems. *Adv. Appl. Prob.*, 26, 498–515.

Thiruvengadam, K. (1963). Queueing with breakdowns. *Operations Research*, 11(1), 62–71.

Thorisson, H. (2000). *Coupling, Stationary and Regeneration*. Springer, New York.

White, H., Christie, L.S. (1958). Queueing with preemptive priorities or with breakdown. *Operations Research*, 6(1), 79–95.

2

Queueing Models in Services – Analytical and Simulation Approach

Srinivas R. CHAKRAVARTHY
Departments of Industrial and Manufacturing Engineering & Mathematics, Kettering University, Flint, MI, USA

Queueing models play a significant role in many walks of life. In this chapter, we will focus on a few selected queueing models that are useful in service sectors. While the literature on queueing models is vast and filled with a variety of applications, the number of models with implementable and useful results needs to be increased. In this chapter, we will use both analytical and simulation approaches to discuss a few selected models that are useful in practice as well as to point out the significant role played by the correlated arrivals. Correlated arrivals occur naturally in service sectors due to the fact that customers/jobs arrive from different sources. We also report some interesting observations based on the analytical models of some well-known queueing models. We believe that some of these, especially the ones dealing with the busy period and the distribution functions of the sojourn times, are not reported in the literature and are very useful for managerial decision-making in service sectors.

2.1. Introduction

On a day-to-day basis we come across items that have either manufacturing aspects or service aspects or both. Manufacturing and service industries invariably use queueing models in one form or another to identify ways of increasing productivity, as well as to optimize the allocation of the limited resources. While the literature on queueing models is vast and filled with a variety of models, the number of papers dealing with practical and implementable solutions definitely needs to be increased. The reason for not having that many practical and implementable queueing models in manufacturing and service industries is not due to choice but rather due to the inherent difficulty present in incorporating proper modeling assumptions coupled with obtaining implementable results when carrying out the analysis. This is especially true in the case when we need to incorporate the correlation that may be present in the interarrival times of jobs/customers. For example, when we look at the output of an assembly line (which will form the input to the next assembly line) ignoring the correlation present will lead to incorrect analysis and conclusions. Similarly when we look at the pooled input from various sources that form the input to a service center, the interarrival times of jobs will invariably be correlated and this fact cannot be ignored. One way to alleviate the challenges posed in analytical modeling is to analyze such queueing models through simulation.

Hence, in this chapter we will also focus on the study of complicated queueing models through simulation using ARENA, a powerful simulation software (Kelton 2010). The main purpose is to motivate and point out the usefulness and the need for simulation, especially when the theoretical analysis from a practical point of view becomes very cumbersome or at times infeasible. Simulation should be considered as an important tool and also a way to get insight into obtaining theoretical results. Hence, a few selected queueing models will be simulated and their results will be analyzed. We also report some interesting observations based on analytical models of some well-known queueing models. To our knowledge, some of these, especially the ones dealing with the busy period and the distribution functions of the sojourn times, are not reported in the literature. These are very useful for managerial decision-making in service sectors. We will briefly review phase-type (PH) distributions and the batch Markovian arrival process (BMAP) as they will be used as building blocks for some of the simulated models.

2.2. Phase-type distributions and the batch Markovian arrival process

In order to make this chapter somewhat self contained, we will give a brief overview of PH distributions and $BMAP$. We refer readers to (Artalejo and Gomez-Correl 2010; Bladt and Nielsen 2017; Chakravarthy 2001, 2010, 2015; Qi-Ming 2014; Lucantoni 1991; Neuts 1975, 1979, 1981, 1989, 1992, 1995) for details on PH distributions and $BMAPs$ including proofs of the results mentioned here. The impact of Poisson processes and exponential distributions in stochastic modeling among other areas in mathematics is very well known and detailed extensively in the literature. However, these have restrictions in many practical applications and that have been documented in the literature. Further, queueing models involving non-exponential assumptions (for arrivals or services or both) had been mostly been analyzed in a form that was not computationally friendly up until 1970s.

Realizing that the computational aspects were no longer a luxury and that they had become both an art and science, Neuts introduced and developed the theory of PH distributions and related point processes during 1970s. Neuts' work made the computational aspects an integral part of the analysis of any stochastic model and it is required to qualitatively study any stochastic model.

We know how much Poisson processes played and still playing an important role in stochastic modeling. Poisson processes are very elegant mathematically and of course they have their limitations in practice due to the memoryless property of the underlying exponential distribution. Note that the PH processes are natural extensions of Poisson processes and enable us to get away from a restricted class of distributions yet preserve computationally tractable solutions in stochastic modeling. Traditionally, Poisson/exponential distributions have been used mainly for mathematical convenience more than for practical purposes.

PH distributions have nice closure properties and a related matrix formalism that make them attractive for algorithmic implementation as well as for use in practice. It is worth pointing out that if Poisson/exponential assumptions in a queueing model do not lead to explicit expressions for quantities like steady-state probability vector, then it is guaranteed that no explicit expressions are possible for more general cases.

For use in sequel, we define the following notation. By $\boldsymbol{e}$, we will denote a column vector (of appropriate dimension which will be clear in the context of usage) of 1's; b y$\boldsymbol{e}_i$, we will denote a unit column vector (of appropriate dimension) with 1 in the ith position and 0 elsewhere; and I an identity matrix (of appropriate dimension). The notation $'$ stands for the transpose of a matrix or a vector. The symbols, $\otimes$ and $\oplus$, respectively, will stand for the Kronecker product and Kronecker sum of matrices. Thus, if G is a matrix of order $m \times n$ and if H is a matrix of order $p \times q$, then $G \otimes H$ will denote a matrix of order $mp \times nq$ whose (i, j)th block matrix is given by $g_{ij}H$; the Kronecker sum of two square matrices, say, A of order a and B of b, is given by $A \otimes I + I \otimes B$, a square matrix of order ab, and is denoted by $A \oplus B$. We refer readers to (Graham 1981; Marcus and Minc 1964; Steeb and Hardy 2011) for details and properties on Kronecker products and Kronecker sums, which play a key role in stochastic models.

2.2.1. *Phase-type distributions*

Neuts (1981) introduced and developed PH distributions, which form the building blocks for $BMAP$. Further, as mentioned earlier, PH distributions are natural extensions of Poisson processes. While PH distributions are defined and developed for both discrete and continuous-time cases, here we will briefly review the definition in the case of continuous-time and list a few key results that will be of great use. Full details can be found in the references cited above.

Consider a continuous-time Markov chain ($CTMC$) with m transient states and one absorbing state with generator

$$Q = \begin{bmatrix} T & \boldsymbol{T^0} \\ \mathbf{0} & 0 \end{bmatrix}. \qquad [2.1]$$

First note that $\boldsymbol{T^0} = -T\boldsymbol{e}$. The eventual absorption is guaranteed when T is non-singular. Let X denote the time until absorption in the $CTMC$ with generator Q. Suppose that the initial probability vector of the process is given by $(\boldsymbol{\alpha}, \alpha_{m+1})$. In most cases, $\alpha_{m+1} = 1 - \boldsymbol{\alpha e}$ will be zero. However, if we need to model X with a positive mass at the origin, then α_{m+1} should to be positive. The knowledge of the matrix T in Q is sufficient to completely determine the other entries of Q. Thus, X is said to follow a PH distribution with representation $(\boldsymbol{\alpha}, T)$ of order m, and we will denote this by indicating that $X \sim PH(\boldsymbol{\alpha}, T)_m$. We will assume that this representation is irreducible unless otherwise indicated. Thus, we assume that T is irreducible.

2.2.2. *Some useful results related to continuous* PH *distributions*

In this section, we briefly state some key results and interpretations assuming that the representation $(\boldsymbol{\alpha}, T)$ is irreducible. For full details, we refer readers to the references on PH distributions listed earlier.

– the probability density function of X is: $f(x) = \boldsymbol{\alpha} e^{Tx}\boldsymbol{T^0}, x \geq 0$;

– the probability distribution function of X is: $F(x) = 1 - \boldsymbol{\alpha} e^{Tx}\boldsymbol{e}, x \geq 0$;

– the Laplace-Stieltje's transform (LST) of X is:$f^*(s) = \boldsymbol{\alpha}(sI - T)^{-1}\boldsymbol{T^0}$, $Re(s) \geq 0$;

– the (i, j)th element of the matrix e^{Tx} gives the probability that the underlying $CTMC$ is in state j at time t given that it started in state i;

– the (i, j)th element of the matrix $(-T)^{-1}$ gives the mean time spent in state j before an absorption given that the underlying $CTMC$ started in state i;

– the nth order raw moments of X are given by: $n!\boldsymbol{\alpha}(-T)^{-n}\boldsymbol{e}, n \geq 1$;

– the class, PH_D, of PH distributions is dense (in the sense of weak convergence) in the class, NND, of all probability distributions on $(0, \infty)$. That is, if $F(.) \in NND$, then $\exists F_n \in PH_D \ni F_n \to F$(weakly). [This result follows on noting that the LST of a PH distribution is a rational function];

– if $-T\boldsymbol{e} = \lambda\boldsymbol{e}$, for some positive λ, then the underlying PH distribution has to be exponential;

– let $\xi < 0$ be an eigenvalue of maximum real part of T with the corresponding left eignevector given by $\boldsymbol{u}$, which is normalized so that $\boldsymbol{ue} = 1$. Then the representation

$(\boldsymbol{u}, T)$ corresponds to an exponential distribution with rate $-\xi$. This is a very useful result when analyzing stochastic models involving PH distributions algorithmically;

– if $X \sim PH(\boldsymbol{\alpha}, T)_m$ and $Y \sim PH(\boldsymbol{\beta}, S)_n$, then $X + Y \sim PH(\boldsymbol{\gamma}, L)_{m+n}$ where

$$\boldsymbol{\gamma} = (\boldsymbol{\alpha}, 0), \quad L = \begin{bmatrix} T & \boldsymbol{T}^0\boldsymbol{\beta} \\ 0 & S \end{bmatrix}.$$

– if $X_i \sim PH(\boldsymbol{\alpha}(i), T(i))_{m_i}$ and $Y = \sum_{i=1}^{n} p_i X_i$, then $Y \sim PH(\boldsymbol{\delta}, K)_m$, where $m = m_1 + \cdots + m_n$, $\boldsymbol{\delta} = (p_1\boldsymbol{\alpha}(1), \cdots, p_n\boldsymbol{\alpha}(n))$ and $K = \Delta(T(1), \cdots, T(n))$, a diagonal matrix with diagonal (block) entries given by $T(i), i = 1, \cdots, n$;

– the forward recurrence time at time t of $X \sim PH(\boldsymbol{\alpha}, T)_m$ (i.e., the waiting time until the next renewal) is a PH distribution with representation $(\boldsymbol{\alpha}e^{Qt}, T)_m$;

– let $X \sim PH(\boldsymbol{\alpha}, T)_m$ and $Y \sim PH(\boldsymbol{\beta}, S)_n$. Define $U = min\{X, Y\}$ and $V = max\{X, Y\}$. Then we have $U \sim PH(\boldsymbol{\alpha} \otimes \boldsymbol{\beta}, T \otimes I + I \otimes S)_{mn}$ and $V \sim PH(\boldsymbol{\phi}, L)_{mn+m+n}$, where

$$\boldsymbol{\phi} = (\boldsymbol{\alpha} \otimes \boldsymbol{\beta}, \boldsymbol{0}), L = \begin{bmatrix} T \oplus S & I \otimes \boldsymbol{S}^0 & \boldsymbol{T}^0 \otimes I \\ 0 & T & 0 \\ 0 & 0 & S \end{bmatrix}.$$

– suppose that $X \sim PH(\boldsymbol{\alpha}, T)_m$. Let $-\eta$ denote the eigenvalue with largest real part of T; $\boldsymbol{u}$ and $\boldsymbol{v}$ be the left and right, respectively, eigenvector of T corresponding to $-\eta$ with $\boldsymbol{ue}$=$\boldsymbol{uv}$=1. The CDF, $F(.)$, of X is asymptotically exponential (from classical Perron–Frobenius theorem). That is, $1 - F(x) = \boldsymbol{\alpha v}e^{-\eta x} + o(e^{-\eta x}),\ x \to \infty$;

– let $\{X_i\}$ be i.i.d with $X_i \sim PH(\boldsymbol{\alpha}, T)_m$ with mean $\frac{1}{\mu}$. Renewals occur when the underlying $CTMC$ enters into the absorbing state. The process, $\{J(t)\}$, keeping track of the phase of the underlying $CTMC$ has the generator, $\hat{Q} = T + \boldsymbol{T}^0\boldsymbol{\alpha}$;

– the vector, $\boldsymbol{\pi}$, satisfying $\boldsymbol{\pi}\hat{Q} = \boldsymbol{0}, \boldsymbol{\pi e} = 1$, is given by $\boldsymbol{\pi} = \mu\boldsymbol{\alpha}(-T)^{-1}$. This result will play an important role as an accuracy check in numerical computation;

– the (i, j)th entry of the matrix $\int_0^t e^{\hat{Q}x}dx$ gives the mean sojourn time in state j during $(0, t)$ given that the process starts in state i at time 0;

– suppose that $X_i \sim PH(\boldsymbol{\alpha}_i, T_i)_{m_i}$, for $1 \le i \le n$. The kth-order statistics, $X_{(k)}$, is also a PH distribution. The representation of this kth-order statistics is quite involved (Bladt and Nielsen 2017; Qi-Ming 2014). For example, if $n = 3$, the second-order statistics is of phase type with representation, say, $(\boldsymbol{\zeta}, L)$ of order N, where

$N = m_1m_2m_3+m_1m_2+m_1m_3+m_2m_3+m_1+m_2+m_3$, $\boldsymbol{\zeta} = (\boldsymbol{\alpha}_1 \otimes \boldsymbol{\alpha}_2 \otimes \boldsymbol{\alpha}_3, \mathbf{0})$, and

$$L = \begin{pmatrix} T_1 \oplus T_2 \oplus T_3 & I \otimes I \otimes \boldsymbol{T_3^0} & I \otimes \boldsymbol{T_2^0} \otimes I & \boldsymbol{T_1^0} \otimes I \otimes I & 0 & 0 & 0 \\ 0 & T_1 \oplus T_2 & 0 & 0 & I \otimes \boldsymbol{T_2^0} & \boldsymbol{T_1^0} \otimes I & 0 \\ 0 & 0 & T_1 \oplus T_3 & 0 & I \otimes \boldsymbol{T_3^0} & 0 & \boldsymbol{T_1^0} \otimes I \\ 0 & 0 & 0 & T_2 \oplus T_3 & 0 & I \otimes \boldsymbol{T_3^0} & \boldsymbol{T_2^0} \otimes I \\ 0 & 0 & 0 & 0 & T_1 & 0 & 0 \\ 0 & 0 & 0 & 0 & 0 & T_2 & 0 \\ 0 & 0 & 0 & 0 & 0 & 0 & T_3 \end{pmatrix}.$$

– suppose that $X \sim PH(\boldsymbol{\alpha}, T)_m$. Let $Y_1 = E[cos(tX)]$ and $Y_2 = E[sin(tX)]$. Then we have (Bladt and Nielsen 2017), $Y_1 = \boldsymbol{\alpha}(-T)[t^2I + T^2]^{-1}\boldsymbol{T^0}$ and $Y_2 = t\boldsymbol{\alpha}[t^2I + T^2]^{-1}\boldsymbol{T^0}$;

– the computation of the PH density can be carried out using uniformization technique, also known as the method of randomization, first described by (Jensens 1953). Let $X \sim PH(\boldsymbol{\alpha}, T)_m$. Suppose that the generator Q is defined as

$$Q = \begin{pmatrix} T & \boldsymbol{T^0} \\ \mathbf{0} & 0 \end{pmatrix}.$$

Let $\tau = -\min\{T_{i,i}\}$ and define $K = I + \frac{1}{\tau}Q$, $P = I + \frac{1}{\tau}T$, and $\boldsymbol{\zeta} = \frac{1}{\tau}\boldsymbol{T^0}$. Note that K is a stochastic matrix. Applying the uniformization technique, we get $f(t) = \tau\boldsymbol{\alpha}\sum_{n=0}^{\infty} e^{-\tau t}(\tau t)^n/n!P^n\boldsymbol{\zeta}$, $t \geq 0$. We can compute the probability density function for any t to any degree of accuracy by truncating the infinite summation using the cumulative Poisson probabilities approach for the truncation. It is worth pointing out that (Bowerman *et al.* 1990) have proposed a stable method to compute cumulative Poisson probabilities;

– let $\{N(t)\}$ denote the counting process associated with the PH-renewal process. The matrix $P(n,t) = (p_{ij}(n,t))$, where $p_{ij}(n,t) = P[N(t) = n, J(t) = j|N(0) = 0, J(0) = i]$, satisfies

$$\begin{aligned} P'(0,t) &= P(0,t)T, \\ P'(n,t) &= P(n,t)T + P(n-1,t)\boldsymbol{T^0\alpha},\ n \geq 1,\ t \geq 0, \end{aligned} \quad [2.2]$$

with $P(n,0) = \delta_{0,n}I$;

– the PH renewal process is a two-dimensional $CTMC$, $\{N(t), J(t)\}$, on the state space $\{(i,j) : i \geq 0, 1 \leq j \leq m\}$ with generator:

$$\tilde{Q} = \begin{bmatrix} T & \boldsymbol{T^0\alpha} & & & \cdots \\ & T & \boldsymbol{T^0\alpha} & & \cdots \\ & & T & \boldsymbol{T^0\alpha} & \cdots \\ & & & \ddots & \ddots \end{bmatrix}.$$

The above form will provide us with an insight into the development of $BMAP$.

$$e^{\tilde{Q}t} = \begin{bmatrix} P(0,t) & P(1,t) & P(2,t) & \cdots \\ & P(0,t) & P(1,t) & \cdots \\ & & P(0,t) & \cdots \\ & & & \ddots \end{bmatrix}.$$

– the computation of $e^{\tilde{Q}t}$ and $P(n,t)$ using the uniformization method is discussed in Neuts and Li (1996);

– the renewal function $M(t) = E[N(t)]$ has a density given by $\boldsymbol{\alpha} e^{Qt}\boldsymbol{T^0}$;

– since $e^{Qt} \to \boldsymbol{e\pi}$ as $t \to \infty$, we note that the renewal density approaches $\boldsymbol{\pi T^0} = \frac{1}{\mu}$, a well-known result from renewal theory;

– the number of (Poisson) arrivals during a random time that has a continuous PH distribution follows a discrete PH distribution. A very powerful result that is useful in analyzing certain queueing and reliability models;

– *special cases of a continuous PH distribution*

1) *Exponential:*

$\boldsymbol{\alpha} = (1), \ T = -\lambda.$

2) *Erlang:*

$$\boldsymbol{\alpha} = (1,0,\cdots,0), \ T = \begin{bmatrix} -\lambda & \lambda & & \\ & -\lambda & \lambda & \\ & & \ddots & \\ & & & -\lambda \end{bmatrix}.$$

3) *Generalized Erlang:*

$$\boldsymbol{\alpha} = (1,0,\cdots,0), \ T = \begin{bmatrix} -\lambda_1 & \lambda_1 & & \\ & -\lambda_2 & \lambda_2 & \\ & & \ddots & \\ & & & -\lambda_m \end{bmatrix}.$$

4) *Hyperexponential:*

$$\boldsymbol{\alpha} = (\alpha_1,\alpha_2,\cdots,\alpha_m), \ T = \begin{bmatrix} -\lambda_1 & & & \\ & -\lambda_2 & & \\ & & \ddots & \\ & & & -\lambda_m \end{bmatrix}.$$

2.2.3. *The batch Markovian arrival process*

Earlier we saw how Poisson processes were generalized to phase-type renewal processes. PH processes are ideal to model point processes whose interarrival times are independent. What happens if that is not the case? For example, suppose that the jobs arrive from different sources and are pooled for further processing. Even if each source generates i.i.d arrivals, the pooled (or combined) will not be i.i.d (unless all sources are Poisson or there are lots of sources that we can approximate with Poisson).

Marked point processes ($MPPs$) play an important role in stochastic modeling. A special class of $MPPs$ is the Markovian arrival process (MAP), which itself is a particular case of batch MAP ($BMAP$). Neuts (1979) introduced $BMAP$ as a versatile Markovian point process initially. Below, we will briefly present key definitions and results pertaining to $BMAP$.

– a versatile point process which was introduced by Neuts in 1979 (with very complicated notation), and later simplified in 1990, addresses this issue;

– a $BMAP$ is characterized by the sequence of matrices $\{D_k\}$ with D_0 governing no arrivals and $D_k, k \geq 1$, arrivals of size k;

– the point process of $BMAP$ is a semi-Markov process with transition probability matrix given by

$$\int_0^x e^{D_0 t} dt D_k = [I - e^{D_0 x}](-D_0)^{-1} D_k,\ k \geq 1.$$

– let $\boldsymbol{\alpha}$ denote the initial probability vector of the $CTMC$ with generator $Q^* = \sum_k D_k$;

– the initial probability vector $\boldsymbol{\alpha}$ can be chosen so as to have the time origin to be (i) an arbitrary arrival point; (ii) the end of an interval during which there are at least k arrivals; (iii) the point at which the system is in specific state such as the busy period ends or busy period begins. The most interesting is $\boldsymbol{\alpha} = \boldsymbol{\pi}$, where $\boldsymbol{\pi} Q^* = \mathbf{0}, \boldsymbol{\pi} \boldsymbol{e} = 1$;

– the fundamental rate λ is defined as $\lambda = \boldsymbol{\pi} \sum_{k=1}^{\infty} k D_k \boldsymbol{e}$;

– the quantity λ gives the rate (per unit of time) at which customers arrive at the system;

– the quantity $\lambda_g = \boldsymbol{\pi}(-D_0)\boldsymbol{e}$ gives the rate (per unit of time) at which batches arrive at the system;

– to have a specific value for λ_g, we multiply the matrices, $D_k, k \geq 0$, by the appropriate common constant;

– let X denote the time to the first arrival in a $BMAP$ process and define $D = \sum_{k=1}^{\infty} D_k$. The CDF, $F(.)$, of X is $F(x) = \boldsymbol{\alpha}[I - e^{D_0 x}](-D_0)^{-1} D\boldsymbol{e},\ x \geq 0$;

– suppose we take $\boldsymbol{\alpha} = (\boldsymbol{\pi} De)^{-1}\boldsymbol{\pi} D$. Then $F(.)$ is the CDF of two successive (group) arrivals in the stationary version. Verify that $\mu' = \frac{\boldsymbol{\pi} D(-D_0)^{-1}\boldsymbol{e}}{\boldsymbol{\pi} De} = \frac{1}{\lambda_g}$, $\sigma^2 = \frac{2}{\lambda_g}\boldsymbol{\pi}(-D_0)^{-1}\boldsymbol{e} - \frac{1}{\lambda_g^2}$;

– the joint density of the first n intervals between (group) arrivals is given by

$$f(x_1, \cdots, x_n) = \boldsymbol{\alpha} e^{D_0 x_1} D \cdots e^{D_0 x_n} D\boldsymbol{e}.$$

– the joint LST of the first n intervals between (group) arrivals is given by

$$\Phi(s_1, \cdots, s_n) = \boldsymbol{\alpha}\left[\prod_{i=1}^{n}(s_i I - D_0)^{-1} D\right]\boldsymbol{e}.$$

let $\rho(X_1, X_n)$ denote the $(n-1)$–lag correlation between the first and nth interarrivals. Then

$$\rho(X_1, X_n) = \frac{\lambda_g \boldsymbol{\pi}[(-D_0)^{-1}D]^{n-1}(-D_0)^{-1}\boldsymbol{e} - 1}{2\lambda_g \boldsymbol{\pi}(-D_0)^{-1}\boldsymbol{e} - 1}. \quad [2.3]$$

– the superposition of a finite number of independent $BMAPs$ is also a $BMAP$. Thus, if we have n independent $BMAPs$ with representation $\{D_k(j)\}$ of order $m_j, 1 \leq j \leq n$, then the superposition of these n $BMAPs$ will be a $BMAP$ with representation $\{D_k(1) \oplus D_k(2) \oplus \cdots \oplus D_k(n)\}$ of dimension $m_1 \times m_2 \times \cdots \times m_n$;

– *special cases of* $BMAP$

1) Batch Poisson process with batch size PMF $\{p_k\}$: $m = 1$, $D_0 = -\eta, D_k = \eta p_k, k \geq 1$.

2) Batch PH-renewal process with batch size PMF $\{p_k\}$: $D_0 = T$, $D_k = p_k \boldsymbol{T^0}\boldsymbol{\alpha}, k \geq 1$. Note that the underlying PH distribution has the representation $(\boldsymbol{\alpha}, T)$.

3) Batch Markov modulated Poisson process with batch size PMF $\{p_k^{(r)}\}, 1 \leq r \leq m$: $D_0 = Q - \Delta(\boldsymbol{\lambda})$, $D_k = \Delta(\boldsymbol{\lambda})\Delta(\boldsymbol{p}_k), k \geq 1$, where $\boldsymbol{p}_k = (p_k^{(1)}, \cdots, p_k^{(m)})$, and $\boldsymbol{\lambda} = (\lambda_1, \cdots, \lambda_m)$. The diagonal matrix, $\Delta(\boldsymbol{a})$, is such that the diagonal elements are given by the entries of the vector $\boldsymbol{a}$.

4) Batch Markov switched Poisson process with batch size PMF $\{p_k\}$; $D_0 = -\Delta(\boldsymbol{\lambda}), D_k = \Delta(\boldsymbol{\lambda})\Delta(\boldsymbol{p})P, k \geq 1$, where $P = (p_{ij})$ is a transition probability matrix.

5) PH semi-Markov process with single arrivals:

$$D_0 = \begin{bmatrix} T(1) & & & \\ & T(2) & & \\ & & \ddots & \\ & & & T(r) \end{bmatrix}, \quad D_1 = \begin{bmatrix} p_{11}\boldsymbol{T}^{\mathbf{0}}(1)\boldsymbol{\alpha}(1) & \cdots & p_{1r}\boldsymbol{T}^{\mathbf{0}}(1)\boldsymbol{\alpha}(r) \\ p_{21}\boldsymbol{T}^{\mathbf{0}}(2)\boldsymbol{\alpha}(1) & \cdots & p_{2r}\boldsymbol{T}^{\mathbf{0}}(2)\boldsymbol{\alpha}(r) \\ \vdots & \cdots & \vdots \\ p_{r1}\boldsymbol{T}^{\mathbf{0}}(r)\boldsymbol{\alpha}(1) & \cdots & p_{rr}\boldsymbol{T}^{\mathbf{0}}(1)\boldsymbol{\alpha}(r) \end{bmatrix}.$$

2.3. Generation of MAP processes for numerical purposes

Illustrative numerical examples dealing with $BMAP$ pose an interesting and challenging problem, namely, the choice of the values for the parameter matrices, $\{D_k\}$. Even in the case of MAP dealing with only two square matrices, D_0 and D_1 we see such challenges. While it is easy to choose a probability distribution to match a given mean and a given standard deviation, it is not the case when it comes to choosing a MAP to have a given 1-lag correlation. However, there is an interesting way to generate $MAPs$ with a variety of values for the 1-lag correlation and here we will present that method (Chakravarthy 2010). This will be helpful when we qualitatively study queueing models of interest to bring out the behavior of correlated arrivals.

Let $(\boldsymbol{\alpha}, T)$ of dimension k denote an Erlang distribution with rate λ_1 in each of k phases. Consider an exponential distribution with parameter λ_2. Let p_i and q_i be such that $0 < p_i < 1, q_i = 1 - p_i,\ i = 1, 2$. We now construct a MAP with parameter matrices D_0 and D_1 of order $k + 1$ as follows.

$$D_0 = \begin{pmatrix} T & 0 \\ 0 & -\lambda_2 \end{pmatrix}, \quad D_1 = \begin{pmatrix} p_1\boldsymbol{T}^{\mathbf{0}}\boldsymbol{\alpha} & q_1\boldsymbol{T}^{\mathbf{0}} \\ q_2\lambda_2\boldsymbol{\alpha} & p_2\lambda_2 \end{pmatrix}. \qquad [2.4]$$

By varying k, p_1, p_2, λ_1 and λ_2 we can generate a variety of $MAPs$ with positive and negative (1-lag) correlated arrivals. To get an explicit expression for the 1-lag correlation coefficient, $\rho = \rho(X_1, X_2)$ (see [2.3]), we need to establish a few results. Suppose that $\boldsymbol{\pi} = (\boldsymbol{\pi}_1, \pi_2)$ of dimension $k + 1$ is the steady-state probability vector of $D_0 + D_1$. Using the equations

$$\begin{aligned} &\boldsymbol{\pi}_1 T + p_1\boldsymbol{\pi}_1\boldsymbol{T}^{\mathbf{0}}\boldsymbol{\alpha} + q_2\pi_2\lambda_2\boldsymbol{\alpha} = \mathbf{0}, \\ &\quad q_1\boldsymbol{\pi}_1\boldsymbol{T}^{\mathbf{0}} - q_2\lambda_2\pi_2 = 0, \end{aligned} \qquad [2.5]$$

along with $\boldsymbol{\pi}_1\boldsymbol{e} + \pi_2 = 1$, it is easy to verify that

$$\boldsymbol{\pi} = \frac{q_1\lambda_1}{q_1\lambda_1 + kq_2\lambda_2}\left(\frac{q_2\lambda_2}{q_1}\boldsymbol{\alpha}(-T)^{-1}, 1\right). \qquad [2.6]$$

Noting that

$$\alpha(-T)^{-1}e = \frac{k}{\lambda_1}, \ \pi(-D_0)^{-1}e = \frac{q_1\lambda_1}{q_1\lambda_1 + kq_2\lambda_2}\left[\frac{q_2k(k+1)\lambda_2}{2q_1\lambda_1^2} + \frac{1}{\lambda_2}\right],$$

$$\alpha(-T)^{-2}e = \frac{k(k+1)}{2\lambda_1^2}, \qquad [2.7]$$

$$\pi(-D_0)^{-1}D_1(-D_0)^{-1}e = \frac{q_1\lambda_1}{q_1\lambda_1 + kq_2\lambda_2}\left[\frac{p_1q_2k^2\lambda_2}{q_1\lambda_1^2} + \frac{2kq_2}{\lambda_1} + \frac{p_2}{\lambda_2}\right],$$

it can readily be seen that

$$\lambda = \frac{\lambda_1\lambda_2(q_1+q_2)}{q_1\lambda_1 + kq_2\lambda_2}, \ \sigma^2 = \frac{\lambda_1^2q_1(q_1+2q_2) - 2\lambda_1\lambda_2kq_1q_2 + k\lambda_2^2q_2[(k+1)q_1+q_2]}{\lambda_1^2\lambda_2^2(q_1+q_2)^2},$$

$$\rho = \frac{q_1q_2(1-q_1-q_2)(\lambda_1-k\lambda_2)^2}{q_1\lambda_1^2(q_1+2q_2) - 2kq_1q_2\lambda_1\lambda_2 + kq_2\lambda_2^2[q_2+(k+1)q_1]}. \qquad [2.8]$$

Fixing $p_1 = p_2 = p$, we see that λ is independent of p and we can get $\rho = \rho(p, k, \lambda_1, \lambda_2)$ in a compact form as

$$\lambda = \frac{2\lambda_1\lambda_2}{\lambda_1 + k\lambda_2}, \ \sigma^2 = \frac{3\lambda_1^2 - 2k\lambda_1\lambda_2 + k(k+2)\lambda_2^2}{4\lambda_1^2\lambda_2^2},$$

$$\rho = \frac{(2p-1)(\lambda_1-k\lambda_2)^2}{3\lambda_1^2 - 2k\lambda_1\lambda_2 + k(k+2)\lambda_2^2}. \qquad [2.9]$$

Fixing $p_1 = p_2 = p$ and $\lambda_2 = r\lambda_1$, we get

$$\lambda = \frac{2r\lambda_1}{1+rk}, \ \sigma^2 = \frac{k^2r^2 + 2(r-1)rk + 3}{4\lambda_1^2r^2}, \ \rho(k,p,r) = \frac{(rk-1)^2(2p-1)}{3 + 2(r-1)rk + r^2k^2}.$$

In Figure 2.1, we display the three-dimensional graphs of $\rho = \rho(k, p, r)$ under various scenarios. Obviously, for smaller values of r, namely when $0 < r < 1$, the values of ρ are restricted between -0.3 and 0.3; however, for $r > 1$, we have a much wider coverage and choice for ρ.

We conclude this section with Table 2.1 that lists a few $MAPs$ that we will use as arrival processes in analyzing various queueing models. Note that these $MAPs$ have interesting properties that they have the same mean (which is 1 for all $MAPs$) and standard deviation (depending on k but the same for both NC_k and PC_k) but a different 1-lag correlation.

k	TaP	$p_1 = p_2$	ρ	σ	TaP	$p_1 = p_2$	ρ	σ
2	NC_1	0.01	-0.3267	1.0392	PC_1	0.99	0.3267	1.0392
3	NC_2	0.01	-0.4804	1.0202	PC_2	0.99	0.4804	1.0202
4	NC_3	0.01	-0.5786	1.0123	PC_3	0.99	0.5786	1.0123
5	NC_4	0.01	-0.6454	1.0082	PC_4	0.99	0.6454	1.0082
6	NC_5	0.01	-0.6935	1.0059	PC_5	0.99	0.6935	1.0059
7	NC_6	0.01	-0.7296	1.0044	PC_6	0.99	0.7296	1.0044
8	NC_7	0.01	-0.7577	1.0035	PC_7	0.99	0.7577	1.0035
9	NC_8	0.01	-0.7802	1.0028	PC_8	0.99	0.7802	1.0028

Table 2.1. *Selected* $MAPs$

2.4. Analysis of selected queueing models of $BMAP/G/c$ type

In this section, we will discuss a few queueing models of $BMAP/G/c$ type through simulation. First, we need to validate the simulated models. In the end, we will discuss analytically two queueing models. This will also make this chapter self contained. However, for more in-depth details we refer readers to the references cited earlier.

2.4.1. $MAP/PH/1$ *queueing model*

Here, we consider a single server queueing model in which the customers arrive according to a MAP with representation (D_0, D_1) of order m. We assume that the matrix $D = D_0 + D_1$ is an irreducible matrix and that $\boldsymbol{\delta}$ is the invariant vector of D. That is,

$$\boldsymbol{\delta} D = \mathbf{0}, \quad \boldsymbol{\delta} \boldsymbol{e} = \mathbf{1}. \qquad [2.10]$$

Recall that the arrival rate, λ, is given by $\lambda = \boldsymbol{\delta} D_1 \boldsymbol{e}$. The single server offers services to customers on a first-come-first-served ($FCFS$) basis and the service times are assumed to follow a PH distribution $(\boldsymbol{\beta}, S)$ of order n. The mean service time is given by $\mu^{-1} = \boldsymbol{\beta}(-S)^{-1}\boldsymbol{e}$. Let $\boldsymbol{S^0}$ be such that $S\boldsymbol{e} + \boldsymbol{S^0} = \mathbf{0}$.

By keeping track of the number of customers, say, $N(t)$, in the system, the phase of the arrival process, say, $J_1(t)$, and the phase of the service process, say, $J_2(t)$ (when the server is busy) at time t, the model can be studied as a quasi-birth-and-death (QBD) process with the state space given by

$$\begin{aligned}\Omega &= \{(0,k), 1 \le k \le m\} \bigcup \{(i,j,k) : i \ge 1, 1 \le j \le n, 1 \le k \le m\} \\ &= \{\underline{\mathbf{0}}\} \bigcup \{\underline{\mathbf{i}}, i \ge 1\}.\end{aligned}$$

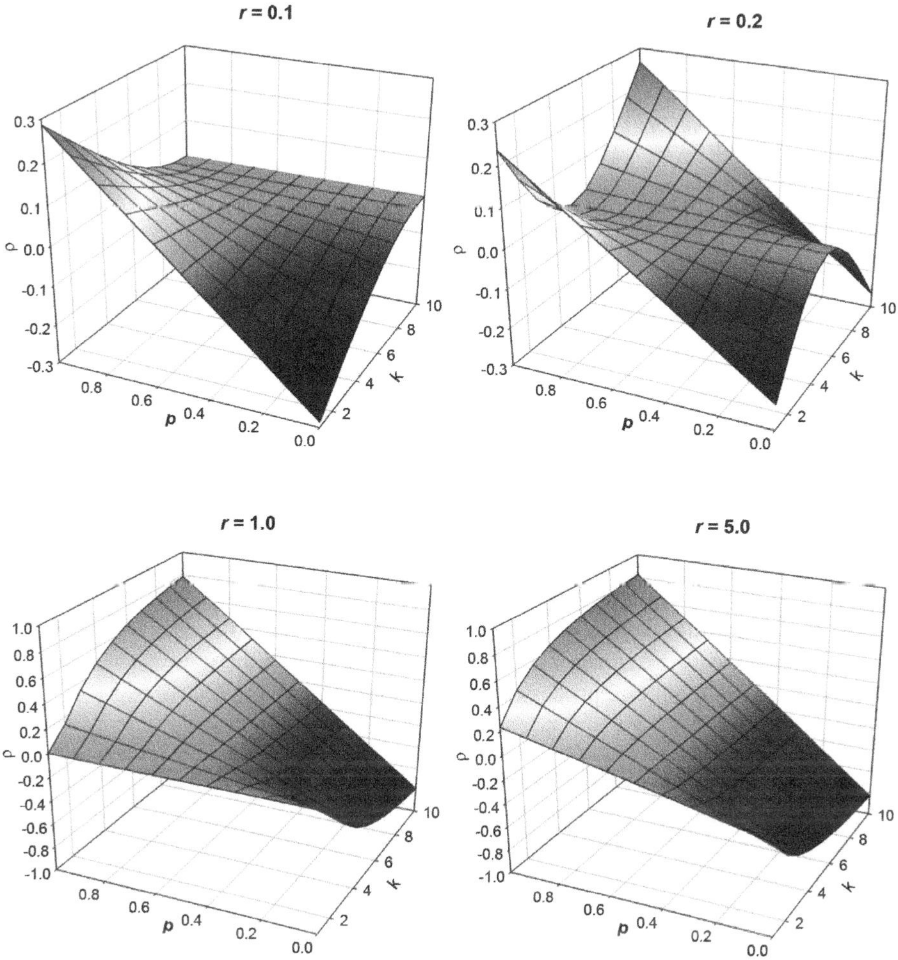

Figure 2.1. *One-lag correlation coefficient of $MAPs$ under different scenarios. For a color version of this figure, see www.iste.co.uk/anisimov/queueing2.zip*

The level, $\underline{\mathbf{0}}$, of dimension m corresponds to the case when the server is idle and the arrival process is in one of m phases. The set of states, $\underline{\boldsymbol{i}}$, of dimension mn corresponds to the case when the server is busy serving in phase j, the arrival process is in phase k and there are $i - 1$ customers waiting in the queue.

The infinitesimal generator of the Markov chain governing the QBD process is given by

$$Q = \begin{pmatrix} D_0 & \boldsymbol{\beta} \otimes D_1 & & & \\ \boldsymbol{S^0} \otimes I & A_1 & A_0 & & \\ & A_2 & A_1 & A_0 & \\ & & A_2 & A_1 & A_0 \\ & & & \ddots & \ddots & \ddots \end{pmatrix}, \quad [2.11]$$

where

$$A_0 = I \otimes D_1, \;\; A_1 = S \oplus D_0, \;\; A_2 = \boldsymbol{S^0}\boldsymbol{\beta} \otimes I. \quad [2.12]$$

Under the stability condition that $\lambda < \mu$ (which can be established through Neuts' results on matrix-geometric solutions), the steady-state probability vector, $\boldsymbol{x}$, of Q is of matrix-geometric type. Suppose that $\boldsymbol{x}$ is partitioned as $\boldsymbol{x} = (\boldsymbol{x}_0, \boldsymbol{x}_1, \boldsymbol{x}_2, \cdots)$ such that the vector $\boldsymbol{x}_0$ of dimension m gives the steady-state probability vector that the system is empty with the arrival process in one of m phases; the vector $\boldsymbol{x}_i$ of dimension mn is further partitioned into vectors of dimension m as $\boldsymbol{x}_i = (\boldsymbol{x}_{i1}, \cdots, \boldsymbol{x}_{in})$ so that the vector $\boldsymbol{x}_{ij}$ gives the steady-state probability vector that there are i customers in the system with the service process in phase j and the arrival process is in one of m phases.

Using Neuts' results on the matrix-geometric method (Neuts 1981), it is known that the steady-state probability vector $\boldsymbol{x}$ is given by

$$\boldsymbol{x}_i = \boldsymbol{x}_1 R^{i-1}, \;\; i \geq 1, \quad [2.13]$$

where the rate matrix R is the minimal non-negative solution to the matrix-quadratic equation given by

$$R^2 A_2 + R A_1 + A_0 = 0, \quad [2.14]$$

and the vectors $\boldsymbol{x}_0$ and $\boldsymbol{x}_1$ are obtained by solving

$$\begin{aligned} &\boldsymbol{x}_0 D_0 + \boldsymbol{x}_1(\boldsymbol{S^0} \otimes I) = \mathbf{0}, \\ &\boldsymbol{x}_0(\boldsymbol{\beta} \otimes D_1) + \boldsymbol{x}_1(A_1 + RA_2) = \mathbf{0}, \end{aligned} \quad [2.15]$$

subject to the normalizing condition

$$\boldsymbol{x}_0 \boldsymbol{e} + \boldsymbol{x}_1 (I - R)^{-1} \boldsymbol{e} = 1. \quad [2.16]$$

A few comments about the rate matrix R. The (j,k)th entry of R matrix gives the mean time spent in state $(i+1,k)$ before the first return to level i, given that the process starts in state (i,j) expressed in the time unit measured in terms of the mean time spent in state (i,j). This probabilistic interpretation enables us to identify the special structure, if any, present in a given model and can be exploited in numerical computation, especially when dealing with large values for m and n.

The computation of the rate matrix can be accomplished in many ways such as (block) Gauss–Siedel iteration by exploiting the special structure of the coefficient matrices A_0, A_1 and A_2, or through a more efficient method, namely, logarithmic reduction, of Latouche and Ramaswami (Latouche and Ramaswami 1999). We will briefly outline the logarithmic reduction the algorithm below.

Logarithmic reduction algorithm: this method, due to Latouche, is a quadratically convergent algorithm for computing both R and G matrices.

Step 0: $H \leftarrow (I - A_1)^{-1}A_0$, $L \leftarrow (I - A_1)^{-1}A_2$, $G = L$ and $T = H$.

Step 1:

$$\begin{array}{l} U = HL + LH, \quad M = H^2, \quad H \leftarrow (I - U)^{-1}M, \\ M \leftarrow L^2, \quad \leftarrow (I - U)^{-1}M, \quad G \leftarrow G + TL, \quad T \leftarrow TH. \end{array}$$

Continue Step 1 until $||\boldsymbol{e} - G\boldsymbol{e}||_\infty < \epsilon$.

Step 2: $R = A_0(I - A_1 - A_0G)^{-1}$.

The stationary distribution of the sojourn time of an arriving customer can be obtained using the technique of (Ozawa 2006). Toward this end, we need to find the steady-state probability vector at arrivals. Let $\boldsymbol{z} = (\boldsymbol{z}_1, \boldsymbol{z}_2, \cdots)$ be such that the vector $\boldsymbol{z}_i$ of dimension mn gives the steady-state probability that soon after an arrival there are i customers in the system with the arrival and the service are, respectively, in one of m and n phases. It is easy to verify that

$$\begin{array}{l} \boldsymbol{z}_1 = \frac{1}{\lambda}\boldsymbol{x}_0(\boldsymbol{\beta} \otimes D_1), \\ \boldsymbol{z}_{i+1} = \frac{1}{\lambda}\boldsymbol{x}_1 R^{i-1}(I \otimes D_1) = \frac{1}{\lambda}\boldsymbol{x}_1(I \otimes D_1)\hat{R}^{i-1}, \; i > 1, \end{array} \quad [2.17]$$

where

$$\hat{R} = -(A_0 + RA_2)^{-1}A_0. \quad [2.18]$$

Defining $\boldsymbol{\eta} = \sum_{i=1}^{\infty} \boldsymbol{z}_i$, it can be verified that

$$\boldsymbol{\eta} = \frac{1}{\lambda}\boldsymbol{x}_0(\boldsymbol{\beta} \otimes D_1)(I - \hat{R})^{-1}. \quad [2.19]$$

Let Y denote the sojourn time in the system of an arriving customer. Then applying Ozawa's technique for the classical $MAP/PH/1$ queueing model we get the following expression for the tail probabilities of Y.

$$P(Y > t) = (\mathbf{e}' \otimes \boldsymbol{\eta})exp\{[(A_0 + A_1)' \otimes I + A_2' \otimes \hat{R}]t\}\boldsymbol{a}, \quad [2.20]$$

where $\boldsymbol{a}$, a column vector of dimension $(mn)^2$, comprises unit vectors given by

$$\boldsymbol{a} = \begin{bmatrix} \mathbf{e}_1(mn) \\ \mathbf{e}_2(mn) \\ \vdots \\ \mathbf{e}_{mn}(mn) \end{bmatrix}. \quad [2.21]$$

NOTE.– The mean, μ_{ST}, time spent in the system by an arriving customer can be obtained from [2.20] as

$$\mu_{ST} = (\mathbf{e}' \otimes \boldsymbol{\eta})\left[-\left((A_0 + A_1)' \otimes I + A_2' \otimes \hat{R}\right)^{-1}\right]\boldsymbol{a}. \quad [2.22]$$

Of course, instead of using the above expression for the mean sojourn time in the system, we can appeal to Little's law to obtain this measure. However, this will also serve as an (internal) accuracy check in numerical computation.

2.4.2. *The system performance measures*

The qualitative analysis of any queueing model requires one or more system performance measures. While we could develop a number of such measures, here we will detail a few and leave the rest to the interested readers.

1) *Probability that the system is idle*: the probability, P_{idle}, that the system is idle (i.e., there are no customers in the system) is given by $P_{idle} = \boldsymbol{x}_0\mathbf{e} = 1 - \frac{\lambda}{\mu}$.

2) *Mean queue length*: the mean, μ_q, queue length is given by $\mu_q = \boldsymbol{x}_1 R(I-R)^{-2}\boldsymbol{e}$.

3) *Mean number in system*: the mean, μ_s, number of customers in the is given by $\mu_s = \boldsymbol{x}_1(I-R)^{-2}\boldsymbol{e}$.

4) *P(an arrival has to wait)*: the probability, P_{wait}, that an arriving customer has to wait to receive service is given by $P_{wait} = \frac{1}{\lambda}\boldsymbol{x}_1(I-R)^{-1}D_1\boldsymbol{e}$.

2.4.3. *Illustrative numerical examples for* $MAP/PH/1$

We will now discuss a few illustrative numerical examples. Toward this end, we consider a number of arrival processes and service times in order to bring out the qualitative aspects of the models.

Arrival processes: for arrival process, we consider one or more of the $MAPs$ listed in Table 2.1. Further, we also consider the following three (renewal) arrivals.

Exponential (ERA):

$$D_0 = \begin{pmatrix} -2 & 2 \\ 0 & -2 \end{pmatrix}, \quad D_1 = \begin{pmatrix} 0 & 0 \\ 2 & 0 \end{pmatrix}$$

Exponential (EXA):

$$D_0 = (-1), \quad D_1 = (1)$$

Hyperexponential (HEA):

$$D_0 = \begin{pmatrix} -8.2 & 0 & 0 \\ 0 & -0.82 & 0 \\ 0 & 0 & -0.082 \end{pmatrix}, \quad D_1 = \begin{pmatrix} 5.74 & 2.05 & 0.41 \\ 0.574 & 0.205 & 0.041 \\ 0.0574 & 0.0205 & 0.0041 \end{pmatrix}$$

All of these arrival processes will be normalized so as to arrive at a specific value for λ.

Service Times: For the service times, we look at the following three. These distributions will be normalized so as to arrive at a desired value for μ.

Erlang (ERS):

$$\beta = (1, 0), \quad S = \begin{pmatrix} -2 & 2 \\ 0 & -2 \end{pmatrix}.$$

Exponential (EXS):

$$\beta = (1), \quad S = \begin{pmatrix} -1 \end{pmatrix}.$$

Hyperexponential (HES):

$$\beta = (0.9, 0.1), \quad S = \begin{pmatrix} -10 & 0 \\ 0 & -1 \end{pmatrix}.$$

The measures, $ln(\mu_q)$ and P_{wait}, are displayed in Figure 2.2, and the tail probabilities, $P(Y > t)$, are displayed in Figure 2.3 under various scenarios. In Figure 2.4, the distribution functions of the sojourn time in the system under a variety of scenarios are given. The label "Ren" corresponds (in that order) to ERA, EXA and HEA. For "Neg correlation", the order is $NC_8, \cdots, NC_1$ and for "Pos correlation", the order is $PC_1, \cdots, PC_8$. This is to show the effect of changing the correlation from -0.7802 to 0.7802 on the selected performance measures. It should be pointed out that for each $i, 1 \leq i \leq 8$, both NC_i and PC_i have the same standard deviation but 1-lag correlation differs only in its sign.

First note that the arrival processes and the service times considered are qualitatively different. While the correlation values for the correlated arrivals are given in Table 2.1, the ratios of the standard deviation of the arrival processes $EXA, HEA, NC_1, PC_1, \cdots, NC_8, PC_8$ with respect to ERA are, respectively, 1.4142, 5.4106, 1.4697, 1.4697, 1.4428, 1.4428, 1.4316, 1.4316, 1.4258, 1.4258, 1.4226, 1.4226, 1.4204, 1.4204, 1.4192, 1.4192, 1.4182 and 1.4182. The ratios of the standard deviations of EXS and HES with respect to ERS are, respectively, 1.4140 and 3.1742.

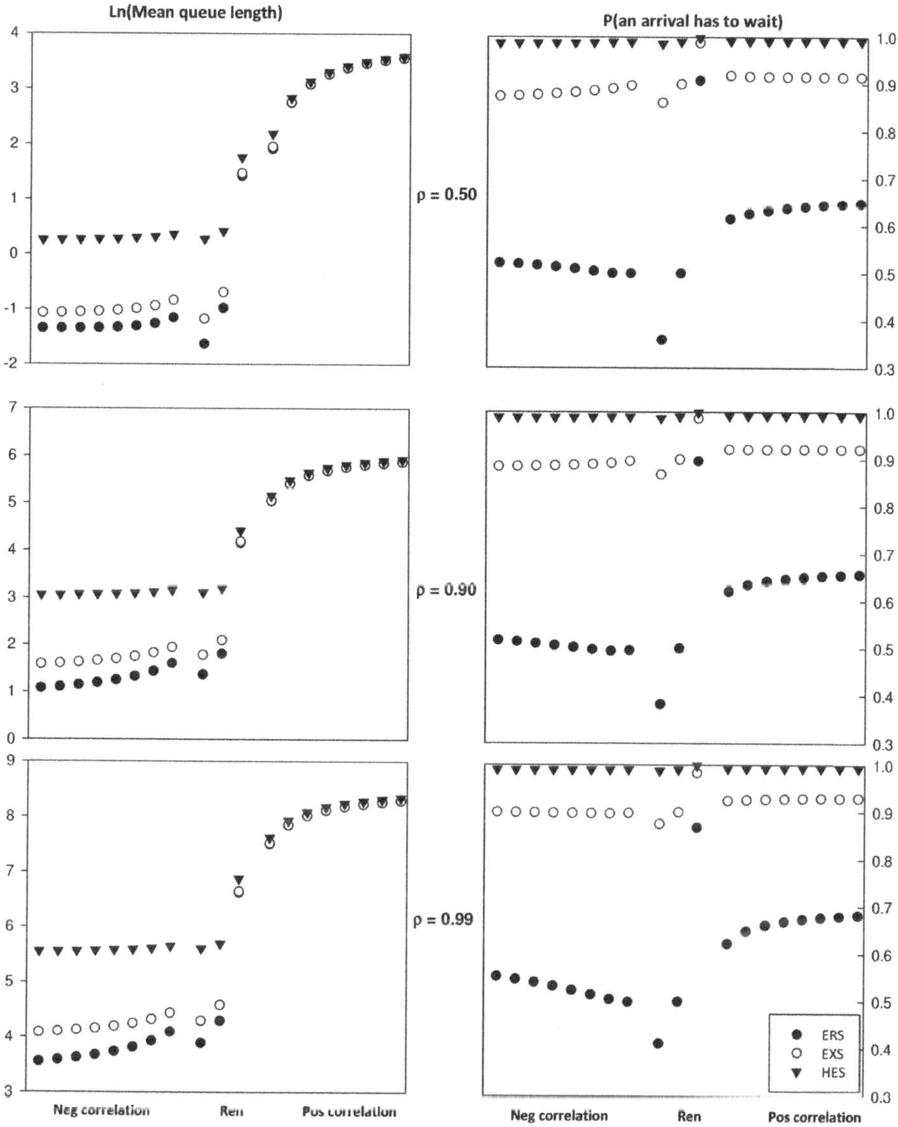

Figure 2.2. *Selected measures for a $MAP/PH/1$ queue under different scenarios*

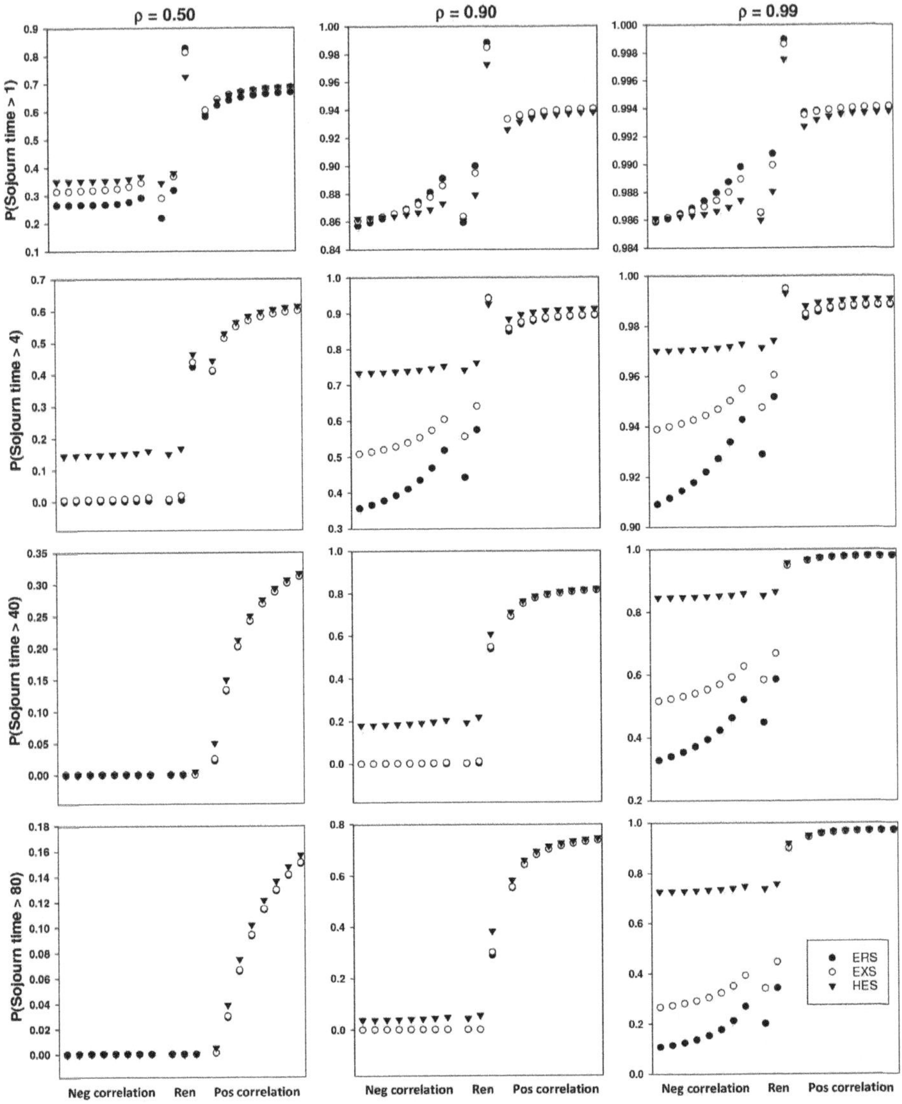

Figure 2.3. *Tail probabilities for the sojourn time for a $MAP/PH/1$ queue under different scenarios*

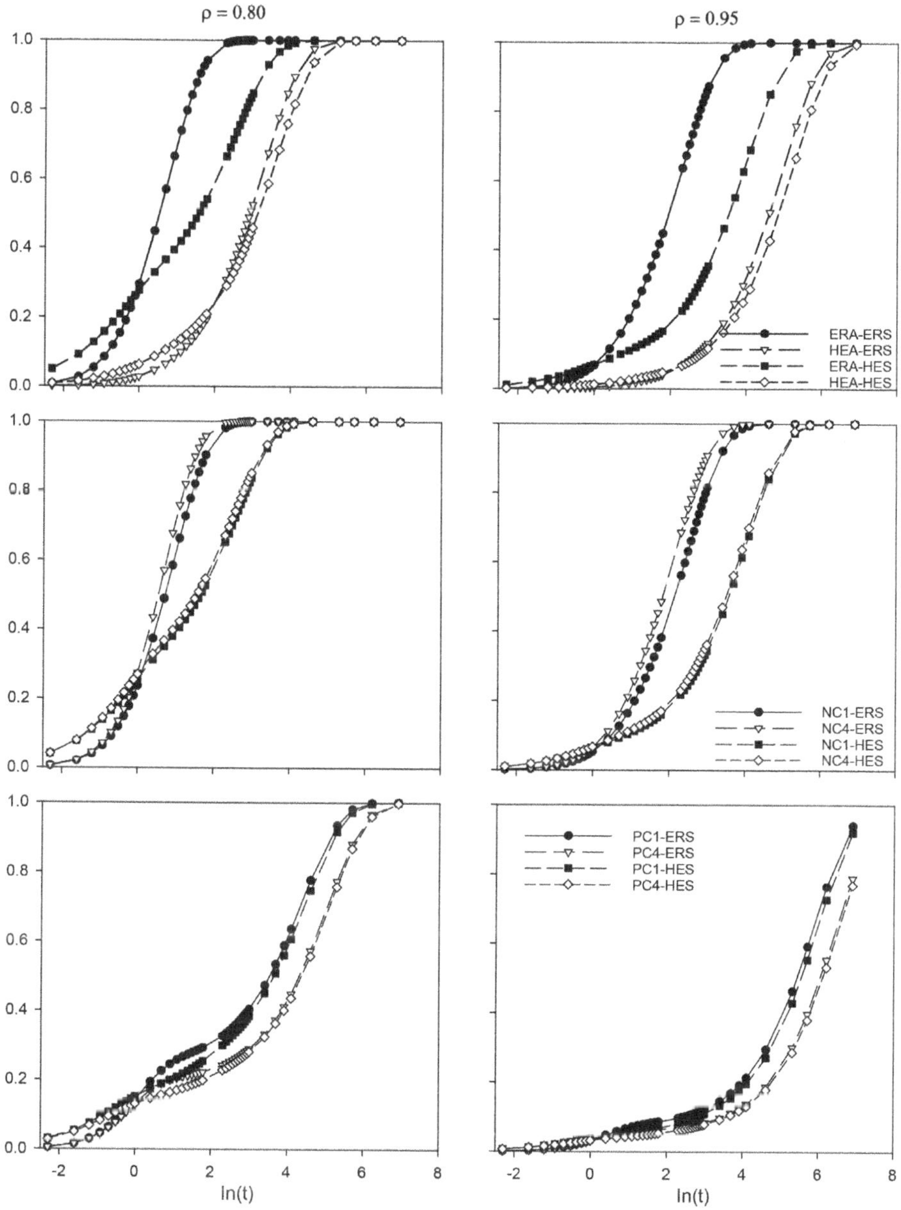

Figure 2.4. *The distribution function of the sojourn time for a $MAP/PH/1$ queue under different scenarios*

Figures 2.2 through 2.4 reveal the following observations:

– in the case of renewal arrivals, we see that the mean queue length appears to increase with increasing variability in the services;

– in the case of negatively correlated arrivals, we notice a significant impact of the service time distribution on μ_q. Further, as the (negative) correlation increases (from a high negative value to zero), μ_q appears to increase only in the case of Erlang and exponential services, whereas for hyperexponential services there appears to be no significant change in the mean queue length;

– in the case of positively correlated arrivals, we do not see a significant impact of the type of service times on μ_q. However, μ_q increases as the (positive) correlation increases;

– all of the above observations appear to be true for the three values of ρ considered;

– now we look at the measure, P_{wait}. For all arrival processes, the type of services considered has a significant impact on this measure. Further, for all arrival processes, this measure appears to increase with increasing variability in the services. It is also easy to see that a higher variability in the services causes a higher probability for a customer to wait before getting into service;

– the measure, P_{wait}, appears to increase as the magnitude of the correlation increases in the case of Erlang services. That is, whether the arrivals are negatively or positively correlated, a higher correlation in absolute value causes a higher value for this probability when services are Erlang. Also, the rate of increase appears to be higher as the traffic intensity becomes larger. However, for the other two services this trend is not seen. In fact, we notice a somewhat insignificant effect as the correlation is changed;

– with regard to the sojourn time of a customer, we notice that the positively correlated arrivals tend to have heavy tails;

– it appears that a higher variability in services leads to a heavy tail in the distribution function of the sojourn time compared to services with less variability;

– suppose we denote by $F(t; a, b)$ the distribution function of the sojourn time of Y when the arrival process is a and the service time distribution is given by b. Below, we will summarize some key observations on $F(t; a, b)$ under a variety of scenarios. To the best of our knowledge, some of these observations have not been reported or available in the literature.

1) $F(t; ERA, ERS) \leq F(t; HEA, ERS)$, $\forall\ t \geq 0$, as well as $F(t; ERA, HES) \leq F(t; HEA, HES)$, $\forall\ t \geq 0$. Thus, we notice the stochastic dominance of the first order for Erlang arrivals compared to hyperexponential arrivals. This is the case irrespective of the type of services used.

2) There exists, say, a $t_0 > 0$, such that for $t \leq t_0, F(t; ERA, ERS) \leq F(t; ERA, HES)$, and $F(t; ERA, HES) < F(t; ERA, ERS)$, $t > t_0$. Similarly, there exists, say, a $t_1 > 0$, such that for $t \leq t_1$, $F(t; HEA, ERS) \leq F(t; HEA, HES)$, and for $t > t_1$, we have $F(t; HEA, HES) < F(t; HEA, ERS)$.

3) Looking at the correlated arrivals, we also notice a stochastic dominance. This dominance has an interesting trend. Specifically in the case of negatively correlated arrivals, for example, we notice that $F(t; NC_4, ERS) \leq F(t; NC_1, ERS)$, $\forall\, t \geq 0$, as well as $F(t; NC_4, HES) \leq F(t; NC_1, HES)$, $\forall\, t \geq 0$. In the case of positively correlated arrivals, we have $F(t; PC_4, ERS) \leq F(t; PC_1, ERS)$ as well as $F(t; PC_4, HES) \leq F(t; PC_1, HES)$, for all $t \geq 0$.

4) Similar to the renewal case, we have the distribution functions cross in the correlated arrivals too. There exists, say, a $t_0 > 0$, such that $F(t; NC_1, ERS) \leq F(t; NC_1, HES)$, $t \leq t_0$ and $F(t; NC_1, HES) < F(t; NC_1, HES)$, $t > t_0$. Similarly, there exists, say, a $t_1 > 0$, such that $F(t; PC_1, ERS) \leq F(t; PC_1, HES)$, $t \leq t_1$ and $F(t; PC_1, HES) < F(t; PC_1, HES)$, $t > t_1$. It is worth pointing out that we noticed this type of behavior in other types of (positively and negatively) correlated arrivals.

5) Finally, when we compare negatively and positively correlated arrivals, we notice that for the positively correlated arrivals the distribution function of the sojourn time appears to be less sensitive to the type of services used. The insensitivity appears to increase as the traffic load becomes higher.

2.4.4. $MAP/M/c$ queueing model

Here, we consider a $c-$server queueing model in which the customers arrive according to a MAP with representation (D_0, D_1) of order m. We assume that the matrix $D = D_0 + D_1$ is an irreducible matrix and that δ is the invariant vector of D satisfying equation [2.10]. Recall that the arrival rate, λ, is given by $\lambda = \boldsymbol{\delta} D_1 \boldsymbol{e}$. The customers are served on an $FCFS$ basis and the service times are assumed to be exponential with parameter μ.

Suppose that $N(t)$ and $J(t)$ denote, respectively, the number of customers and the phase of the arrival process at time t. Suppose that the level, $\underline{\boldsymbol{i}}$, $i \geq 0$, of dimension m corresponds to the case when the i customers are in the system with at most c servers busy serving customers. The state space of the queueing system, $MAP/M/c$, is given by

$$\Omega = \{(i, j) : i \geq 0, 1 \leq j \leq m\} = \{\underline{\boldsymbol{i}},\ i \geq 0\}.$$

The infinitesimal generator of the Markov chain governing the $QBD-$process for $MAP/M/c$ queueing model is given by

$$Q = \begin{pmatrix} D_0 & D_1 & & & & & & \\ \mu I & D_0 - \mu I & D_1 & & & & & \\ & 2\mu I & D_0 - 2\mu I & D_1 & & & & \\ & & \ddots & \ddots & \ddots & & & \\ & & & (c-1)\mu I & D_0 - (c-1)\mu I & D_1 & & \\ & & & & c\mu I & D_0 - c\mu I & D_1 & \\ & & & & & c\mu I & D_0 - c\mu I & D_1 \\ & & & & & \ddots & \ddots & \ddots \end{pmatrix}. \quad [2.23]$$

Under the stability condition that $\lambda < c\mu$ (which can be established through Neuts' results on the matrix-geometric solution), the steady-state probability vector, $\boldsymbol{x}$, of Q is of matrix-geometric type. Note that $\boldsymbol{x}_i$ of dimension m gives the steady-state probability vector that there are i customers in the system and the arrival process in one of m phases. Using Neuts' results on the matrix-geometric method (Neuts 1981), it is known that the steady-state probability vector $\boldsymbol{x}$ is given by

$$\boldsymbol{x}_{c-1+i} = \boldsymbol{x}_{c-1} R^i, \ \ i \geq 0, \quad [2.24]$$

where the rate matrix R is the minimal non-negative solution to the matrix-quadratic equation given by

$$c\mu R^2 + R(D_0 - c\mu I) + D_1 = 0, \quad [2.25]$$

and the vectors $\boldsymbol{x}_i$, $0 \leq i \leq c$ are obtained by solving

$$\begin{aligned} &\boldsymbol{x}_0 D_0 + \mu \boldsymbol{x}_1 = \mathbf{0}, \\ &\boldsymbol{x}_{i-1} D_1 + \boldsymbol{x}_i (D_0 - i\mu I) + (i+1)\mu \boldsymbol{x}_{i+1} = \mathbf{0}, \ 1 \leq i \leq c-2, \\ &\boldsymbol{x}_{c-2} D_1 + \boldsymbol{x}_{c-1}[D_0 - c\mu I + c\mu R] = \mathbf{0}, \end{aligned} \quad [2.26]$$

subject to the normalizing condition: $\sum_{i=0}^{c-2} \boldsymbol{x}_i \boldsymbol{e} + \boldsymbol{x}_{c-1}(I - R)^{-1}\boldsymbol{e} = 1$.

2.4.5. *The system performance measures*

The qualitative analysis of any queueing model requires one or more system performance measures. While we could develop a number of such measures, here we will detail a few and leave the rest to the interested readers.

1) *Probability that the system is idle*: the probability, P_{idle}, that the system is idle (i.e., there are no customers in the system) is given by $P_{idle} = \boldsymbol{x}_0\mathbf{e} = 1 - \frac{\lambda}{\mu}$.

2) *Mean queue length*: the mean, μ_q, queue length is given by $\mu_q = \boldsymbol{x}_{c-1}R^2(I - R)^{-2}\mathbf{e}$.

3) *P(an arrival has to wait)*: the probability, P_{wait}, that an arriving customer has to wait to receive service is given by $P_{wait} = \frac{1}{\lambda}\boldsymbol{x}_{c-1}R(I - R)^{-1}D_1\mathbf{e}$.

4) *Mean number in the system*: the mean, μ_s, number of customers in the system is given by $\mu_s = \boldsymbol{x}_{c-1}R^2(I - R)^{-2}\mathbf{e} + \frac{\lambda}{\mu}$.

2.4.6. *Illustrative numerical examples for* $MAP/M/c$

Here, we will discuss a few examples using the same set of arrival processes listed in section 2.4.3. When we vary c we modify the value of μ so that we get a given value for ρ. Similarly, when we vary ρ, we find the value of μ for a fixed c so as to arrive at the given value of ρ. In Figure 2.5, we display the two measures, $ln(\mu_q)$ and P_{wait}. Figure 2.5 reveals the following observations:

– With regard to the measure, $ln(\mu_q)$, we notice that for a large value of ρ, the effect of increasing c (recall that as c is increased, μ is decreased to arrive at a given ρ) is not seen in all arrival processes. However, for small to reasonably large ρ, we notice the effect of changing c in all but positively correlated arrivals and hyperexponential arrivals. This indicates the significant role played by the high variability as well as positively correlated arrivals.

– Looking at P_{wait} we notice that while $c = 1$ produces a largest value, $c = 10$ produces the smallest value. This is the case for all types of arrival processes. However, for positively correlated arrivals the differences (between $c = 1$ and $c = 10$ cases) are relatively small compared to other arrival processes.

– As c increases, the sensitivity to the magnitude of the (positive or negative) correlation with regard to P_{wait} can be seen.

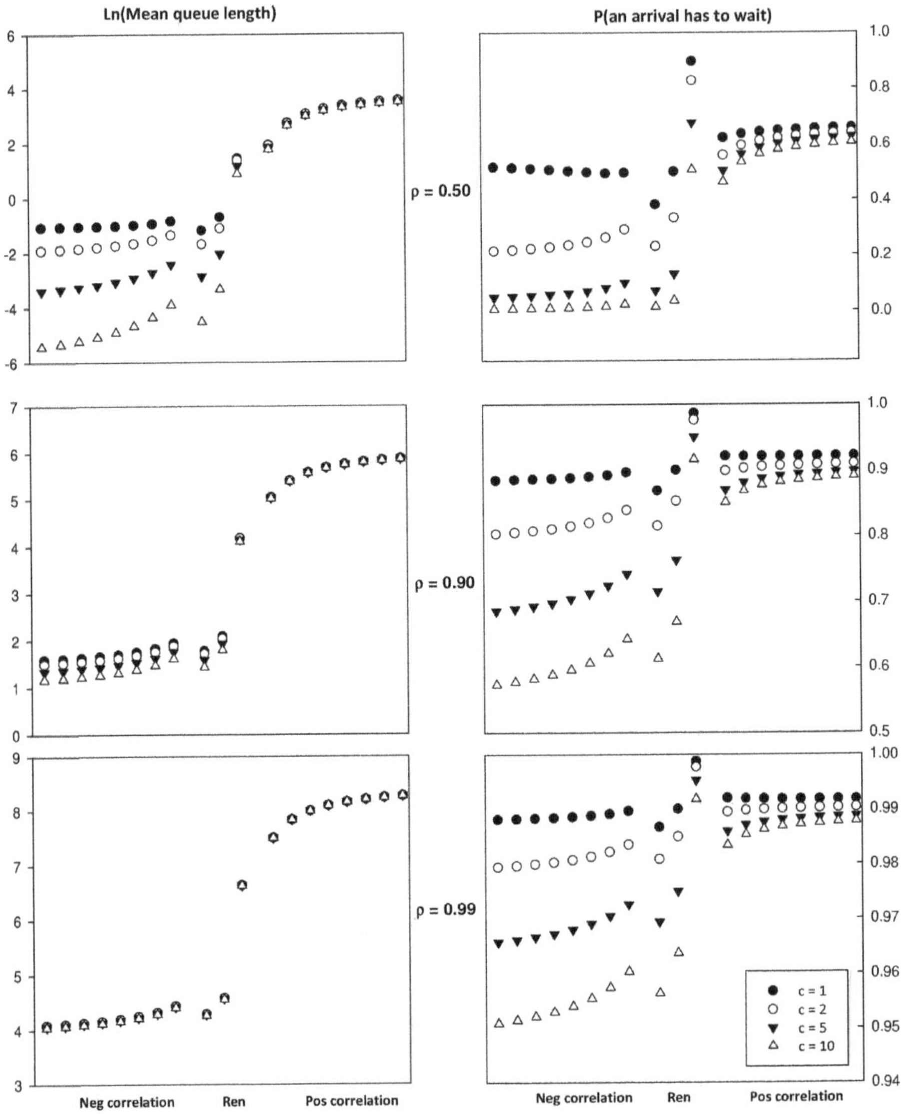

Figure 2.5. *Selected measures for a $MAP/M/c$ queue under different scenarios*

2.5. Simulated models of $BMAP/G/c$ type queues

In this section, we will present the analysis of selected queueing models using simulation. We used ARENA, a powerful simulation software, to simulate the queueing models presented here. We used 3 to 10 replicates and each replicate is taken to be of length 1,000,000 or 10,000,000 units (in minutes) depending on the type of distribution. For example, when dealing with hyperexponential services and or positively correlated arrivals, a longer replication length and a higher number of replications are required to reduce the error percentages.

2.5.1. *Simulated model validation using* $MAP/M/c$ *type queues*

Here, we validate our simulated models using $MAP/M/c$ type queues for which analytical results are available (see section 2.4.4). In Table 2.2, we display the error percentages for the two measures μ_q and P_{wait}. The error percentages for the measure P_{wait} are displayed within parentheses.

	$\rho = 0.80$			$\rho = 0.95$		
ToA	$c = 1$	$c = 2$	$c = 5$	$c = 1$	$c = 2$	$c = 5$
ERA	0.48 (0.09)	0.24 (0.01)	0.12 (0.01)	2.71 (0.20)	2.94 (0.16)	1.61 (0.14)
EXA	0.08 (0.11)	1.03 (0.31)	3.52 (0.54)	4.46 (0.17)	4.32 (0.02)	0.63 (0.41)
HEA	0.38 (0.01)	3.32 (0.09)	0.55 (0.02)	1.10 (0.01)	2.90 (0.08)	2.17 (0.05)
NC_1	0.11 (0.00)	0.23 (0.03)	0.28 (0.05)	4.65 (0.01)	4.75 (0.00)	4.96 (0.03)
NC_4	0.07 (0.19)	0.30 (1.08)	0.41 (2.08)	3.39 (0.02)	3.53 (0.01)	3.83 (0.01)
NC_8	0.02 (0.04)	0.22 (0.06)	0.35 (0.03)	2.98 (0.03)	3.12 (0.02)	3.42 (0.02)
PC_1	8.57 (0.58)	8.58 (0.79)	8.65 (1.07)	9.20 (0.22)	9.20 (0.30)	9.24 (0.41)
PC_4	6.82 (0.41)	6.83 (0.69)	6.87 (1.01)	5.79 (0.29)	5.79 (0.36)	5.81 (0.43)
PC_8	6.80 (0.37)	6.81 (0.42)	6.85 (0.44)	6.18 (0.34)	6.18 (0.39)	6.20 (0.45)

Table 2.2. *Error percentage for a* $MAP/M/c$ *queue under selected scenarios*

It is clear from the entries in Table 2.2 that the simulated results agree very much with those corresponding to the analytical models.

2.5.2. *Simulated model validation using* $MAP/PH/1$ *type queues*

Here, we validate our simulated models using $MAP/PH/1$ type queues for which analytical results are available (see section 2.4.1).

Again, from the entries in Table 2.3, it is clear that the simulated results agree very much with the corresponding analytical results.

2.5.3. *Selected simulated models of* $BMAP/G/c$ *type queues*

In this section, we will illustrate a few examples (among many) simulated. In addition to the arrival processes and the services mentioned in section 2.4.3 we also consider two Weibull distributions for services and one Weibull for arrivals. These are mentioned below.

		$\rho = 0.80$					$\rho = 0.95$				
ToS	TaP	μ_Q	P_{wait}	$P_{ST}^{(1)}$	$P_{ST}^{(4)}$	$P_{ST}^{(7)}$	μ_Q	P_{wait}	$P_{ST}^{(1)}$	$P_{ST}^{(4)}$	$P_{ST}^{(7)}$
ERS	ERA	0.87	0.27	0.40	0.00	0.00	5.39	0.17	0.17	3.25	0.00
	EXA	3.14	0.47	0.52	2.03	0.00	0.22	0.32	0.30	0.97	5.38
	HEA	1.22	0.19	0.28	2.25	0.00	0.59	0.00	0.00	0.00	0.00
	NC_1	2.08	0.00	0.00	1.19	0.00	3.54	0.25	0.28	0.80	4.01
	NC_4	2.12	0.00	0.00	0.51	4.43	6.31	0.00	0.00	0.15	2.15
	NC_8	3.82	0.00	0.12	0.68	6.68	3.47	0.00	0.00	0.49	3.78
	PC_1	0.98	0.00	0.00	4.22	0.00	0.79	0.00	0.00	0.26	0.00
	PC_4	6.49	0.52	0.48	1.08	4.14	4.83	0.17	0.16	0.50	1.28
	PC_8	0.87	0.00	0.00	7.88	0.00	1.78	0.00	0.00	0.00	0.00
HES	ERA	4.69	0.35	0.22	0.62	2.88	1.25	0.25	0.21	0.52	1.22
	EXA	0.44	0.00	0.00	0.00	0.00	1.74	0.00	0.00	0.36	0.00
	HEA	3.78	0.43	0.36	0.62	0.90	2.23	0.30	0.22	0.52	1.30
	NC_1	0.15	0.00	0.00	0.00	0.00	2.12	0.00	0.00	0.36	6.77
	NC_4	3.08	0.32	0.32	0.86	3.87	6.64	0.00	0.00	0.17	0.73
	NC_8	0.14	0.00	0.00	0.32	0.00	2.28	0.00	0.11	0.52	6.52
	PC_1	1.44	0.19	0.13	0.34	1.31	6.05	0.11	0.15	0.27	0.79
	PC_4	1.85	0.15	0.19	1.08	0.00	2.57	0.00	0.13	0.56	6.53
	PC_8	1.69	0.93	0.88	1.39	1.85	4.14	0.13	0.16	0.29	0.74

Table 2.3. *Error percentage for a $MAP/PH/1$ queue under selected scenarios*

Weibull (WBS). The two-parameter Weibull considered here has the probability density function

$$f(t) = \begin{cases} \frac{\beta}{\alpha}\Big(\frac{t}{\alpha}\Big)^{\beta-1} e^{-(\frac{t}{\alpha})^{\beta}}, & t \geq 0, \\ 0, & t < 0, \end{cases}$$

where $\alpha > 0$ and $\beta > 0$ are, respectively, scale and shape parameters. For our numerical examples, we consider the following.

Weibull (WBA): Here, we take the scale and shape parameters to be $\alpha = \frac{1.128360749411992}{\lambda}$ and $\beta = 2$.

Weibull ($W1S$): Here, we take the scale and shape parameters to be $\alpha = \frac{1.128360749411992}{\mu}$ and $\beta = 2$.

Weibull ($W2S$): Here, we take the scale and shape parameters to be $\alpha = \frac{0.5}{\mu}$ and $\beta = 0.5$.

It should be pointed that $W2S$ corresponds to a heavy-tailed distribution since its shape parameter is less than 1. Note that the means of $WBA, W1S$, and $W2S$ are, respectively, $\frac{1}{\lambda}, \frac{1}{\mu}$, and $\frac{1}{\mu}$. Their respective standard deviations are $\frac{0.522715}{\lambda}$, $\frac{0.522715}{\mu}$, and $\frac{2.23607}{\mu}$. The parameters λ and μ will be chosen so as to arrive at specific arrival and service rates.

The ratio of the standard deviation of the arrival process WBA with respect to ERA is 0.7392. The ratios of the standard deviations of $W1S$ and $W2S$ with respect to ERS are, respectively, 0.7392 and 3.1619.

In our examples below, we consider three distributions for the batch sizes for the arrivals. These are as follows.

– *Constant batch size* (CoN): here, we assume that the arriving batch is constant and is given by N.

– *Poisson batch size* (PbS): here, we assume that the arriving batch is of size k with probability given by $e^{-\theta}\left(\frac{\theta^{k-1}}{(k-1)!}\right)$, $k \geq 1$. Note that the mean batch size is given by $\theta + 1$.

– *Geometric batch size* (GbS): here, the arriving batch is of size k with probability given by $(1-p)p^{k-1}$, $k \geq 1$. Note that the mean batch size is given by $\frac{1}{1-p}$.

For a meaningful comparison of various types of batch arrival processes, the mean batch size will remain the same. That is, given the mean batch size, say N, we take $1 + \theta = \frac{1}{1-p} = N$.

Under a variety of scenarios obtained through varying the arrival processes, the service times (listed earlier), the batch size distributionsand the number of servers, we simulated the multiserver queuing system of the type $BMAP/G/c$. Since the simulated results agreed with the qualitative findings of the analytical results for their respective scenarios, here we will highlight a few findings involving scenarios for which the analytical results are not reported as well as scenarios for which certain measures such as busy period are not reported. A sample figure is displayed in Figure 2.6.

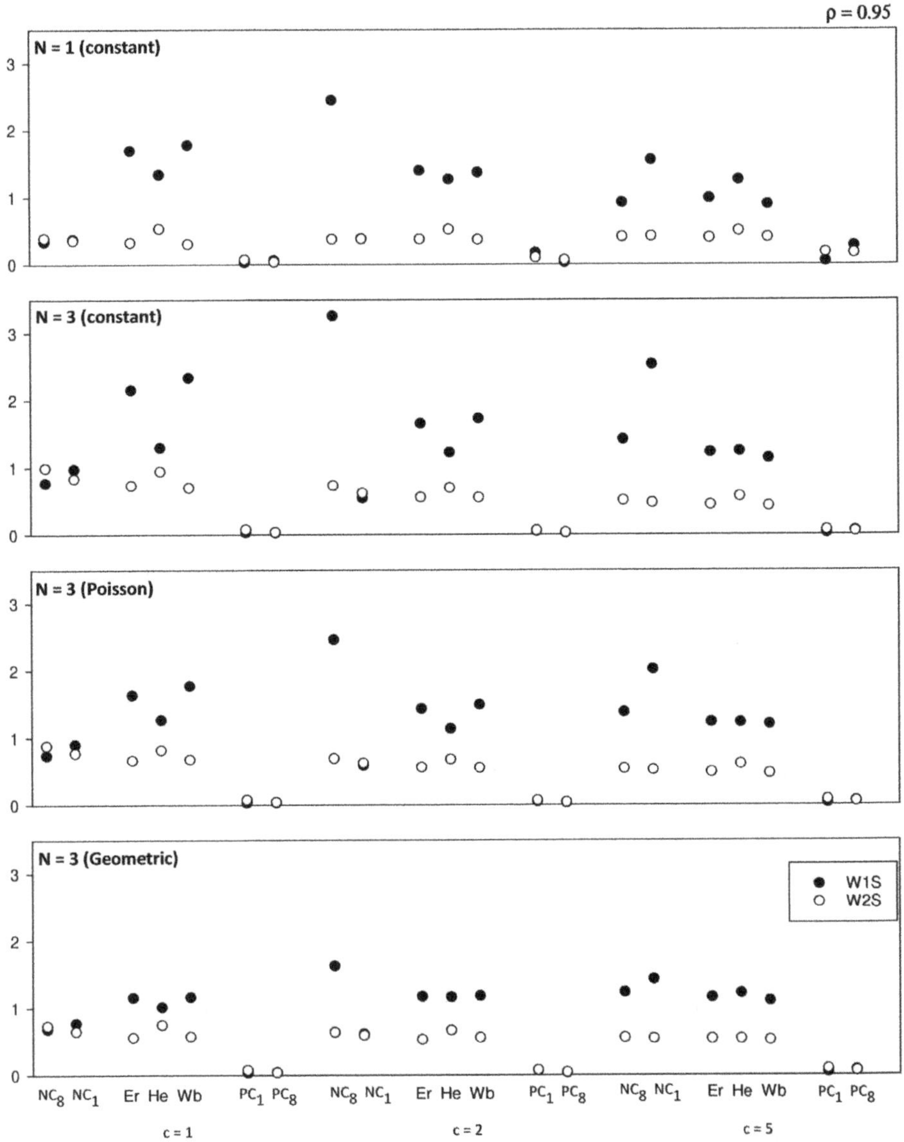

Figure 2.6. *Ratio of the mean busy period to the mean sojourn time for (simulated) a* $BMAP/G/c$ *queue under various scenarios*

– Poisson and Geometric distributions for batch sizes appear to yield similar results.

– Among all arrival processes considered, HEA and PC_8 appear to yield higher tail probabilities for the busy period. Note that HEA has the largest variability while PC_8 has the largest (positive) correlation. This again points out the significant role played by high variability as well as high positive correlation in stochastic modeling.

– The ratio of the mean busy period to the mean sojourn time appears to be smaller for services with a high variability and or a heavy tail. This ratio is the smallest when the arrival process has the highest positive correlation. Further, as the correlation in the arrival process increases this ratio appears to decrease.

– Now we compare the selected system performance measures when service times are modeled using the two Weibull, namely, $W1S$ and $W2S$, distributions for the services. As seen earlier, a highly positively correlated arrival process yields the largest value for the measure μ_q for the cases: $c = 1, 2, 5$, $\rho = 0.80, 0.95$, and for all batch size distributions considered here. When ρ is large, the insensitivity to the service times used is seen for in the case of HEA and PC_8 arrivals for all c and N cases considered. We notice an interesting trend in the case of NC_1 and NC_8 arrivals. As c is increased, the sensitivity of the type of services is seen for these two negatively correlated arrivals. The latter phenomenon appears to be the case for all batch size distributions considered here.

– When ρ is not that large, say, $\rho = 0.8$, the sensitivity to the type of services is seen in all but HEA and PC_8 arrivals. This indicates a somewhat interesting interaction of negatively correlated arrivals and the traffic intensity of the system. Further investigation is required to fully understand this behavior.

– We note that a heavy-tailed distribution like $W2S$ for services results in a heavy tailed distribution for the sojourn time in the system as well as for the busy period.

– A heavy-tailed distribution like $W2S$ results in the mean busy period to be less than the mean sojourn time in the system in all cases considered here.

– It appears that the mean busy period is less than the mean sojourn time in the system in the case of PC_8 arrivals for all c, N and the ρ values considered here.

– It is worth pointing out that (Chakravarthy 2019) that the mean busy period maybe less than or greater than the mean sojourn time in the system.

– The best fits for the sojourn time and for the busy period based on ARENA's analyzer are displayed in Tables 2.4 and 2.5 for selected scenarios. It is clear from these fits that the distributions fitted are all skewed, which are typical for the sojourn times and the busy periods. It is worth pointing out that in almost all cases, the busy period is fitted with a heavy tailed distribution like Weibull with the shape parameter less than 1.

N	TAP	ERS	EXS	HES	$W1S$	$W2S$
1	ERA	GAMM(8.39, 1.16)	EXPO(14)	333*BETA(0.717, 4.62)	GAMM(5.82, 1.32)	GAMM(66.3, 0.721)
	EXA	GAMM(12.1, 1.15)	EXPO(18.6)	351*BETA(0.695, 3.8)	GAMM(10.4, 1.19)	376*BETA(0.765, 5.23)
	HEX	EXPO(154)	GAMM(174, 0.948)	GAMM(175, 0.934)	GAMM(127, 1.09)	EXPO(153)
	WBA	GAMM(6.26, 1.23)	EXPO(12.3)	285*BETA(0.704, 3.69)	LOGN(6.12, 7.48)	GAMM(63.8, 0.697)
	NC_1	GAMM(10.4, 1.16)	GAMM(17.3, 0.997)	GAMM(69.2, 0.755)	GAMM(8.28, 1.24)	545*BETA(0.802, 8.82)
	PC_1	1940*BETA(0.629, 3.08)	1820*BETA(0.674, 3.27)	2080*BETA(0.625, 2.79)	1880*BETA(0.627, 2.94)	1970*BETA(0.682, 3.38)
	NC_8	GAMM(5.57, 1.33)	GAMM(11.7, 1.05)	GAMM(64.5, 0.741)	LOGN(5.71, 6.17)	423*BETA(0.707, 6.42)
	PC_8	GAMM(1030, 0.695)	2950*BETA(0.65, 2.15)	3420*BETA(0.652, 2.22)	3190*BETA(0.623, 2.11)	3170*BETA(0.646, 2.24)
3 (PbS)	ERA	GAMM(7.49, 1.14)	EXPO(10.5)	173*BETA(0.856, 5.67)	WEIB(2.09, 0.435)	193*BETA(0.857, 7.17)
	EXA	EXPO(13.3)	EXPO(10.5)	GAMM(34.8, 0.833)	EXPO(13)	GAMM(32.1, 0.836)
	HEX	EXPO(139)	EXPO(137)	964*BETA(0.924, 4.76)	945*BETA(0.995, 5.98)	GAMM(148, 1.03)
	WBA	GAMM(5.87, 1.14)	GAMM(7.33, 1.07)	GAMM(28.9, 0.796)	GAMM(5.01, 1.21)	158*BETA(0.854, 6.33)
	NC_1	GAMM(10.1, 1.11)	GAMM(11.7, 1.06)	GAMM(31.2, 0.831)	GAMM(9, 1.15)	198*BETA(0.81, 5.93)
	PC_1	1690*BETA(0.643, 2.81)	1560*BETA(0.647, 2.69)	1940*BETA(0.631, 3.13)	1690*BETA(0.645, 2.92)	1610*BETA(0.69, 2.85)
	NC_8	GAMM(5.05, 1.28)	GAMM(6.81, 1.15)	GAMM(24.8, 0.836)	GAMM(4.18, 1.35)	GAMM(22.9, 0.848)
	PC_8	3050*BETA(0.725, 2.67)	2920*BETA(0.724, 2.65)	3600*BETA(0.724, 3.18)	2970*BETA(0.719, 2.66)	2930*BETA(0.769, 2.74)

Table 2.4. *Best fit probability distribution for the sojourn time when* $c = 1, \rho = 0.95$

N	TAP	ERS	EXS	HES	$W1S$	$W2S$
1	ERA	WEIB(4.5, 0.478)	WEIB(3.63, 0.409)	WEIB(1.56, 0.445)	WEIB(4.52, 0.525)	WEIB(0.965, 0.324)
	EXA	WEIB(5.2, 0.492)	WEIB(3.85, 0.434)	WEIB(1.7, 0.45)	WEIB(6.63, 0.521)	WEIB(0.988, 0.357)
	HEX	WEIB(27.6, 0.408)	WEIB(20.8, 0.37)	WEIB(8.2, 0.371)	WEIB(31.2, 0.42)	WEIB(3.24, 0.299)
	WBA	WEIB(3.19, 0.497)	WEIB(2.87, 0.435)	WEIB(1.45, 0.428)	WEIB(3.62, 0.606)	WEIB(0.924, 0.332)
	NC_1	WEIB(5.57, 0.535)	WEIB(3.57, 0.443)	WEIB(1.85, 0.39)	WEIB(6.25, 0.586)	WEIB(1.3, 0.312)
	PC_1	EXPO(21.6)	WEIB(1.75, 0.551)	WEIB(1.04, 0.471)	EXPO(21.1)	WEIB(0.697, 0.41)
	NC_8	WEIB(6.06, 0.606)	WEIB(4.38, 0.477)	WEIB(2.43, 0.401)	WEIB(6.33, 0.643)	WEIB(1.38, 0.315)
	PC_8	EXPO(19.7)	EXPO(21.6)	EXPO(24.1)	EXPO(18.6)	WEIB(1.38, 0.315)
3 (PbS)	ERA	WEIB(3.9, 0.517)	WEIB(3.92, 0.485)	WEIB(2.29, 0.424)	WEIB(5.2, 0.477)	WEIB(1.97, 0.394)
	EXA	WEIB(5.65, 0.47)	WEIB(4.94, 0.462)	WEIB(3.74, 0.425)	WEIB(5.06, 0.517)	WEIB(1.73, 0.406)
	HEX	WEIB(27.4, 0.407)	WEIB(24.2, 0.375)	WEIB(15.5, 0.329)	WEIB(30.7, 0.421)	WEIB(14, 0.33)
	WBA	WEIB(3.43, 0.549)	WEIB(3.45, 0.513)	WEIB(2.09, 0.435)	EXPO(10.5)	WEIB(1.88, 0.417)
	NC_1	WEIB(5.25, 0.494)	WEIB(5.84, 0.475)	WEIB(3.36, 0.416)	WEIB(6.56, 0.504)	WEIB(2.03, 0.415)
	PC_1	WEIB(2.2, 0.58	WEIB(2.19, 0.558)	WEIB(3.36, 0.416)	WEIB(2.2, 0.601)	WEIB(1.37, 0.465)
	NC_8	WEIB(6.42, 0.572)	WEIB(6.4, 0.52)	WEIB(4.56, 0.441)	WEIB(7.24, 0.588)	WEIB(2.88, 0.439)
	PC_8	EXPO(21.8)	EXPO(22.2)	EXPO(24.3)	EXPO(20.9)	WEIB(1.1, 0.49)

Table 2.5. *Best fit probability distribution for the (full) busy pericd when* $c = 1, \rho = 0.95$

2.6. Analysis of selected queueing models of $BMAP/G/c$ type with a vacation

In this section, we will discuss a few queueing models of $BMAP/G/c$ type in which the servers can go on a vacation whenever the system has no customers waiting in the queue at a service completion through simulation. In order to validate the simulated models, we will look at the classical vacation model. For more in-depth details, we refer readers to references mentioned.

2.6.1. $MAP/PH/1$ *queueing model with a vacation*

This model is a variation to the classical $MAP/PH/1$ queueing model in which the server goes on a vacation whenever the system becomes empty. It is possible for the server to go on multiple vacations when the vacationing server returns to find the system to be empty. Queueing models with vacations (and working vacations) have many applications in practice and have been studied extensively by several authors (Chakravarthy 2013; Chakravarthy *et al.* 2020; Chandrasekaran *et al.* 2016; Doshi 1986; Tian and Zhang 2006). In the case of a single server system, it is clear how the vacation models work; however, in the case of multiserver systems, the vacation models need more clarification such as whether the vacations are synchronous (i.e., all servers take a vacation simultaneously at the moment the only server who is busy finishes service), asynchronous (i.e., servers take a vacation as and when they are eligible to take a vacation), and others. We refer readers to (Tian and Zhang 2006) for more details.

In the model under consideration, the customers arrive at a single server system according to a MAP with representation (D_0, D_1) of order m and with $D = D_0 + D_1$ being irreducible and that δ is the invariant vector of D. The arrival rate, λ, is given by $\lambda = \boldsymbol{\delta} D_1 \boldsymbol{e}$. The server offers services to customers on a $FCFS$ basis and the service times are assumed to follow a PH distribution with representation $(\boldsymbol{\beta}, S)$ of order n. The mean service time is given by $\mu^{-1} = \boldsymbol{\beta}(-S)^{-1}\boldsymbol{e}$. Let $\boldsymbol{S^0}$ be such that $S\boldsymbol{e} + \boldsymbol{S^0} = \mathbf{0}$. Whenever the system becomes empty upon a completion of a service, the server goes on a vacation. The vacation times are assumed to be exponentially distributed with parameter θ.

Like in the classical $MAP/PH/1$ queueing model, the model can be studied as a QBD by keeping track of the number of customers, say, $N(t)$, in the system, the phase of the arrival process, say $J_1(t)$, the phase of the service process, say, $J_2(t)$ (when the server is busy) at time t, and the status of the server (busy or on vacation). The state space is given by

$$\Omega = \{(0,k), 1 \le k \le m\} \bigcup \{(i,0,k) : i \ge 1, 1 \le k \le m\}$$
$$\bigcup \{(i,j,k) : i \ge 1, 1 \le j \le n, 1 \le k \le m\} = \{\underline{\mathbf{0}}\} \bigcup \{\underline{\mathbf{i}}, i \ge 1\}.$$

The level, $\underline{\mathbf{0}}$, of dimension m corresponds to the case when the server is on vacation with no customers in the system and the arrival process is in one of m phases. The set of states, $\underline{\boldsymbol{i}} = \{(i,0,k) : 1 \leq k \leq m\} \bigcup \{(i,j,k) : 1 \leq j \leq n, 1 \leq k \leq m\}$, of dimension $m + mn$, corresponds to the case when there are $i, i \geq 1$, customers in the system with the server either on vacation or busy serving in phase j, and the arrival process is in phase k.

The infinitesimal generator of the Markov chain governing the QBD−process is given by

$$Q = \begin{pmatrix} D_0 & e_1' \otimes D_1 & & & \\ e_1 \otimes \boldsymbol{S}^0 \otimes I & A_1 & A_0 & & \\ & A_2 & A_1 & A_0 & \\ & & A_2 & A_1 & A_0 \\ & & & \ddots & \ddots & \ddots \end{pmatrix}, \qquad [2.27]$$

where

$$A_0 = \begin{bmatrix} D_1 & 0 \\ 0 & I \otimes D_1 \end{bmatrix}, A_1 = \begin{bmatrix} D_0 - \theta I & \theta(\boldsymbol{\beta} \otimes I) \\ 0 & S \oplus D_0 \end{bmatrix}, A_2 = \begin{bmatrix} 0 & 0 \\ 0 & \boldsymbol{S}^0 \boldsymbol{\beta} \otimes I \end{bmatrix}. \qquad [2.28]$$

Under the stability condition that $\lambda < \mu$ (which can be established through Neuts' results on the matrix-geometric solution), the steady-state probability vector, $\boldsymbol{x}$, of Q is of matrix-geometric type. Suppose that $\boldsymbol{x}$ is partitioned as $\boldsymbol{x} = (\boldsymbol{x}_0, \boldsymbol{x}_1, \boldsymbol{x}_2, \cdots)$ such that the vector $\boldsymbol{x}_0$ of dimension m gives the steady-state probability vector that the system is empty with the arrival process in one of m phases; the vector $\boldsymbol{x}_i$ of dimension mn is further partitioned into vectors of dimension m as $\boldsymbol{x}_i = (\boldsymbol{x}_{i1}, \cdots, \boldsymbol{x}_{in})$ so that the vector $\boldsymbol{x}_{ij}$ gives the steady-state probability vector that there are i customers in the system with the service process in phase j and the arrival process is in one of m phases.

Neuts' results on the matrix-geometric method (Neuts 1981) indicate that the steady-state probability vector $\boldsymbol{x}$ is given by

$$\boldsymbol{x}_i = \boldsymbol{x}_1 R^{i-1}, \; i \geq 1, \qquad [2.29]$$

where the rate matrix R is the minimal non-negative solution to the matrix-quadratic equation given by

$$R^2 A_2 + R A_1 + A_0 = 0, \qquad [2.30]$$

and the vectors $\boldsymbol{x}_0$ and $\boldsymbol{x}_1$ are obtained by solving

$$\begin{aligned} \boldsymbol{x}_0 D_0 + \boldsymbol{x}_1(\boldsymbol{e}_1 \otimes \boldsymbol{S}^\mathbf{0} \otimes I) &= \mathbf{0}, \\ \boldsymbol{x}_0(\boldsymbol{e}_1' \otimes D_1) + \boldsymbol{x}_1(A_1 + RA_2) &= \mathbf{0}, \end{aligned} \quad [2.31]$$

subject to the normalizing condition: $\boldsymbol{x}_0\boldsymbol{e} + \boldsymbol{x}_1(I - R)^{-1}\boldsymbol{e} = 1.$

From the probabilistic interpretation of the rate matrix, R, we note that starting in the set of states, $\{(i, j, k) : 1 \leq j \leq n, 1 \leq k \leq m\}$, that corresponds to the server being busy with i customers in the system, the mean time (expressed in units of the mean time spent in $(i, j, k), 0 \leq j \leq n, 1 \leq i \leq m)$ in state $(i + 1, 0, k), 1 \leq k \leq m$, before hitting level $\boldsymbol{i}$ has to be zero. This is clear from the fact that once the server starts serving, the server will go on vacation only by hitting the boundary states. This interpretation results in the fact that R is of the form:

$$R = \begin{pmatrix} R_1 & R_2 \\ 0 & R_3 \end{pmatrix}. \quad [2.32]$$

Due to the special structure of R, we see that equation [2.30] can be rewritten as

$$\begin{aligned} &R_1 = D_1(\theta I - D_0)^{-1}, \\ &R_3^2(\boldsymbol{S}^\mathbf{0}\boldsymbol{\beta} \otimes I) + R_3(S \oplus D_0) + (I \otimes D_1) = 0, \\ &(R_1R_2 + R_2R_3)(\boldsymbol{S}^\mathbf{0}\boldsymbol{\beta} \otimes I) + \theta R_1(\boldsymbol{\beta} \otimes I) + R_2(S \oplus D_0) = 0. \end{aligned} \quad [2.33]$$

NOTE.– (a) The second equation in [2.33] is the matrix-quadratic equation corresponding to the classical $MAP/PH/1$ queue (see [2.14]). Thus, we can see the decomposition result for the rate matrix R for the vacation model. It is worth pointing out a similar decomposition result for $MAP/M/c$ with a vacation proved in Chakravarthy (2007). (b) Once R_1 and R_3 are obtained, R_2 can be obtained explicitly in terms of R_1 and R_3. This is accomplished through recognizing that the third equation in [2.33] is a Sylvester matrix equation of the form $BXC + EXF + H = 0$, where the matrices B, C, E, F, and H are known and X is unknown. Suppose $\tau(B)$ denotes the direct sum of the rows of B. Then, using the fact that (Horn and Johnson 1991) $\tau(BXC) = \tau(X)(B' \otimes C)$, it is easy to verify that

$$\begin{aligned} \tau(R_2) &= \theta\tau(R_1(\boldsymbol{\beta} \otimes I)) \\ &\times \left\{ -\left[[R_1' \otimes (\boldsymbol{S}^\mathbf{0}\boldsymbol{\beta} \otimes I)] + [I \otimes R_3(\boldsymbol{S}^\mathbf{0}\boldsymbol{\beta} \otimes I)] + [I \otimes (S \oplus D_0)] \right]^{-1} \right\}. \end{aligned}$$

2.6.2. *The system performance measures*

The qualitative analysis of any queueing model requires one or more system performance measures. While we could develop a number of such measures, here we will detail a few and leave the rest to the interested readers.

1) *Probability that the server is on vacation*: the probability, P_{vac}, that the server is on vacation is given by $P_{vac} = \boldsymbol{x}_0\boldsymbol{e} + \boldsymbol{x}_1(I - R)^{-1}(\boldsymbol{e}_1 \otimes \boldsymbol{e}) = 1 - \dfrac{\lambda}{\mu}$.

2) *Probability that the server is on vacation with at least one customer waiting in the queue*: the probability, P_{vwqp}, that the server is on vacation with at least one customer waiting in the queue is $P_{vwqp} = \boldsymbol{x}_1(I - R)^{-1}(\boldsymbol{e}_1 \otimes \boldsymbol{e})$.

3) *Mean queue length*: the mean, μ_q, queue length is given by $\mu_q = \sum_{i=1}^{\infty} i\boldsymbol{x}_i(\boldsymbol{e}_1 \otimes \boldsymbol{e}) + \sum_{i=1}^{\infty}(i - 1)\boldsymbol{x}_i(\boldsymbol{e}_2 \otimes \boldsymbol{e}) = \boldsymbol{x}_1(I - R)^{-2}\boldsymbol{e} - \boldsymbol{x}_1(I - R)^{-2}(\boldsymbol{e}_2 \otimes \boldsymbol{e})$.

4) *Mean number in the system*: the mean, μ_s, queue length is given by $\mu_s = \boldsymbol{x}_1(I - R)^{-2}\boldsymbol{e}$.

2.6.3. *Illustrative numerical examples for* $MAP/PH/1$ *with a vacation*

The purpose of this section is to discuss a few illustrative examples for the vacation model. We will use the same set of arrival processes listed in section 2.4.3. In Figures 2.7-2.10, we display, respectively, the ratio of the mean queue lengths for a $MAP/PH/1$ vacation queue to a $MAP/PH/1$ classical queue, the ratio of the system idle probability for vacation and classical $MAP/PH/1$ queues, the distribution functions of the sojourn time for $MAP/PH/1$ classical and vacation queues (when $\rho = 0.80$), and the distribution functions of the sojourn time for $MAP/PH/1$ classical and vacation queues (when $\rho = 0.95$) under different scenarios. Note that, we added a "v" in the legend to indicate that it represents for the vacation model. Now, we briefly summarize some key observations below.

– Before we compare the vacation model to the corresponding classical $MAP/PH/1$ queue, it is worth pointing out the observations summarized in item (ix) in section 4.3 holds good for this case.

– As is to be expected the ratio of the mean queue length approaches to 1 as θ increases. The interesting thing is that for the services with a large variability (HES) the rate of approach is faster. Further, the ratio is much higher for ERS compared to HES. This appears to be the case for renewal as well as for correlated arrivals. We also notice that positively correlated arrivals appear to yield a smaller ratio compared to those of negatively correlated ones.

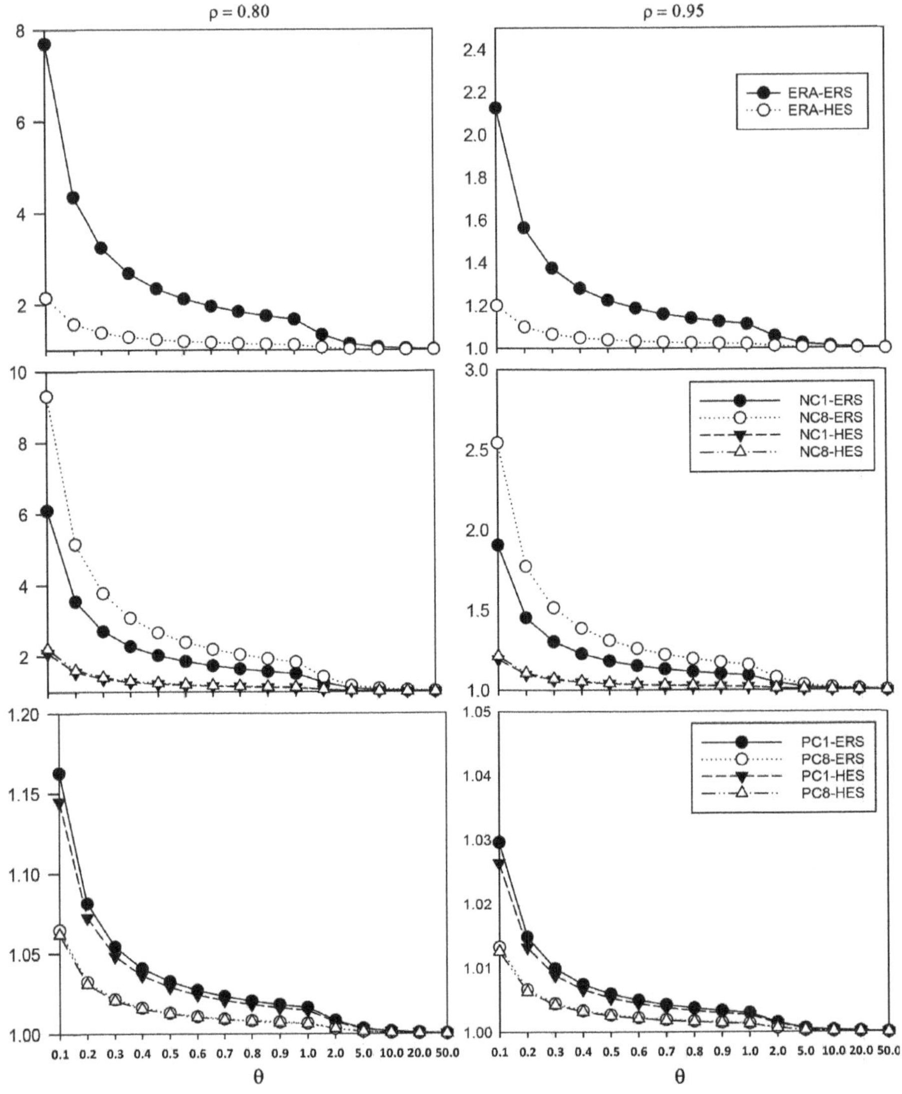

Figure 2.7. *Ratio of the mean queue length for $MAP/PH/1$ vacation to $MAP/PH/1$ classical queues under different scenarios*

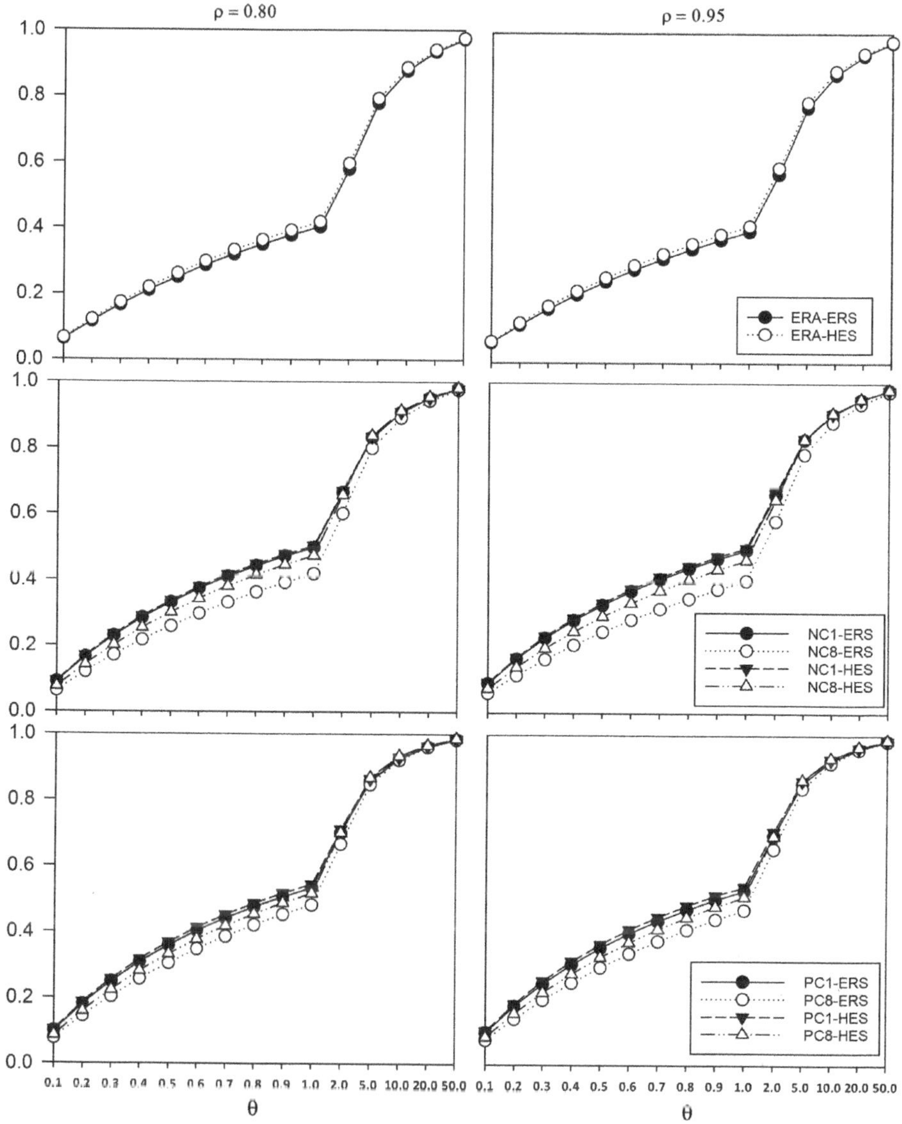

Figure 2.8. *Ratio of the system idle probability for vacation and classical $MAP/PH/1$ queues*

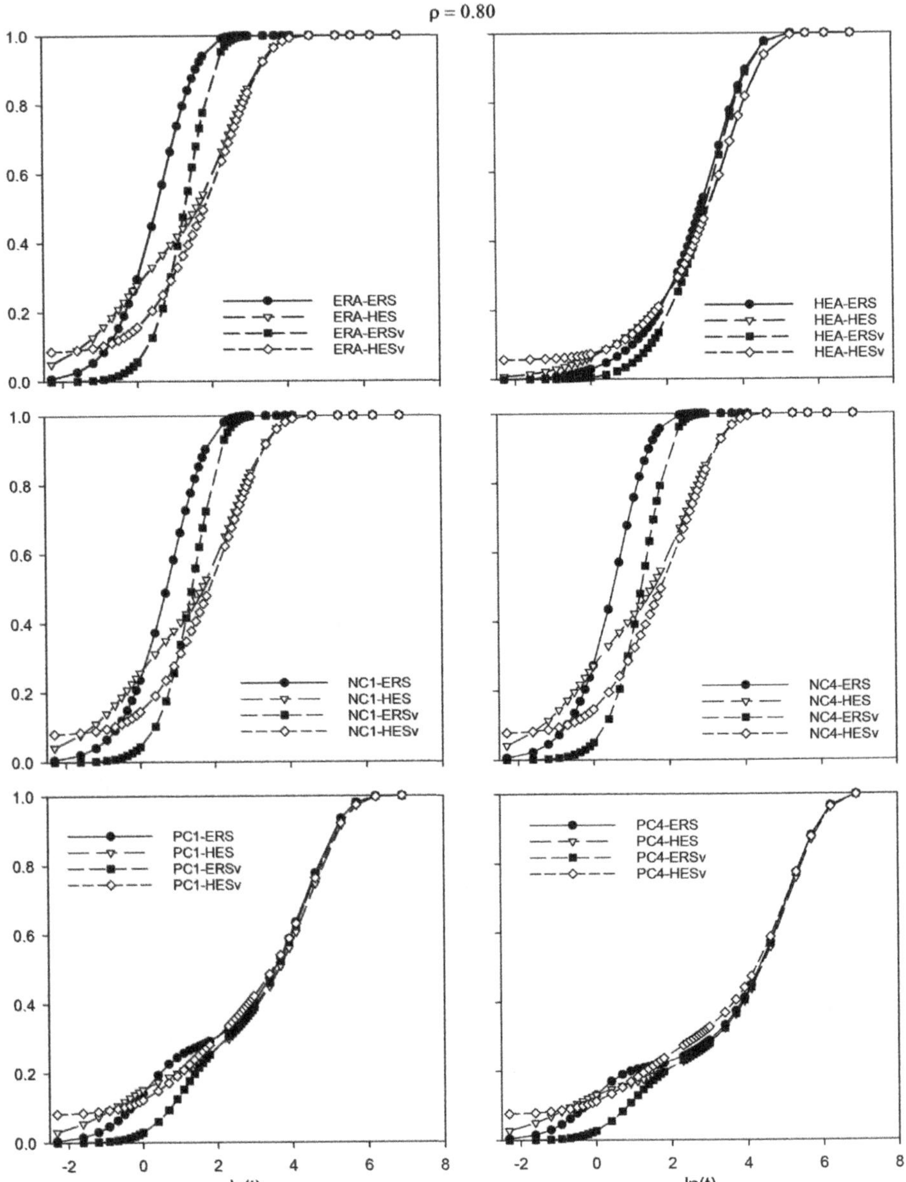

Figure 2.9. *The distribution functions of the sojourn time for $MAP/PH/1$ classical and vacation queues under different scenarios*

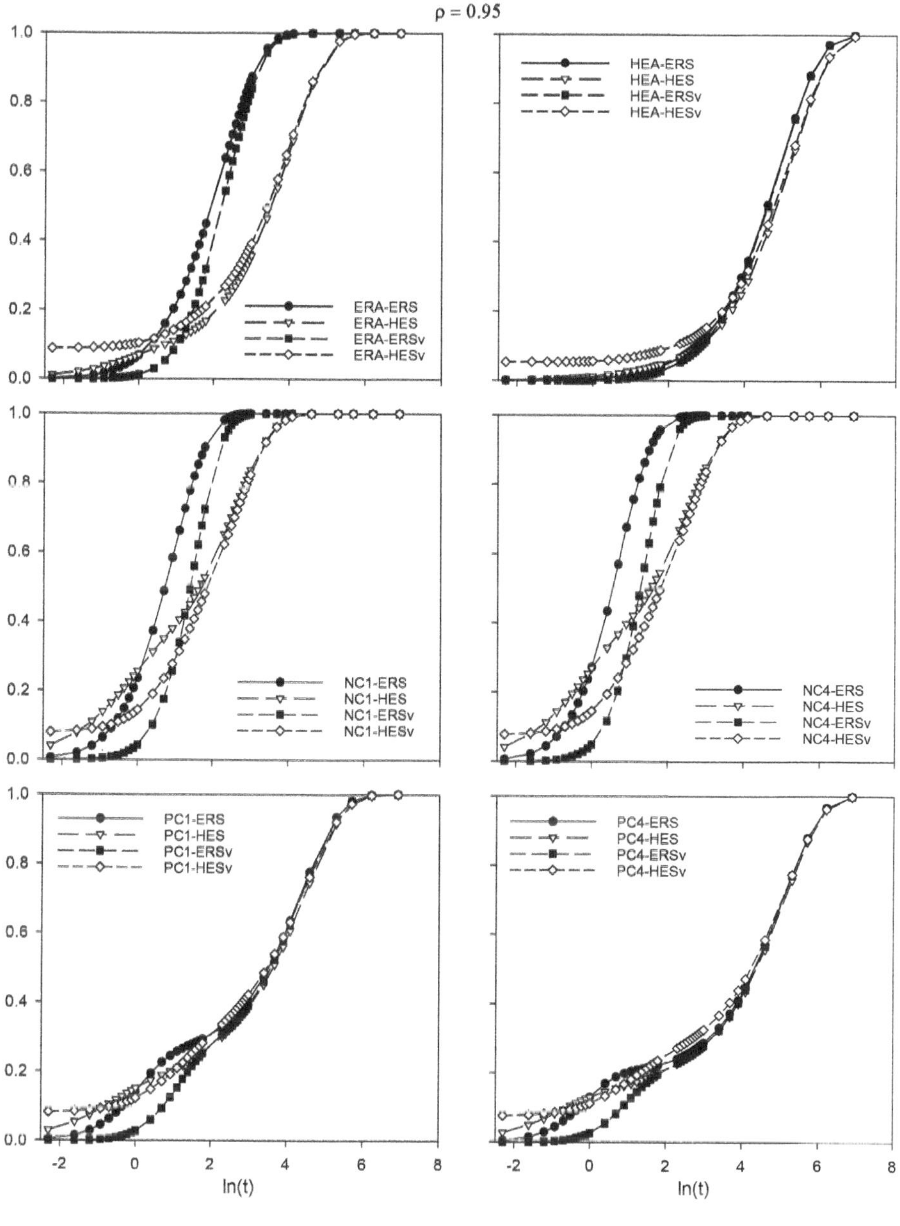

Figure 2.10. *The distribution functions of the sojourn time for $MAP/PH/1$ classical and vacation queues under different scenarios*

– When we look at the ratio of the probability that the system has no customers, we see an interesting behavior for the correlated arrivals. While for ERA we notice an insignificant role played by the service time distribution, we see a different behavior for the correlated arrivals. When the absolute value of the correlation is large, we notice the significance of the type of services used, especially for smaller values of θ. However, for large values of θ this significance appears to become less, even though the ratio is still well below 1.

– Comparing the distribution functions of the sojourn time in the system of $MAP/PH/1$ with a vacation and the corresponding classical queueing models, we notice the stochastic dominance of the classical model over the vacation model when the services are modeled with less variable distribution like ERS. That is, if Y_c and Y_v denote, respectively, the sojourn time in the system of a customer for the classical and vacation models, then we have $P(Y_v \leq t) \leq P(Y_c \leq t), t \geq 0$ for all types of arrival processes considered here. However, when the services have a high variability like HES, we notice (in all types of arrival processes considered here) that there exists, say, a $t_0 \geq 0$ (which depends on the type of arrival process considered) such that $P(Y_c \leq t) \leq P(Y_v \leq t), 0 \leq t \leq t_0$, and $P(Y_v \leq t) \leq P(Y_c \leq t), t \geq t_0$.

2.6.4. *Validation of simulated model for vacation type queues*

Before we discuss a few simulated vacation models, we validate the simulated models against the vacation models of the type $MAP/PH/1$ for which analytical results are available (see section 2.6.1). Toward this end, we display the error percentages of the two measures, μ_q and P_{vwqp}, in Tables 2.6 and 2.7, respectively.

		$\rho = 0.80$				$\rho = 0.95$			
ToA	ToS	$\theta = 0.1$	$\theta = 0.5$	$\theta = 1.0$	$\theta = 10.0$	$\theta = 0.1$	$\theta = 0.5$	$\theta = 1.0$	$\theta = 10.0$
ERA	ERS	0.21	0.14	0.44	0.26	1.82	1.25	1.08	4.13
	HES	1.77	0.28	1.01	1.11	3.67	2.74	1.88	4.55
NC_1	ERS	0.42	0.12	0.18	1.36	0.48	0.06	1.30	1.32
	HES	0.40	0.15	0.26	0.15	5.08	4.82	2.26	2.06
NC_8	ERS	0.01	0.43	0.01	0.06	0.43	2.03	3.13	0.33
	HES	0.61	0.45	1.60	0.25	1.04	0.95	2.14	3.89
PC_1	ERS	0.35	0.12	1.23	3.74	3.37	3.07	4.65	1.87
	HES	0.29	1.93	2.33	0.94	7.48	2.92	7.04	0.42
PC_8	ERS	2.79	0.18	0.76	1.27	1.71	1.54	1.65	5.32
	HES	1.67	1.85	1.30	2.59	7.72	0.53	5.92	5.69

Table 2.6. *Error percentage for a $MAP/PH/1$ queue with a vacation under selected scenarios for μ_q*

		$\rho = 0.80$				$\rho = 0.95$			
ToA	ToS	$\theta = 0.1$	$\theta = 0.5$	$\theta = 1.0$	$\theta = 10.0$	$\theta = 0.1$	$\theta = 0.5$	$\theta = 1.0$	$\theta = 10.0$
ERA	ERS	0.11	0.07	0.01	0.27	0.76	0.49	1.27	2.19
	HES	0.32	0.22	0.25	0.09	3.15	3.46	1.00	2.70
NCA_1	ERS	0.24	0.05	0.34	0.43	0.76	0.18	0.54	3.18
	HES	0.12	0.13	0.31	0.54	1.89	3.73	2.39	0.57
NCA_8	ERS	0.20	0.06	0.13	0.06	0.23	0.18	1.95	1.16
	HES	0.01	0.22	0.24	0.69	2.24	1.82	0.63	3.36
PCA_1	ERS	0.61	0.40	0.09	1.33	6.04	0.71	0.26	0.75
	HES	0.07	0.01	1.40	0.35	6.33	0.92	0.94	0.50
PCA_8	ERS	1.37	0.08	0.53	0.56	3.89	0.77	1.20	0.17
	HES	0.13	0.98	0.76	0.94	5.53	3.66	6.54	0.37

Table 2.7. *Error percentage for a $MAP/PH/1$ queue with a vacation under selected scenarios for P_{vwqp}*

2.6.5. *Selected simulated models of $BMAP/G/c$ type queues with a vacation*

In this section, which is similar to section 2.5.3, we will discuss some simulated results. We consider the same input distributions (both arrival and service times), we use Poisson batch size distribution with a mean of 3 (i.e., $N = 3$ and hence $\theta = 2$) and vary the number of servers. Also, we look at the case when services are offered in batches of fixed size, say, 5. We denote the service batch size as (SbS) in the figures below. The best fits for the sojourn time distribution under a selected set of scenarios are given in Table 2.8 and a few sample figures on selected measures are displayed in Figures 2.11 and 2.12. Some key observations based on our simulated results are summarized below.

TaP	ToV	ERS	HES	$W1S$	$W2S$
ERA	ERV	LOGN(1.67, 1.87)	LOGN(1.95, 2.92)	LOGN(1.64, 1.74)	WEIB(1.7, 0.988)
	HEV	LOGN(4.51, 10.7)	LOGN(5.34, 15.7)	LOGN(4.42, 10.2)	LOGN(5.8, 19)
HEX	ERV	GAMM(1.42, 1.61)	GAMM(2.26, 1.18)	LOGN(2.35, 2.41)	GAMM(2.28, 1.16)
	HEV	96*BETA(0.384, 5.82)	LOGN(7.43, 19.6)	102*BETA(0.384, 6.25)	LOGN(7.69, 21.5)
WBA	ERV	LOGN(1.62, 1.8)	LOGN(1.87, 2.77)	LOGN(1.6, 1.68)	LOGN(1.97, 3.25)
	HEV	LOGN(4.46, 10.6)	LOGN(5.27, 15.6)	LOGN(4.28, 9.75)	LOGN(5.56, 18.1)
NC_1	ERV	LOGN(1.45, 1.39)	WEIB(1.49, 1.18)	LOGN(1.38, 1.33)	GAMM(1.29, 1.22)
	HEV	LOGN(4.77, 11.2)	98*BETA(0.322, 5.83)	115*BETA(0.31, 6.93)	LOGN(6.11, 19.7)
PC_1	ERV	LOGN(1.48, 1.42)	LOGN(1.86, 2.56)	LOGN(1.45, 1.31)	GAMM(1.5, 1.17)
	HEV	LOGN(4.82, 11.2)	LOGN(5.99, 17.2)	LOGN(4.69, 10.5)	GAMM(1.5, 1.17)
NC_8	ERV	GAMM(0.768, 1.96)	GAMM(1.15, 1.4)	LOGN(1.54, 1.33)	GAMM(1.22, 1.32)
	HEV	118*BETA(0.32, 6.99)	103*BETA(0.324, 5.97)	LOGN(4.91, 10.4)	LOGN(6.18, 18.8)
PC_8	ERV	96*BETA(0.403, 7.02)	99*BETA(0.322, 5.38)	84*BETA(0.396, 6.03)	89*BETA(0.341, 5.17)
	HEV	131*BETA(0.523, 7.04)	140*BETA(0.482, 6.83)	133*BETA(0.519, 7.08)	156*BETA(0.476, 7.36)

Table 2.8. *Best fit probability distribution for the sojourn time when $c = 1, \rho = 0.95, N = 3, SbS = 5$*

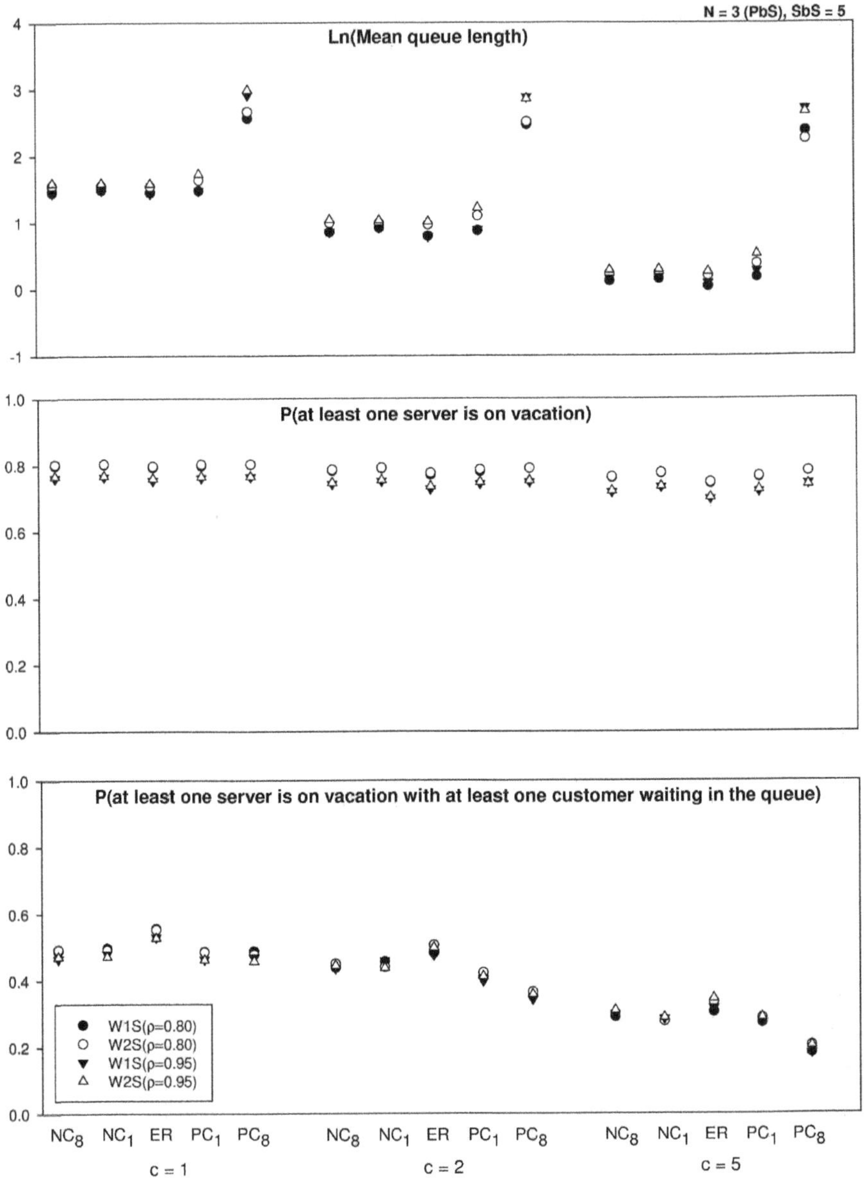

Figure 2.11. *Selected measures for (simulated) a $BMAP/PH/c$ queue with a vacation under different scenarios*

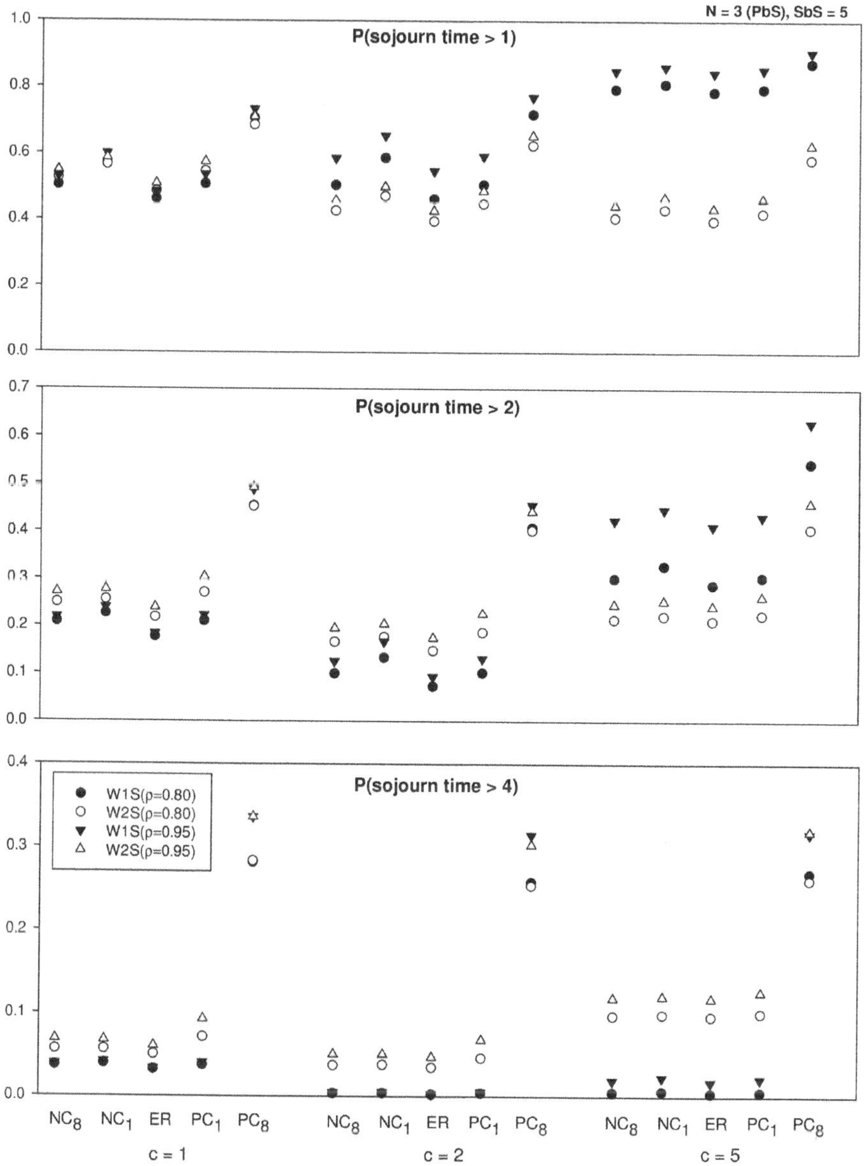

Figure 2.12. *The tail probabilities of the sojourn time for (simulated) a $BMAP/G/c$ queue with a vacation under different scenarios*

– A high variability service distribution like HES appears to yield the largest value for μ_q, P_{vac}, P_{vwqp} in all cases.

– As c is increased (note that μ is chosen so as to fix the system load, ρ, to a given value), all three measures, μ_q, P_{vac}, P_{vwqp}, appear to decrease for all types of arrival and service times considered.

– When c is increased, we notice some distinct patterns highlighting the impact of high variability as well as heavy-tailed services on the sojourn time tail probabilities.

– As pointed out earlier a high variable or a heavy-tailed services appear to produce a heavy-tailed sojourn time distribution.

– In the case of highly positively correlated arrivals, the tail probabilities (beyond a certain time point) appear to be insensitive to the type of services used, whereas this phenomenon is not seen in the negatively correlated arrivals as well as the renewal arrivals considered.

– As is to be expected, we need to use a heavy-tailed distribution or a long-tailed distribution as the best fit for the sojourn time depending on the combination of arrival and service times.

2.7. Acknowledgment

This chapter is dedicated to the memory of Prof. Marcel F. Neuts for his pioneering work on queues.

2.8. References

Artalejo, J.R., Gomez-Correl, A., He, Q.M. (2010). Markovian arrivals in stochastic modelling: a survey and some new results. *SORT*, 34(2), 101–144.

Bhat, U.N. (2015). *An Introduction to Queueing Theory*, 2nd edition. Springer, New York.

Bladt, M., Nielsen, B.F. (2017). *Matrix-Exponential Distributions in Applied Probability*. Springer LLC, New York.

Bowerman, P.N., Nolty, R.G., Scheuer, E.M. (1990). Calculation of Poisson cumulative distribution function. IEEE Transactions on Reliability. *IEEE Trans. Rel.*, 39, 158–161.

Chakravarthy, S.R. (2001). The batch Markovian arrival process: A review and future work. In *Advances in Probability Theory and Stochastic Processes*, Krishnamoorthy, A., Raju, N., Ramaswami, V. (eds). Notable Publications Inc, Neshanic Station, NJ, 21–39.

Chakravarthy, S.R. (2007). A multi-server synchronous vacation model with thresholds and a probabilistic decision rule. *European Journal of Operational Research*, 182, 305–320.

Chakravarthy, S.R. (2013). Analysis of MAP/PH$_1$/PH$_2$/1 queue with vacations and optional secondary services. *Applied Mathematical Modeling*, 37, 8886–8902.

Chakravarthy, S.R. (2015). Matrix-analytic queueing models. In *An Introduction to Queueing Theory*, Narayan Bhat, U., (ed.). 2nd edition. Birkhauser, Springer Science + Business Media, New York, 177–197.

Chakravarthy, S.R. (2019). Busy period analysis of $GI/G/c$ and $MAP/G/c$ queues. In *Performance Prediction and Analysis of Fuzzy, Reliability and Queueing Models, Asset Analytics*. Deep, K., Jain, M., Salhi, S. (eds). Springer Nature Singapore Pte Ltd., 1–31. https://doi.org/10.1007/978-981-13-0857-4_1.

Chakravarthy, S.R., James, J.C., Jr. Louis, A.C., Pinar, K., Jeffrey, P.K., Cole Smith, J. (2010). Markovian arrival processes. *Wiley Encyclopedia of Operations Research and Management Science*. https://doi.org/10.1002/9780470400531.eorms0499.

Chakravarthy, S.R., Shruti Rakhee, K. (2020). A queueing model with server breakdowns, repairs, vacations, and backup server, *Operations Research Perspectives*, 7, 1–13, 100131. https://doi.org/10.1016/j.orp.2019.100131.

Chandrasekaran. V.M., Indhira, K., Saravanarajan, M.C., Rajadurai, P. (2016). A survey on working vacation queueing models. *International Journal of Pure and Applied Mathematics*, 106(6), 33–41.

Doshi, B.T. (1986). Queueing systems with vacations: A survey. Queueing System, 1, 29–66.

Graham, A. (1981). *Kronecker Products and Matrix Calculus with Applications*. Ellis Horwood, Chichester, UK.

Horn, R.A., Johnson, C.R. (1991). *Topics in Matrix Analysis*. Cambridge University Press, Cambridge.

Jensens, A. (1953). Markoff chains as an aid in the study of Markoff processes. Skand. Aktuarietidskr, 36, 87–91.

Kelton, W.D., Sadowski, R.P., Swets, N.B. (2010). *Simulation with ARENA*, 5th edition. McGraw-Hill, New York.

Latouche, G., Ramaswami, V. (1999). Introduction to matrix analytic methods in stochastic modeling. SIAM, Philadelphia, PA 19104-2688.

Lucantoni, D.M. (1991). New results on the single server queue with a batch Markovian arrival process. Stochastic Models, 7, 1–46.

Marcus, M., Minc, H. (1964). *A Survey of Matrix Theory and Matrix Inequalities*. Allyn and Bacon, Boston, MA.

Neuts, M.F. (1975). Probability distributions of phase type. In Liber Amicorum Prof. Emeritus H. Florin. Department of Mathematics, University of Louvain, 173–206.

Neuts, M.F. (1979). A versatile Markovian point process. *J. Appl. Prob.*, 16, 764–779.

Neuts, M.F. (1981). *Matrix-geometric solutions in stochastic models: an algorithmic approach*. The Johns Hopkins University Press, Baltimore, MD. [1994 version is Dover Edition].

Neuts, M.F. (1989). *Structured Stochastic Matrices of $M/G/1$ Type and Their Applications*. Marcel Dekker, Inc., New York.

Neuts, M.F. (1992). Models based on the Markovian arrival process. IEICE Transactions on Communications, E75B, 1255–1265.

Neuts, M.F. (1995). *Algorithmic Probability: A Collection of Problems*. Chapman and Hall, NY.

Ozawa, T. (2006). Sojourn time distributions in the queue defined by a general QBD process. *Queueing Systems*, 53, 203–211.

Qi-Ming, H. (2014). *Fundamentals of Matrix-Analytic Methods*. Springer, New York.

Steeb, W-H., Hardy, Y. (2011). *Matrix Calculus and Kronecker Product*. World Scientific Publishing, Singapore.

Tian, N., Zhang, Z.G. (2006). *Vacation Queueing Models Theory and Applications*. Springer, New York.

3

Distributions and Random Processes Related to Queueing and Reliability Models

Boyan DIMITROV
Department of Mathematics, Kettering University, Flint, MI, USA

In this chapter, we discuss the similarity between probability distributions and random processes related to queueing and reliability models. Analysis of the properties of these distributions helps us to understand their structure, and extend the area of their applications and modeling power. We see how these properties are used to model various processes in economics, industry, demography and environmental studies, and how their use reflects on insurance, risk and cost analysis.

The chapter is mostly based on results obtained by the author and many collaborators and followers. It is designed to give incentives, analytic skills and inspiration for students and researchers. Some open and challenging new problems for further research directions are marked, as well as various possible areas of practical application of mathematics.

3.1. Some useful notations, relationships and interpretations

Most variables in queueing and reliability are non-negative, since they describe the duration of activities, or counting a random number of events. Therefore, they can be treated as lifetimes.

Queueing Theory 2,
coordinated by Vladimir ANISIMOV, Nikolaos LIMNIOS.

We use capital letters for notation of random variables and the same subscripts for their characteristics. So, if X is a random variable (r.v.), its cumulative distribution function (cdf) is defined as

$$F_X(t) = P(X \leq t).$$

Equivalent to the cdf is the knowledge of the *survival function* $\bar{F}_X(t) = P(X > t) = 1 - F_X(t)$. If X is continuous r.v., its probability density function (pdf) is $f_X(t) = \frac{dF_X(t)}{dt}$. If X is discrete, then its probability mass function (pmf) is $P(X = k) = p_X(k)$. Everyone knows that a r.v. is well known if either its cdf, pdf or pmf is known.

However, there are also some other function, which also uniquely define the distributions. For lifetimes, these are as follows:

– The failure rate function (FRF)

$$\lambda_X(t)dt = \frac{dF_X(t)}{1 - F_X(t)} = \frac{f_X(t)dt}{1 - F_X(t)}.$$

The name FRF comes from reliability theory (Barlow and Proshan 1966). It represents the conditional probability the lifetime to end immediately after the time t, given that it did not ended until the moment t. For this reason, the FRF $\lambda_X(t)$ sometimes is called the *mortality rate* of individuals at age t. According to us, another appropriate name is *risk function* since it also represents the risk of failure at this age. Given the FRF, the cdf is found from the equation

$$F_X(t) = 1 - e^{-\int_0^t \lambda_X(u)du}, t \geq 0.$$

We believe that readers are able to define FRF for discrete distribution, since $\lambda_X(k) = P(X = k|X > k - 1)$. Hence, we omit details.

– Another useful and interesting characteristic that also uniquely identifies the distribution of a positive r.v. is the *hazard distribution function* (HDF) of X, defined by

$$\Lambda_X(x,t) = P\{X - x \leq t \mid X > x\} = \frac{F_X(x+t) - F_X(x)}{1 - F_X(x)} = \int_x^{x+t} \lambda_x(u)du.$$

This probability measure represents the conditional probability of the event (end of life) occurring within the interval $[x, x+t)$, given that it did not happen before time x. It is called *hazard measure relative to* $F_X(x)$ (Kotz and Shanbhag 1980). Originally, the HDF was introduced by Barlow and Proshan (1966) when $x = 0$. This gives us the reason to call $\Lambda_X(t)$ as well as $\Lambda_X(x,t)$ accumulated stress during the respective intervals

Our main assumptions in lifetime modeling here and further are that: *The sensors of the r.v.'s recording the impact of the surrounding random environment are that these are followed either by the HDF $\Lambda_X(x,t)$ or by the failure rate function $\lambda_X(x)$, if these exist.*

In other words, if a r.v. X is embedded into an environment of periodic nature, or other varying stress environment, these two measures would reflect these periodic properties and will have a periodic structure.

– *Transformations also uniquely define probability distributions.* For positive r.v. and integer valued counting variables, two functional transformations are usually used. These are the Laplace–Stiltjes transform (LST)

$$\phi_X(s) = \mathbf{E}(e^{-sX}) = \int_0^\infty e^{-st}\, dF_X(t),\ s \geq 0, \qquad [3.1]$$

and the *probability generating function* (pgf)

$$P_X(z) = \mathbf{E}(z^X) = \sum_{k=0}^{\infty} z^k P(X = k),\ |z| \leq 1. \qquad [3.2]$$

LST and pgf are interesting, because they allow to determine any moments of order r of the r.v. X if it exists (without integration or summation) due to the relationships

$$\mathbf{E}(X^r) = (-1)^r \frac{d^r \phi_X(s)}{ds^r}|_{s=0}\,;\ \mathbf{E}[X(X-1)\cdots(X-r+1)] = \frac{d^r P_X(z)}{dz^r}|_{z=1}.$$

Therefore, a r.v. X is uniquely and sufficiently well defined by either its cdf, pdf, pmf, FRF, HDF, LST or pgf.

In addition, we refer to the following also well known facts:

– The exponential and geometric distributions of a r.v. X is uniquely defined by either the equality

$$\bar{F}_X(x+y) = \bar{F}_X(x)\bar{F}_X(y),\ \text{valid for any values } x \geq 0, y \geq 0,$$

or

$$\phi_X(s) = \frac{\mu}{\mu + s}\,;\ P_X(z) = \frac{1-\beta}{1-\beta z}.$$

Also, $\lambda(t) = \mu$ =const, and $\Lambda(0,t) = \mu t$.

We conclude this section with two very interesting, useful and intensively used in numerous queueing studies interpretations of the LST and pgf (Gnedenko *et al.* 1972). This approach is known as *method of complimentary events*, Originally it was proposed by Kesten and Runnenburg (1957) and popularized by Klimov (1966). The method was intensively developed by Klimov's students E. Danielyan, V. Matveev, G. Mishkoy and many others, including the author.

For $s > 0$, the term e^{-st} in the LST [3.1] can be treated as the probability that a r.v. S, independent of X, with exponential distribution of parameter $s > 0$ is exceeding the time t when X terminates. Then the function $\phi_X(s) = P(S > X)$. If we conditionally call the occurrence of S a "catastrophe", then $\phi_X(s)$ can be interpreted as the probability of the event A = "during the time of duration X no catastrophes occurred". If we can find two different approaches to derive the probability $P(A)$ of this event A, it will come as a functional equation about $\phi_X(s)$. When an equation for analytic function is true for all values in an interval, the same is true elsewhere where all these functions are analytic.

Similar interpretation can apply to the pgf $P_X(z)$ when $z \in (0,1)$. Introduce a complementary event independent of X which declares event $(X = k)$ "green" with probability z^k. Then, $P_X(z)$ is the probability that all events counted by X are green. Similar derivations of probability of such events in two different ways may lead to equations for pgf's and algebraic functional equations, from which many useful conclusions can be made.

A combination of the two interpretations also holds, assuming all events are independent. The function

$$P(s,z) = \mathbf{E}[e^{-sX_t} z^{X_t}] = \sum_{k=0}^{\infty} \int_0^{\infty} e^{-sx} z^k P(X_t = k) dF_{X_t}(x)$$

can be interpreted as the probability of event A = {until a counting process X_t lasts there will be no "catastrophes" and all arrived events will be "green"}.

As an illustration we show the derivation of the LST $\pi(s)$ of the total occupation time caused by one call in the system $M/G/1$, where the service time has a cdf $B(x)$, and the input is a Poisson process of intensity λ. (a) $\pi(s)$ for $s > 0$ is the probability that during this occupation time, there will be no "catastrophes". (b) Meanwhile during the first service, other calls will arrive, and each new call will generate its own occupation time, during which no catastrophes should occur, and the probability of this is $\pi(s)$. The chance of a call coming which does not satisfy this requirement has the probability 1 - $\pi(s)$. We name such calls "bad". The input of "bad" calls is a screened Poisson with parameter $\lambda(1-\pi(s))$. Therefore, there will be no "catastrophes" during occupation time if during the first service there will be no "catastrophes" and no "bad calls" will arrive. This is a combined Poisson process of parameter $s+\lambda(1-\pi(s))$, and the probability for this event is $\beta(s+\lambda(1-\pi(s)))$. Hence

$$\pi(s)=\beta(s+\lambda(1-\pi(s))).$$

Similarly, the pgf $L(z)$ of the number L of serviced calls during the total occupation time in the same service system $M/G/1$. (a) For $z\in(0,1)$, the pgf $L(z)$ is probability that all served calls are "green". (b) Also this $L(z)$ is probability that the calls that come during occupation periods generated by any new arriving call during the first service also will be "green". "Bad calls" are those that do not fulfill this property, and they form a Poisson input of parameter $\lambda(1-L(z))$. All served calls will be green if "the first one is green" (probability of it is z, and during the first service "no bad calls" will arrive (probability of it is $\beta(\lambda(1-L(z)))$, and both events are independent. Hence

$$L(z)=z\beta(\lambda(1-L(z))).$$

We leave as an elementary exercise for our reader's pleasure and to show the validity of the equation

$$\pi(z,s)=z\beta(s+\lambda(1-\pi(z,s)))$$

for the combined above situation.

In this way, a long detailed probability analysis of situations is replaced by "tricky" auxiliary constructions, very similar to some approaches in derivation of identities in combinatorics. The good side of this approach is it allows multidimensional interpretations and generalizations. Many of these can be found in the cited above sources. The approach (it seems to me) is less explored in reliability studies, and still has great potential for further use. For instance, many asymptotic approximations for the service characteristics can be obtained from their transformations using Abelian and Tauberian theorems for Laplace transforms of functions (see, e.g. Omey and Willenkens 1989).

3.2. Unreliable service model and reliability maintenance

Service models with unreliable servers are used as basic models in services with priorities. Calls with priority may interrupt current services of calls with lower priorities. Such interruptions may be considered as breakdowns of the server, and interest to such models is understandable. However, these models are in full scale equivalent to many maintenance models for operating reliability items. Our goals here are to point out these similarities.

Consider a single server queueing system in which, while the jobs are processed, the server is subject to failures. The failures are repaired during some independent "down times" that form a sequence $\{Y_n\}_{n=1}^{\infty}$, following the interrupted services. We have instantaneous repairs when $P\{Y_n = 0\} = 1$ for all n.

We consider mainly three service disciplines as follows:

– Pre-emptive repeat different

The model assumes that a job enters service at $t_0 = 0$ on a ready to serve server. The server is able to serve during times $\{X_n\}_{n=0}^{\infty}$, usually assumed to be iid with some distribution $F_X(x)$. When in service the lifetime expires, the ongoing service is interrupted, and starts anew for a similar new service time T_n. If $T_n < X_n$, service is complete.

– Pre-emptive repeat identical

The model assumes that a job enters service at $t_0 = 0$ on a ready to serve server and requires service time $T = t$. The server is able to serve during times $\{X_n\}_{n=0}^{\infty}$, usually assumed to be iid with some distribution $F_X(x)$. When in service the lifetime expires, the ongoing service is interrupted, and starts anew for the same service time duration $T = t$. If $t < X_n$, the service is complete.

– Pre-emptive repeat resume

The model assumes that a job enters service at $t_0 = 0$ on a ready to serve server and required service time is $T = t$. The server is able to serve during times $\{X_n\}_{n=0}^{\infty}$, usually assumed to be iid with some distribution $F_X(x)$. When in service the lifetime expires, the ongoing service is interrupted, and starts from the same service status as at the time of interruption. When $T < X_1 + \cdots + X_n$, service is completed. Service is completed when the sum of all earlier interrupted times plus the last uninterrupted time becomes equal to the originally required service time T.

If interrupted, continuation of service proceeds according to the considered policy. Unreliable service means that an interrupted job is restarted according to adopted policy, and will be ended when declared accomplished. A reliable server never fails, i.e. $P\{X_1 = \infty\} = 1$. Then service is never interrupted, and is naturally ended, like a death.

Service models on an unreliable server with instantaneous repairs assume no times for repair. After an interruption, server is immediately staring the service again according to the specified policy. Other models consider some times for repair $\{Y_n\}_{n]1}^{\infty}$ and these prolong the occupation time. All of these models are used in reliability maintenance too.

– Relative priority disciplines

These are specific in priority considerations only. Instead of interruptions all arrived calls of higher priority are waiting for the current service to be completed, and will be served next. In reliability models, it could be equivalent to something like trials to "hack" the ongoing process, and then trials to recover the situation. To the best of our knowledge, such models have not been discussed in reliability. But in queueing theory they are analyzed for specific purposes.

In addition to these "classic" service disciplines, we offer here some that have not been discussed yet.

– Pre-emptive repeat resume with relevation

The concept of relevation is introduced by Krakowski (1973). It is easily understood in terms of reliability, as the total lifetime of a system is formed by two components. The first component has a lifetime T with distribution function $F_T(t)$, and the second component has a lifetime X. As soon as the first component fails, it is immediately replaced by the second component, but of the same age, as the first when it failed. The system lifetime is the total life when the second component also fails. The relevation of the two distributions is the distribution of the total lifetime Z of this system, and is denoted by the symbol "#", i.e.

$$F_Z(t) = F_T \# F_X(t)$$

The distribution of the r.v. Z is determined by its survival function

$$P(Z > t) = P(T > t) + \int_0^t \frac{1 - F_X(t)}{1 - F_X(u)} dF_T(u). \qquad [3.3]$$

And the r.v. Z is called a result of the relevation of the r.v.'s T and X. Its symbolic notation is

$$Z = T \# X.$$

How relevation may work in modeling a service on an unreliable server? It could be the following. The required service time is T, and the live times are X_n. If the service is not interrupted at a certain attempt, the service of that call is complete. If it is interrupted at a certain accumulation $X_n = u$, then at the next restarted attempt it is continued as a service time survived duration u from many unfinished services at the same "age". If this "residual" service time is finished before the next lifetime of the server X_{n+1} then the service will be completed. If the server fails earlier, then the residual service will be "relevated" from the age of the last interrupted stage. We call this service discipline *pre-emptive repeat resume with relevation.*

If the accumulated service at the interruption equals u, then the "relevated" residual time T_u has distribution as the conditional residual lifetime of an item, which is at age u, i.e.

$$P(T_u > t) = P(T > t\,|\,T > u) = \frac{1 - F_T(t)}{1 - F_T(u)}. \tag{3.4}$$

Take into account that at the kth interruption, the accumulated service work is $X_1 + \cdots + X_k$ and we introduce the random index

$$\nu = \min\{k;\ T_{X_1+\cdots+X_{k-1}} < X_k\} \tag{3.5}$$

Then the total occupation time of the server by a call under this service discipline will be presented by the random sum

$$Z = \sum_{k=1}^{\nu-1}[X_1 + \cdots + X_k] + T_{X_1+\cdots+X_{\nu-1}}. \tag{3.6}$$

Properties of queueing systems under pre-emptive repeat resume with relevation have not been studied yet. We just noted some to challenge researchers in this area. There are some useful approaches provided by Baxter (1982). Models could be made even more interesting, by taking repair times into account. It is also challenging to consider restarts from some earlier, or later stage than the interruptions, in analogy to maintenance with improvements, or sloppy repair in reliability. But, let me stop here.

3.3. Characterizations of exponential and geometric distributions via properties of service times

Since Galambos and Kotz (1978), new properties based on processes and sampling features have been shown to be equivalent to the well-known lack of memory property (noticed in section 3.1), which uniquely characterize the exponential distribution. Dimitrov and Khalil (1990) derived a new characterization using properties of a queueing model with an unreliable server. We give here the most important components that participate in such processes. Afterward, we consider those coincidence in distributions of variables that lead to characterization of the exponential distribution.

The model considerations usually assume that a job enters service at $t_0 = 0$ on a ready to serve server. The server is able to serve during times $\{X_n\}_{n=0}^{\infty}$, usually assumed to be iid with some distribution $F_X(x)$. When in service the lifetime expires, the ongoing service is interrupted. If interrupted, continuation of service starts according to the considered policy. Unreliable service means that an interrupted job is restarted according to adopted policy, and will be ended when it is declared accomplished (see section 3.2). Reliable server never fails, i.e. $P\{X_1 = \infty\} = 1$. Then service is never interrupted and is naturally ended, like a death.

Consider the service discipline "pre-emptive repeat identical".

Since the service time is a r.v. T, it will be completed accordingly, when the required service time is interruption. Similar thing go with the policy "pre emptive repeat different", and every interrupted service starts again with a new service time with the same distribution as T. The service will be finished if no interruption occurs. Our focus here is on characterization of probability distributions via properties of service times. We do not consider service characteristics. And we present only those results that contribute to various techniques of proofs or approaches.

3.3.1. *Instant repairs: characterization of geometric distribution*

In this section, we present that the consecutive service times T_n have just some discrete distribution over the non-negative integers with $P\{T_n = k\} = p_k \geq 0$ for $k = 0, 1, \ldots$. An ongoing service T_n starts on a working server and it is interrupted if the server's lifetime ends. Also the server lifetimes X_n are iid r.v. with discrete distribution over the non-negative integers with $P\{X_n = k\} = q_k \geq 0$ for $k = 0, 1, \ldots$. We call $\{X_n\}_{n=1}^{\infty}$ *truncating process* to the process $\{T_n\}_{n=1}^{\infty}$. Let $p = P\{X_n > T_n\} \in (0, 1)$.

We introduce the r.v.

$$N = \inf\{k \geq 1 \, : \, \{T_k \leq X_k\}. \qquad [3.7]$$

Clearly, $P\{N = k\} = p^{k-1}p, k = 1, 2, \ldots$, i.e. N has a geometric distribution with parameter p. We mark such a fact with notation $N \sim Ge(p)$

Next we define

$$Z = \sum_{k=1}^{N-1} X_k + T_N \qquad [3.8]$$

where an empty sum in convenience is assumed equal to 0. The r.v. Z represents the total occupation time of the server to complete one service under this pre-emptive repeat different service discipline on a server with instanced repairs. A service is completed when the lifetime of the server is longer or equal to required service time at this attempt. When server's lifetime is shorter, service is interrupted, the used time for service is counted to the total service time, and anew service starts when the server is ready to start again. If a service is not completed during the first (as well as after any other repeated attempt), then each interruption is point of regeneration for this service process. This fact is used in most of the considerations in the sequel, when we analyze probabilities of events at the start and their possible parts later.

Introduce the pgf

$$P_Z(z) = \mathbf{E}[z^Z] = \textstyle\sum_{k=0}^{\infty} z^k P\{Z = k\};$$
$$P_T(z) = \mathbf{E}[z^{T_n}] = \textstyle\sum_{k=0}^{\infty} z^k p_k\};$$
$$P_X(z) = \mathbf{E}[z^{X_n}] = \textstyle\sum_{k=0}^{\infty} z^k q_k,$$

Denote the joint pgf

$$P_1(z) = \mathbf{E}[z^{T_1} \cap \{T_1 \leq X_1\}] = \sum_{k=0}^{\infty} z^k p_k \sum_{m=k+1}^{\infty} q_m,$$

and

$$Q_1(z) = \mathbf{E}[z^{X_1} \cap \{T_1 > X_1\}] = \sum_{k=0}^{\infty} z^k q_k \sum_{m=k+1}^{\infty} p_m.$$

LEMMA 3.1.– The pgf of the occupation time Z is given by the equation

$$P_Z(z) = P_1(z)[1 - Q_1(z)]^{-1}.$$

PROOF.– Let Z_1 be the conditional occupation time after the first interruption. Since Z_1 is independent, then on the variables T_1 and X_1 we have

$$\begin{aligned} P_Z(z) &= \mathbf{E}[z^Z \cap \{T_1 \leq X_1\}] + \mathbf{E}[z^Z \cap \{T_1 > X_1\}] \\ &= P_1(z) + \mathbf{E}[z^{X_1+Z_1} \cap \{T_1 > X_1\}] \\ &= P_1(z) + Q_1(z)P_Z(z). \end{aligned}$$

Solution of this equation determines $P_Z(z)$ as stated.

COROLLARY 3.1.– The expected occupation time is determined by the expression

$$\mathbf{E}Z = \frac{\sum_{k=0}^{\infty} kp_k \sum_{m=k+1}^{\infty} q_m}{1 - \sum_{k=0}^{\infty} kq_k \sum_{m=k+1}^{\infty} p_m}$$

The proof is immediate, since $\mathbf{E}Z - P_Z'((1)$

COROLLARY 3.2.– If the truncating process $\{X_n\}$ has some geometric distribution, i.e. $X_n \sim Ge(\alpha)$ with some $\alpha \in (0, 1)$, then

$$P_Z(z) = \frac{1 - \alpha P_T(\alpha z)}{1 - \frac{1-\alpha}{1-\alpha z}(1 - \alpha z P_T(\alpha z))} \qquad [3.9]$$

and

$$\mathbf{E}Z = [1 - P_T(\alpha)]/[(1 - \alpha)P_T(\alpha)] \qquad [3.10]$$

PROOF.– Since $q_m = \alpha^m(1 - \alpha)$, we have

$$P_1(z) = \sum_{k=0}^{\infty} z^x p_k \alpha^{k+1} = \alpha P_T(\alpha z)$$

and

$$\begin{aligned} Q_1(z) &= \sum_{k=0}^{\infty} z^k \alpha^k (1 - \alpha) \sum_{m=k}^{\infty} p_m \\ &= (1 - \alpha) \sum_{m=0}^{\infty} p_m \sum_{k=0}^{m} \alpha^k z^k \\ &= \tfrac{1-\alpha}{1-\alpha z} \sum_{m=0}^{\infty} p_m [1 - (\alpha z)^{m+1}] \\ &= \tfrac{1-\alpha}{1-\alpha z]} [1 - \alpha z P_T(\alpha z)] \end{aligned}$$

Using these expression, we get [3.9].

COROLLARY 3.3.– If $T_n \sim Ge(\beta)$, and truncating sequence has an arbitrary distribution with $q_0 = P\{X_n = 0\} < 1$, then Z has the same distribution as T_n

PROOF.– Proceeding as in the proof of Corollary 3.2, we obtain

$$P_1(z) = (1 - \beta)(1 - P_X(\beta z))/(1 - \beta z).$$

and

$$Q_1(z) = P_T(\beta z)$$

After substituting these expressions in the equation of Lemma 3.1, we get

$$P_Z(z) = \frac{1-\beta}{1-\beta z} = P_T(z)$$

This fact that for any truncating process $\{X_n\}$, the service times T_n and the total blocking time Z have the same distribution (are stochastically equivalent) appears as a new property of the geometric distribution. It leads us to the following characterizations of the geometric distribution.

THEOREM 3.1.– Let the truncating process X_n have a geometric distribution $Ge(\alpha)$ with some parameter $\alpha \in (0.1)$ and the service times satisfy $P\{T_n = 0\} < 1$. Then X and T are stochastically equivalent (the coincidence in distribution $Z =^d T_n$ holds for the total blocking time and the required service times) is fulfilled if and only if T_n have a geometric distribution.

PROOF.– The necessity part is actually proven in corollary 3.3.

Sufficiency. Let $Z =^d T_n$. Then it is also valid $P_Z(z) = P_T(z)$. Define

$$h(z) = \frac{1 - P_T(z)}{1 - zP_T(z)|}.$$

Then the following is also valid

$$P_T(z) = \frac{1-h(z)}{1-zh(z)} \qquad [3.11]$$

Substitute $P_Z(z) = P_T(z)$ and [3.11] into [3.10], and obtain

$$h(z) = h(\alpha z).$$

Hence, by iteration we obtain that for any integer n the following is fulfilled

$$h(z) = h(\alpha^n z).$$

Therefore, taking the limit in $n \to \infty$, and use of continuity

$$h(z) = h(0).$$

If $h(0) = 1$, then $P\{T_n = 0\} = 1$ – a case to be excluded. If $h(0) > 0$, then [3.11] becomes the pgf of a geometric function of parameter $h(0)$.

Another interesting characterization can be obtained just from observation of the expected blocking time Z.

THEOREM 3.2.– The blocking time Z has a constant expectation $\mathbf{E}Z = c$ if and only if for any truncating sequence $X_n \sim Ge(\alpha)$ with any $\alpha \in (0,\ 1)$ the service times $T_n \sim Ge(\beta)$, and $\beta = c/(1+c)$.

PROOF.– Necessity is already part of the statement of theorem 3.1.

Sufficiency. Let $\mathbf{E}Z = c$ for any $\alpha \in (0,\ 1)$, then from the statement of Corollary 3.2 we get

$$P_T(\alpha) = \frac{1}{1+c(1+\alpha)} = \frac{1-\frac{c}{1+c}}{1-\frac{c}{1+c}\alpha}$$

If we denote here $\beta = c/(1+c)$, we will see that for all $\alpha \in (0,\ 1)$ the following is fulfilled

$$P_T(\alpha) = \frac{1-\beta}{1-\alpha\beta} = (1-\beta)\sum_{k=0}^{\infty}(\alpha\beta)^k.$$

Since the series $P_T(\alpha)$ coincides with the geometric series $(1-\beta)\sum_{k=0}^{\infty}(\alpha\beta)^k$ for a continuous set of arguments α, the coefficients in front of the corresponding powers α^k also coincide. Thus

$$p_k = P\{T_n = k\} = (1-\beta)\beta^k.$$

Hence, $T_n \sim Ge(\beta)$.

Referring to corollary 3.3, it seems natural to conjecture that blocking time and the service times coincide in distribution only when this is the geometric distribution. However, the following counterexample (due to S. Chukova) shows something different.

Let the truncating process (lifetimes of the server) has constant duration $K > 1$, and let the service times satisfy $0 < P\{T_n > K\} < 1$. Then the pgf $P_1(z)$ and $Q_1(z)$ get the forms

$$P_1(z) = \sum_{k=1}^{K} p_k z^k; \;\; Q_1(z) = z^{K+1}(1 - \sum_{k=1}^{K} p_k)$$

Now the equation in Lemma 3.1 for

$$P_Z(z) = P_T(z) = \sum_{k=1}^{\infty} p_k z^k$$

gives the identity

$$\sum_{k=1}^{\infty} z^k p_k \left[1 - z^{K+1}(1 - \sum_{k=1}^{K} p_k)\right] = \sum_{k=1}^{K} p_k z^k. \qquad [3.12]$$

If we set $p_k = 1/K$ for k = 1, 2, ..., K and

$$p_{nK+k} = \frac{1}{K+1}\left(\frac{1}{K+1}\right)^n \quad \text{for } n = 0, 1, 2, \ldots; \, k = 1, 2, \ldots, K,$$

then this distribution is not geometric, but satisfies [3.12]. This is a key example for the discussions and consequences we use in section 3.2 that generated the idea of the ALM probability distributions.

Results here are based on the work of Khalil *et al.* (1991).

3.3.2. *Instant repairs: characterizations of the exponential distribution*

Here, we consider the same model as in previous section, in the case of continuously distributed components. The service times $\{T_n\}$ are iid with continuous distribution $F_T(x) = P\{T_n \le x\}$ and the LST $\phi_T(s) = \mathbf{E}[e^{-sT_n}]$. The server lifetimes $\{X_n\}$ (the truncating process intervals) also are iid r.v. with distributions $F_X(x) = P\{X_n \le x\}$ and LST $\phi_X(s) = \mathbf{E}[e^{-sX_n}]$. The total blocking (processing time of a job) Z under instant repairs is presented by the random sum as in the discrete distribution case. This time is defined as the time from the first start at $t_0 = 0$ until the job is completed, i.e.

$$Z = X_1 + X_2 + \cdots + X_{\nu-1} + T_\nu,$$

where the random index in the sum is defined as

$$\nu = \min\{n: \ [T_1 \geq X_1] \cap \cdots \cap [T_{n-1} \geq X_{n-1}] \cap [T_n < X_n]\}$$

THEOREM 3.3.– Under the assumption that $\{X_n\}$ have an exponential distribution with any parameter $\lambda > 0$, than Z and X_n have the same probability distribution if and only if this distribution is the exponential.

PROOF.– During the first attempt for service here may be no interruption, if service ends before the breakdown, and $Z = T_1$. If the interruption occurs, the service will be started again as new, and the process is restarted as at the beginning with duration Z. Therefore, the first interruption during a service, if occurring, is the point of regeneration for the service process. Considering the two occasions we find the relationships

$$F_Z(x) = \int_0^x e^{-\lambda u} dF_{T_1}(u) + \int_0^x [1 - F_{T_1}(u)] F_Z(x-u) \lambda e^{-\lambda u} du,$$

valid for any $x > 0$. The second integral above is convolution of the functions $F_Z(x)$ and $\lambda e^{-\lambda x}[1 - F_{T_a}(x)]$. This observation helps us obtain the following relationship between the corresponding LST:

$$\phi_Z(s) = \phi_T(s+\lambda) + \frac{\lambda}{s+\lambda}[1 - \phi_T(s+\lambda)] \qquad [3.13]$$

Necessity. If the required service times are exponential and $F_T(x) = 1 - e^{-\mu x}$, then $\phi_T(s) = \frac{\mu}{\mu+s}$. Substitute this expression in

$$\phi_Z(s) = \frac{\phi_T(s+\lambda)}{1 - \frac{\lambda}{s+\lambda}[1 - \phi_T(s+\lambda)]}$$

and after some algebra get that $\phi_Z(s) = \frac{\mu}{\mu+s}$. Therefore, the total service time has the same exponential distribution as the required service duration.

Sufficiency. Now assume that $F_Z(x) = F_T(x)$ for any exponentially distributed truncating sequence $\{X_n\}$. Using that $\phi_Z(s) = \phi_T(s)$, let us rewrite equation [3.13] in the form (we drop temporarily the subscripts for these equal LST)

$$\frac{\phi(s+\lambda) - \phi(s)}{\lambda} - \frac{1}{s+\lambda}[1 - \psi(s+\lambda)]\psi(s).$$

This is a functional equation for the LST $\phi(s)$, valid for any positive values of the arguments s and λ. Taking the limit in it $\lambda \to 0$, we obtain the differential equation

$$\phi'(s) = \frac{1}{s}[\phi(s) - 1]\phi(s).$$

This is a Bernoulli differential equation for $\phi(s)$ with initial condition $\phi(0) = 1$. Its solution is

$$\phi(s) = \frac{1}{1+cs} \text{ with some } c > 0. \qquad [3.14]$$

Since $\phi'(0) = \mathbf{E}Z = \mathbf{E}T_n = 1/c$, we get from [3.14]

$$\phi_Z(s) = \phi_t(s) = \frac{1/c}{1/c + s}.$$

The therefore, the following is fulfilled

$$F_Z(x) = F_T(x) = 1 - e^{-x/c}.$$

The theorem is proved.

On many occasions the methods of proof determine the conditions when your statement is true. There were Van Harn and Steuiel (1991) who gave me the hint, that only for one value $\lambda > 0$ in the exponential distribution of the truncating process the statement of the above theorem is true, then $Z =^d T_n$ holds if and only if Z and T_n are equally exponentially distributed. We use the approach in the proof of Van Harn and Steuiel and get to our own proof of the following simplifying conditions for the same statement:

THEOREM 3.4.– The equality in distributions $Z =^d T_n$ between the total blocking time Z and the required service times X_n in the pre-emptive repeat different service discipline holds if and only if for some $\lambda > 0$ for the truncating process $X_n \sim Exp(\lambda)$ it holds that $T_n \sim Exp(\mu)$ with some $\mu > 0$.

PROOF.– The necessity part is already proved in the above theorem.

The sufficiency. Note that if we use the transformation

$$\phi_Z(s) = \phi_T(s) = \phi(s) = \frac{\varphi(s)}{s + \varphi(s)} \qquad [3.15]$$

in equation [3.13], we will see that this function $\varphi(s)$ satisfies the functional equation

$$\varphi(s + \lambda) = \varphi(s)$$

for any $s \geq 0$ and the given $\lambda > 0$. Therefore, the function $\varphi(s)$ is periodic with period $\lambda > 0$, and satisfies the equation

$$\varphi(s + n\lambda) = \varphi(s) = \varphi(0).$$

Thus, $\varphi(s)$ is periodic with period $\lambda > 0$, and $\lim_{n\to\infty} \varphi(s + n\lambda) = \varphi(0)$. But from [3.15] we obtain

$$\varphi(0) = \lim_{s\to 0} \frac{\phi(s)}{\frac{1-\phi(s)}{s}} = \frac{1}{\mathbf{E}Z} = \mu.$$

Now put $\varphi(s) = \varphi(0) = \mu$ in [3.15] and get the statement.

The most interesting things in these characterizations is that these are characterizations *via the past history* of the process. The well-known characterization called "Lack of Memory property" (LM property, see section 3.1) $P\{Z > t + x|T \geq x\} = P\{Z > t\}$ is a characterization *via the future*. In our opinion, this is the reason for the huge research interest into the topic. The moral of the here described characterizations is that the requirements are too complicated. The researchers started looking for simple practical conditions for characterization of certain classes of probability distributions, which could be practically recognized, and possibly statistically confirmed.

Huang and Shoung (1993) have shown that this characterization is still true under the assumption that the truncating process $\{X_n\}$ forms a general renewal process (not necessarily Poisson), with a continuous cdf $F_X(x)$. Their statement is:

THEOREM 3.5.– Let the distributions $F_Z(x), F_{X_n}(x)$ and $F_{T_n}(x)$ are all continuous. If $F'_X(0+) \neq 0$ exists and $F_T(x)$ has points of increase at any $x > 0$, then the stochastic equivalence $Z =^d T$ holds if and only if T_n have an exponential distribution.

Use the fact that X_1 is a point of regeneration, and some elementary probabilistic analysis of the process obtains the relationship

$$F_Z(x) = \int_0^x (1 - F_T(u))dF_X(u) + \int_0^x (1 - F_T(u))F_Z(x - u)dF_X(u). \quad [3.16]$$

Necessity. If $T_n \sim Exp(\mu)$, this equation yields the relation between corresponding LST:

$$\phi_Z(s) = \frac{\mu}{s(\mu + s)} (1 - (\mu + s)\phi_T(\mu + s) + (\mu + s)\phi_T(\mu + s))\, \phi_Z(s).$$

Solving for $\phi_Z(s)$ get $\phi_Z(s) = \mu/(\mu + s)$.

Sufficiency. Let $F_Z(x) = F_T(x) = G(x)$, where $G(x)$ is temporary notation for the assumed equal distribution. After some manipulations from [3.16] can be rewritten in the form

$$\int_0^x (G(x) - G(t))dF_X(t) = \int_0^x (1 - G(t))G(x - t)dF_X(t),$$

or equivalently

$$\int_0^x (1 - G(t))(1 - G(x - t))dF_X(t) = \int_0^x (1 - G(x))dF_X(t).$$

Introduce the notation

$$dQ(t) = (1 - G(t))dF_X(t)$$

in the last equation, and obtain the functional equation

$$\int_0^x (1 - G(x - t))dQ(t) = \int_0^x \frac{1 - G(x)}{1 - G(t)} dQ(t). \qquad [3.17]$$

The following lemma, due to Lau and Rao (1990), proves that the only solution of this functional equation is valid for all $x \geq 0$, where $G(x) \neq 1$ is the exponential function $G(x) = 1 - e^{-\mu x}$ for some $\mu > 0$.

LEMMA 3.2.– Let $G(x)$ and $Q(x)$ be two lifetime distributions that satisfy the equation [3.17] for all x where $G(x) \neq 1$. If the survival functions $1 - G(x)$ and $1 - Q(x)$ are continuous, the derivative $G'(0)$ exists, and $Q(x)$ has a non-zero change at any argument $x > 0$, then the survival function $1 - G(x)$ is exponential, i.e. $1 - G(x) = e^{-\mu x}$ for some $\mu > 0$.

PROOF.– This proof (according to me) is a brilliant illustration of the application of calculus on specific proofs.

Necessity. When $1 - G(x) = e^{-\mu x}$, then [3.17] holds since

$$\int_0^x e^{-\mu(x-t)} dQ(t) = \int_0^x \frac{e^{-\mu x}}{e^{-\mu t}} dQ(t)$$

always holds.

Sufficiency. Let $c = \sup\{x;\, G(x) < 1\}$. Let h be the non-negative function such that $h(0) = 0$ and $e^{-h(x)} = 1 - G(x)$ for all $x \in [0,\, c)$. Then $h'(0+)$ exists according to the assumptions, and

$$\int_0^x \left[e^{-h(x-t)} - e^{-h(x)+h(t)}\right] dQ(t) = 0 \text{ for all } x \in [0,\, c)\,,$$

according to [3.17]. Note that $m(t) = [e^{-h(x-t)} - e^{-h(x)+h(t)}$ and by the mean value theorem in calculus, there exists a point $\eta \in [0,\, x)$ such that the following is fulfilled

$$\int_0^x \left[e^{-h(x-t)} - e^{-h(x)+h(t)}\right] dQ(t) = m'(\eta)[O)(x) - Q(0)].$$

The only factor here that can be equal to 0 is $m'(\eta) = 0$. This is equivalent to

$$h(x) - h(x-\eta) - h(\eta) = 0, \text{i.e. } h(x) = h(x-\eta) - h(\eta)$$

for any $x \in [0, c)$ and for some $0 < \eta < x$. And the last equation above is equivalent to

$$\frac{h(x) - h(\eta)}{x - \eta} = \frac{h(x-\eta)}{x-\eta}.$$

Here if we allow $x \to 0$, then also $\eta \to 0$ and $x - \eta \to 0$ and the limit above is $h'(0+) = \mu$. Hence, $h(x)$ is a linear function $h(x) = \mu x + b$, where $b = 0$ since $h(0) = 0$, and $h'(0+) = \mu \geq 0$. Therefore $1 - G(x) = e^{-\mu x}$ and the service times T_n and the occupation time Z are exponentially distributed, since it was just a notation $F_Z(x) = F_T(x) = G(x)$. The statement is proved.

Moreover, Huang and Shoung proved in the same article, that if the service times and the truncating times have strictly increasing cdf's, and in addition it is known that the class of service time distributions is either from NBU (New Better than Used), or from NWU (New Worst than Used), then the equality $\mathbf{E}Z = \mathbf{E}T_n$ implies that both distributions are exponential. The NBU and NWU are well-known classes of probability distributions in reliability theory, roughly expressing physical properties of the lifetimes of technical items with these distributions.

Lin (1993) remarked that our characterization of the exponential distribution in the exponential truncating process is also valid either when $\mathbf{E}(Z) = \mathbf{E}(T_n) = \mathbf{const}$ or when $\mathbf{E}(Z^2) = 2[\mathbf{E}(Z)]^2$. Here we give a simplified proof of Lin's statement.

THEOREM 3.6.– For a sequence of exponentially distributed truncating process $\{X_n\}$ with parameters $\lambda_n \in (0, \lambda_0)$, $\lambda_0 > 0$, either the equality

$$\mathbf{E}Z = c, \quad \text{where } c > 0 \text{ is a fixed constant,}$$

is fulfilled or the equality $\mathbf{E}Z^2 = 2[\mathbf{E}Z]$ if and only if the service times T_n are exponentially distributed.

PROOF.– Sufficiency part follows from already proven above facts. *Necessity.* From Theorem 3.3, we know that (we drop the subscript n in λ_n)

$$\phi_Z(s) = \frac{(s+\lambda)\phi_T(s+\lambda)}{s+\lambda\phi_T(s+\lambda)} \text{ for } s > 0,$$

Solve this equation for $\phi_T(s)$ and obtain that

$$\phi_T(s) = \frac{(s-\lambda)\phi_Z(s-\lambda)}{s-\lambda\phi_Z(s-\lambda)} \text{ for all } s > \lambda. \qquad [3.18]$$

So, the distribution of T_n is uniquely determined by the distribution of Z for any λ. It is well known that the moments of order r for a positive r.v. are determined by its LST derivatives at 0. So, accordingly, we have

$$\mathbf{E}Z^r = \lim_{s\to 0}(-1)^r \phi_Z^{(r)}(s),$$

where $\phi_Z^{(r)}(s)$ is the rth derivative of the function $\phi_Z(s)$. Hence, if $\mathbf{E}Z = c$ and relation [3.18] holds, then

$$\begin{aligned}\mathbf{E}Z &= \lim_{s\to 0}(-1)\phi_Z'(s)\\ &= \frac{1}{\lambda}\left(\frac{1}{\phi_T(\lambda)} - 1\right) = c \text{ for any } \lambda \in (0, \lambda_0).\end{aligned}$$

Solve this equation for $\phi_T(\lambda)$ and get

$$\phi_T(\lambda) = \frac{1}{1+\lambda c}.$$

Thus, by the uniqueness theorem from the theory of analytic, functions, Tn are exponentially distributed as Z with parameter $1/c$. The first part of the statement is proved. Let $\mathbf{E}Z^2 = 2[\mathbf{E}Z]$. Introduce for a while the function

$$h(\lambda) = \frac{1}{\lambda}\left(\frac{1}{\phi_T(\lambda) - 1}\right).$$

Then, again by using equation [3.18] and evaluating two derivatives of $\phi_Z(s)$ at 0, for the moments of Z we get

$$\mathbf{E}Z = h(\lambda); \ \mathbf{E}Z^2 = 2[h^2(\lambda) - h'(\lambda)].$$

The assumed equation $\mathbf{E}Z^2 = 2[\mathbf{E}Z]$ tells us that in the equation

$$2[h^2(\lambda) - h'(\lambda)] = 2[h(\lambda)]^2$$

the quantity $h'(\lambda) = 0$ for all $\lambda \in (0, \lambda_0)$. Hence $h(\lambda) = c$, and therefore, T_n are exponentially distributed. The fact that $Z =^d T$ follows from the necessity. Theorem is proved.

3.3.3. *Various simplifying conditions*

Galambos and Hagwood (1994) established a similar characterization of the exponential distribution under the following assumptions, named failure scenarios:

(Fs1) The lifetimes X_n are non-random, infinitely many, taken from the set of all non-random sequences (where $X_i = t_1$ may be arbitrarily close to 0), and $t_k \to \infty$ for each sequence in use.

(Fs2) The times $t_k = X_1 + ..., X_k$ are random and, infinitely many, do not depend on service; X_1 may be arbitrarily close to 0 with positive probability, and the number of interruptions N_t for each finite interval $(0, t)$ satisfies $P(N_t < \infty) = 1$.

An additional assumption that the service time $F_{T_n}(t)$ has the right derivative at zero is also made. In other words, they assume that the process N_t is generated by any arbitrary sequence $\{X_n\}_{n=0}^{\infty}$, non-random for (Fs1) and random for (Fs2), which may have infinitely many points located arbitrarily close to zero (no need for $\{N_t\}$ to be a renewal process). These conditions imply that the server may have arbitrarily small lifetimes, and this is the key in their proofs. In our opinion, these are too artificial mathematical requirements, far away of any practical sense. We try to give it some vitality below.

In what follows we show that exponential characterization of the exponential distribution is true by relaxing the assumptions of Galambos and Hagwood. It is enough to assume that in scenario (Fs1), the truncating point process consists of just two non-random points of specific nature $0 < t_1 < t_2 < \infty$ or at one random point (one continuous variable, X_1) in the scenario (Fs2), where $P(X_2 = \infty) = 1$. Our additional assumption is that the function $F_{X_1}(t)$ is right continuous at zero.The results below have been published in the work of Dimitrov *et al.* (2001).

To avoid possible misinterpretation for the involved time points, we will use notations a and b instead of t_1 and t_2, and specify the models of the breakdown process N_t in the following cases:

Case 1. The process $\{N_t\}$ contains just two points, formed by the deterministic variables $X_1 = a$, $X_2 = b - a$ (since $X_3 = \infty$), where a and b are two incommensurable numbers. That is, a and b are two positive numbers such that the ratio a/b is an irrational number, and these are the only points of the process

Case 2. The process $\{N_t\}$ contains only one random point X_1, since the second variable is $X_2 = \infty$ with probability 1.

More specifically, we assume that the service process for any new job obeys one of the following two failure scenarios:

FS1: the job processing starts at $t_0 = 0$. The server failures (or interruptions) occurring on $(0, \infty)$ are only at the two non-random times, $t_1 = a$ and $t_2 = b$ after the process starts, where $0 < a < b < \infty$, with the property that a and b are incommensurable numbers. If an interruption occurs, the service is immediately restarted as a new job. One job processing may have at most two interruptions. Occupation time ends when the job is successfully completed without interruption.

FS2: the job processing starts at $xzt_0 = 0$. The server may fail just once during the total processing time. The time to failure $X_1 > 0$ has a continuous distribution function $H(x)$ strictly increasing on $[0, \infty)$, and X_1 is independent of the service times before and after the failure. If interrupted, the service is restarted as for a new job and no more interruptions will occur for this particular service. Occupation time ends when the service is successfully done.

The practical motivation for the above two scenarios is as follows: in the first case, we pick two time points, a and b, with the property to be incommensurable and give two chances for the job to be finished. At these two time points, the server will be refreshed instantaneously (as another identical server) and the service (if not completed prior to this failure) will start anew. After the second interruption, the service will be completed without any interruptions (no more hazard). In the second scenario, at the start of processing a new job, a clock with a random lifetime (given by X_1) is also started. If the clock fails before the processing is over, the ongoing service is interrupted, the server is refreshed instantaneously, and the job processing is restarted with no further interruptions.

We show that the total processing time is not affected by the possible service interruptions under either one of the two failure scenarios if and only if the originally requested service time T is exponentially distributed.

Our analysis is based on a long-acknowledged fact from Carl G. Jacobi (1804–1851), concerning real-valued periodic functions. (Its original formulation and proof can be found in Jacobi (1882), II, 25–26.) Here, we state this main proposition in the terms required for our purposes.

PROPOSITION 3.1.– Let $a > 0$, $b > 0$ be two incommensurable real numbers. If a real-valued function $g(t)$ defined on $[0, \infty)$ is right-continuous at zero and is a periodic of period a as well as of period b, then $g(t)$ is a constant function $g(t) = g(0)$ for all $t \geq 0$.

PROOF.– It is worth noting some steps of the proof. First, observe that if $g(t)$ is a periodic of period a, it is also a periodic of period ka for any choice of the integer $k \geq 1$. Second, observe that if $g(t)$ is a periodic of period b, then it is periodic with a period mb. If $g(t)$ is periodic in both a and b, it is periodic with periods $ka - mb$ with an arbitrary choice of integers $k \geq 1, m \geq 1$. The later difference could be made arbitrarily small due to the properties of the incommensurable real numbers.

Now we turn to our specific considerations. We write $T =^d Z$ to state that two r.v.'s T and Z have the same probability distribution, i.e., that $P(T \leq t) = P(Z \leq t)$ for any $t \in 0, \infty)$.

Case 1: Consider the first scenario FS1. During a service at most two failures may occur at certain non-random times a and b, where $a < b$.

THEOREM 3.7.– Let a single server queueing system have a job requiring service time T, whose probability distribution is right continuous at 0. The server is unreliable according to FS1. The server's occupation time Z by the job coincides in distribution with the required processing time T (i.e., the equality $T =^d Z$ holds) if and only if one of the following two cases holds:

i) either $P(T \leq a) = 1$ or $P(T \leq a) = 0$, and then $P(T = \infty) = 1$;

ii) $0 < P(T \leq a) < 1$, and then t is exponentially distributed.

PROOF.– *Necessity:* First, we prove that when (i) or (ii) holds, then we have $T =^d Z$.

Denote by T_0, T_1 and T_2 the requested processing times at the start, and the possible first and second interruptions correspondingly. The value $a > 0$ is fixed in FS1. Thus, if $P(T \leq a) = 1$, then the service will always be completed before the first breakdown of the server occurs, and hence $T =^d Z$. If $P(T = \infty) = 1$, then T_0, T_1 and T_2 are the same, any service will need an infinite amount of time. The first two processes will be interrupted and the third process, started at point b, will not exist. Adding infinite amount of time T_2 to the already used time, namely b, we get the total service time $Z = T_2 + b = \infty$. Thus, we have $P(Z > t) = P(T > t) = 1$ for all $t \geq 0$, without the exponential assumption.

Let us now assume that the required service time has exponential distribution. Then the service times of any other attempt for service T_0, T_1 and T_2 are also exponentially distributed with the same parameter, say μ. Therefore, we have

$$P(T_i > t) = e^{-\mu t}, \quad t \geq 0, \quad i = 0, 1, 2. \qquad [3.19]$$

We are interested in the distribution of the r.v. Z, the total processing time of the job, under the FS1 working conditions. When there are no failures in $(0, t]$, $(t \leq a)$, we simply have

$$P(Z > t) = P(T > t) = exp(-\mu t), \quad 0 < t \leq a. \qquad [3.20]$$

When server failures may occur in $(0, t]$ for the failure scenario FS1, the non-random failure mode, we consequently have

$$P(Z > t) = \begin{cases} P(T_0 > t) = e^{-\mu t}, & if\ t \leq a; \\ P(T_0 > a, T_1 > t - a) = e^{-\mu t}, & if\ t \in (a, b], \\ P(T_0 > a, T_1 > b - a, T_2 > t - b) = e^{-\mu t}, & if\ t > b. \end{cases} \qquad [3.21]$$

We show the third case of the above equation. We use [3.19] and the independence. If $t > b$, then

$$\begin{aligned} P(Z > t) &= P(T_0 > a, T_1 > b - a, T_2 > t - b) \\ &= P(T_0 > a)P(T_1 > b - a)P(T_2 > t - b) \\ &= e^{-\mu a} e^{-\mu(b-a)} e^{-\mu(t-b)} = e^{-\mu t}. \end{aligned}$$

The other proof is similar. That is, for any $t \geq 0$, we have $P(Z > t) = P(T > t)$ for whichever of the cases (i) or (ii) is valid.

Sufficiency: Let $T =^d Z$ be fulfilled under the FS1 model. The inequalities

$$0 < P(T > a) < 1 \qquad [3.22]$$

mean that with positive probability at least one interruption will be observed in this case, and also that there is a positive chance of the service being completed on the first (or second) attempt. Then this will correspond to case (ii) in our statement.

In view of [3.22], we will prove that any service duration also can be observed, and also we show what the two extreme cases $P(T \leq a) = 1$, and $P(T > a) = 1$ produce case (i) as a consequence.

Since T_i's are independent, and the interruptions do not prolong the service, we have $Z =^d T_i,\ i = 0, 1, 2$. Detailed probability analysis of the "what may happen beyond first interruption" shows that we also have the equation

$$P(Z > t) = P(T_0 > a, T_1 > t - a) = P(T_0 > a)P(T_1 > t - a). \qquad [3.23]$$

It holds for any $t > 0$. Notice that equation [3.23] is equivalent to

$$G(a)G(t - a) = G(t) \quad \text{for} \quad \text{any} \quad t > 0. \qquad [3.24]$$

By setting in [3.24] $t = 2a, 3a, ...$, we obtain (by induction, and in view of [3.22]) that it is true

$$0 \leq G(na) = [G(a)]^n \leq 1. \qquad [3.25]$$

It follows from the equations in [3.25] that if $G(a) = 0$, then $P(T \leq a) = 1$ and the first case of statement (i) holds. If $G(a) = 1 = P(T > a)$, then also $P(T > na) = 1$ for any arbitrary n. This implies that $P(T = \infty) = 1$, and the second case of statement (i) is then true.

To avoid these trivial cases in our further considerations, we assume that strict inequalities in [3.22] hold. Then we show that the service time T may have arbitrarily large values with positive probability, since always $G(na) = P(Z > na) = P(T > na) > 0$. Moreover, from equations [3.24] and [3.25], we obtain the equalities

$$G(t) = G\left(t - \left[\frac{t}{a}\right] a\right) [G(a)]^{\left[\frac{t}{a}\right]a}, \quad \text{for} \quad \text{any} \quad t > 0, \qquad [3.26]$$

where $[x]$ denotes the integer part of number x.

Now we consider the case when two breakdowns occur. In this case, from the last equality in [3.21] we see that the function $G(t)$ also satisfies the equations

$$G(t) = G(t - b)G(b - a)G(a), \quad \text{for any} \quad t > b. \qquad [3.27]$$

If we put $t = b$ in [3.24], we get $G(b) = G(b - a)G(a)$. Substitute this expression into the right-hand side of [3.27] and obtain that the following equation is true

$$G(t) = G(t - b)G(b) \quad \text{for} \quad \text{any} \quad t > b. \qquad [3.28]$$

Similarly as in derivation of [3.26] the following is fulfilled

$$G(t) = G\left(t - \left[\frac{t}{b}\right] b\right) []G(b)]^{[\frac{t}{b}]b}, \quad \text{for any} \quad t > 0/,. \qquad [3.29]$$

From both equations [3.26] and [3.29], we conclude that the equalities

$$G(ka) = [G(a)]^k, \quad \text{and} \quad G(mb) = [G(b)]^m$$

are true for any choice of the positive integers k and m. From the later, we obtain

$$\frac{\ln G(a)}{a} = \frac{\ln G(ka)}{ka}; \qquad \frac{\ln G(b)}{b} = \frac{\ln G(mb)}{mb}. \qquad [3.30]$$

For any $t > 0$, there exist two numbers k and m such that the following is fulfilled

$$ka \leq t < (k+1)t, \quad \text{and} \quad mb \leq t < (m+1)b.$$

Since the left-hand sides for each of the equations in [3.30] are finite, and $G(t) > 0$ for any t is monotonic (therefore $\ln G(t)$ is monotonic), we get the limits

$$\lim_{t\to\infty} \frac{\ln G(t)}{t} = \lim_{k\to\infty} \frac{\ln G(ka)}{ka} = \lim_{m\to\infty} \frac{\ln G(mb)}{mb}.$$

Moreover, in view of [3.30]

$$\lambda = \frac{\ln G(a)}{a} = \frac{\ln G(b)}{b} < 0. \qquad [3.31]$$

Consider the function

$$\alpha(t) = \ln G(t) - \lambda t.$$

Using equations [3.24], it can easily be verified that

$$\begin{aligned} \alpha(t+a) &= \ln G(t) + \ln G(a) - \lambda t - \lambda a \\ &= \alpha(t) + a\left(\frac{\ln G(a)}{a} - \lambda\right) = \alpha(t), \end{aligned}$$

for any $t > 0$. That is, the function $\alpha(t)$ is periodic with period a. Similarly, using equation [3.28] we obtain

$$\begin{aligned}\alpha(t+b) &= \ln\ G(t) - \ln\ G(b) - \lambda t - \lambda b \\ &= \alpha(t) + b\left(\frac{\ln\ G(b)}{b} - \lambda\right) = \alpha(t).\end{aligned}$$

Therefore, $\alpha(t)$ is also periodic with period b. The two numbers a and b are incommensurable, and $G(t)$ is supposed right continuous at 0. Thus, from Proposition 1 it follows that $\alpha(t)$ is a constant function. That is,

$$\alpha(t) = \alpha(a) = \alpha(b) = \alpha(0) = 0,$$

implying that $\ln\ G(t) = \lambda t$, or $G(t) = e^{\lambda t}$, $\lambda < 0$, i.e. T has an exponential distribution.

REMARK.– The following example shows that the requirement a and b to be incommensurable is essential. We will show that there are continuous non-exponential distributions of T, for which $P(Z > t) = P(T > t)$ for the model of scenario FS1.

EXAMPLE.– Let the p.d.f. of T be given by the expression

$$g(t) = \beta^{[\frac{t}{a}]}(1-\beta)g_0(t - \left[\frac{t}{a}\right] a), \quad 0 < \beta < 1,$$

where $g_0(t)$ is the p.d.f. of any random variable T_0 satisfying the condition $P(0 \leq T_0 < a) = 1$. Then the survival function of Y is

$$\begin{aligned}G(t) = 1 - \int_0^t g(t)\ dt &= 1 - \sum_{k=0}^{[\frac{t}{a}]-1} \int_{ka}^{(k+1)a} \beta^k(1-\beta)g(t-ka)\ dt \\ &+ \int_{[\frac{t}{a}]a}^{t} \beta^{[\frac{t}{a}]}(1-\beta)g_0(t - \left[\frac{t}{a}\right] a)\ dt \\ &= 1 - \sum_{k=0}^{([\frac{t}{a}]-1)a} \beta^k(1-\beta) + \beta^{[\frac{t}{a}]}(1-\beta)\int_0^{t-[\frac{t}{a}]a} g_0(u)\ du \\ &= \beta^{[\frac{t}{a}]}\left(\beta + (1-\beta)\left(1 - G_0(t - \left[\frac{t}{a}\right] a)\right)\right) \\ &= \beta^{[\frac{t}{a}]}\left(G_0(t - \left[\frac{t}{a}\right] a)\right), \quad 0 < \beta < 1.\end{aligned}$$

Here, $G_0(t) = P(T_0 > t)$ stays for the survival function of T_0 that satisfies the conditions $G_0(0) = 1$, $G_0(a) = 0$, and may be continuous on $[0, a)$.

Let the failures occur at times $t_1 = a$ and $t_2 = b = 3a$. Then, using the fact that

$$k + \left[\frac{t-ka}{a}\right] = \left[\frac{t}{a}\right], \quad \text{for any} \quad t \geq ka, \quad k = 1, 2, \ldots,$$

it can readily be verified that equation [3.21] is also satisfied by $G(t)$.

Case 2: Now consider the second scenario FS2. Suppose we have one failure that occurs at random time X. It holds the following.

THEOREM 3.8.– Let us consider in a single server queueing system a job that requires service time T with continuous probability distribution. The service is unreliable and follows FS2. The total occupation time by the job, Z, will have the same probability distribution as the required processing time, T, (i.e., $Z =^d T$) if and only if T has an exponential distribution.

PROOF.– *Necessity.* Denote by T_1 and T_2 the requested service times in the first, and the possible second start of the job processing. Both variables are iid, with exponential distribution of some parameter $\mu > 0$ as the distribution of T, and are independent of the time to failure, X. Conditioning on the values of X, by the total probability rules we get

$$\begin{aligned} P(T > t) &= P(Y_1 > t, X > t) + P(Y_1 > X, X + Y_2 > t) \\ &= P(Y_1 > t)P(X > t) + \int_0^t P(Y_1 > x)P(Y_2 > t - x)dH(x) \\ &= e^{-\mu t}P(X > t) + \int_0^t e^{-\mu x}e^{-\mu(t-x)}dH(x) \\ &= e^{-\mu t}P(X > t) + e^{-\mu t}P(X \leq t) = e^{-\mu t}. \end{aligned} \quad [3.32]$$

Sufficiency: Let it be true that $T =^d Z$, i.e., $G(t) = P(T > t) = P(Z > t)$ for arbitrary $t > 0$. With the assumption that X has an absolutely continuous c.d.f. on $[0, \infty)$, we can easily see, from the first of the equations in [3.32], that the following relation

$$\int_0^t [G(u)G(t-u) - G(t)]dH(u) = \int_0^t [G(u)G(t-u) - G(t)]h(u)\, du = 0 \quad [3.33]$$

must be true for any $t > 0$. With the assumption that X has a continuous distribution which is strictly increasing on the interval $[0, \infty)$, $h(t) = H'(t) > 0$ holds for all $t > 0$. We will prove that [3.33] implies

$$G(u)G(t-u) = G(t) \quad \text{for all} \quad 0 \leq u \leq t, \quad \text{and all} \quad t > 0.$$

This is the usual *lack of memory* property that characterizes the exponential distribution.

Let

$$\theta = inf\{t;\ t \geq 0,\ G(t) \neq G(t-u)G(u), \quad \text{for all } u,\ 0 \leq u \leq t\},$$

and assume that $\theta < \infty$. Then, by the continuity of $G(t)$, there exists some $\varepsilon > 0$ such that $r(t,u) := G(t) - G(t-u)G(u)$ is either constantly positive, or constantly negative for any $t \in (\theta, \theta+\varepsilon)$. Thus, for any value of $t \in (\theta, \theta+\varepsilon)$ we have (since there is at least one u for which $h(u) > 0$)

$$\begin{aligned} \textstyle\int_0^t r(t,u)h(u)du &= \textstyle\int_0^\theta r(t,u)h(u)du + \int_\theta^t r(t,u)h(u)du \\ &= \textstyle\int_\theta^t r(t,u)h(u)du \neq 0, \end{aligned}$$

which contradicts equation [3.33].

We conclude with a result, which is in some sense a refinement of Theorem 3.3 in Huang and Shoung (1993). They proved that if the required service time Y belongs to the class of NBUE (or NWUE) distributions, i.e. when $\int_t^\infty (1 - F_T(u))\, du \leq (\geq)$ $\mathbf{E}(T)(1 - F_T(t)$ for any $t \geq 0$, then under a scenario of failures that form a renewal process the equality $\mathbf{E}(Z) = \mathbf{E}(T)$ implies T has an exponential distribution. We drop the requirement for possibly infinitely many breakdowns during a service, but narrow the class to NBU (or NWU) distributions, where the property

$$P(T \geq x+y) \leq (\geq) P(T \geq x)P(T \geq y)$$

holds for any non-negative values of x and y.

THEOREM 3.9.– Assume that the service follows FS2, and the server's lifetime distribution function $F_X(x)$ contains at least two incommensurable points of increase, $a > 0, b > 0$, i.e. $dF_X(a) > 0$ and $dH(b) > 0$ hold. Furthermore, assume that the survival function $G(y)$ of the required service time T is continuous with $\mathbf{E}(T) < \infty$ and belongs to either one of the classes NBU, or NWU, of probability distributions. Then the total occupation time Z by the job has the same average as the originally requested service time T, i.e. $\mathbf{E}(Z) = \mathbf{E}(T)$ if and only if T is exponentially distributed.

PROOF.– The sufficiency portion is already proved in Theorem 3.8 since when T has exponential distribution, then the same distribution has Z. We prove the sufficiency that $\mathbf{E}(Z) = \mathbf{E}(T)$ implies exponential distribution of T. From the definition of Z in the FS2 mode, we obtain (please, refer to the second line of equation [3.32])

$$\begin{aligned} P(Z > t) = G(t)(1 - F_X(t)) \\ + \textstyle\int_0^t [G(u)G(t-u) - G(t)]dF_X(u) = 0 \end{aligned}, \qquad [3.34]$$

Since T and Z are non-negative random variables, using the expressions

$$\mathbf{E}(Z) = \int_0^\infty P(Z > t)\, dt$$

and

$$\mathbf{E}(T) = \int_0^\infty P(T > t)\, dt = \int_0^\infty G(t)\, dt$$

in an integration of equation [3.34] from 0 to ∞, we get

$$\mathbf{E}(T) = \mathbf{E}(Z) + \int_0^\infty \int_0^t G(t)G(t-u)dF_X(u)dt - \int_0^\infty \int_0^t G(t)dF_X(u)dt.$$

If $\mathbf{E}(Z) = \mathbf{E}(T)$, then the above equality implies that the equation

$$\int_0^\infty \left(\int_0^t (G(t)G(t-u) - G(t))dF_X(u) \right) dt = 0 \qquad [3.35]$$

holds. Note that the difference $(G(u)G(t-u) - G(t))$ is either non-negative (for $G(t)$ in NBU the class) or non-positive (for $G(t)$ in the NWU class). Thus, equation [3.35] implies that

$$G(t) = G(u)G(t-u) \text{ for any } u \text{ with } dF_X(u) > 0, \text{ for all } t > u.$$

According to our assumption, there are at least two incommensurable points $a > 0$ and $b > 0$, where

$$G(t) = G(a)G(t-a) \quad \text{and} \quad G(t) = G(b)G(t-b), \text{ for any } t > \max(a, b).$$

By using the statement of Theorem 3.7 for the FS1 scenario, the stated result follows. Statement is proved.

REMARK.– Characteristic properties of the exponentially distributed r.v. with instantaneous repairs may have many faces. The unreliable service under specific conditions provides simple feasible features that can be used for recognition of exponentially distributed random variables. In this section, by using a single server queue with simple failure modes (with one, or at most two failures), we obtain another characterization of the exponent. Similar characterization can be carried out for the geometric distribution also. To the best of our knowledge, such studies have not been published.

3.3.4. *Unreliable service, repair times included*

We keep the notation T_n for the service times, X_n for the server lifetimes, Y_n for the repair times and Z for the total service time of the server from servicing one job. All involved random sequences are supposed independent. We exclude the situations, where $P\{T_n \leq X_n\} = 1$ or $P\{T_n > X_n\} = 1$. These immediately lead to the conclusions that $Z \stackrel{d}{=} T) = 1$ or $P\{Z = \infty\} = 1\}$. Therefore, consider the case $0 < P\{T_n \leq X_n\} < 1$.

Introduce the random variable

$$\nu_1 = \inf\{n;\ n \geq 1,\ T_n < X_n\}.$$

Then the total blocking time Z_1 of the service is presented by the equation

$$Z_1 = \sum_{k=1}^{\nu_1 - 1} (X_k + Y_k) + T_{\nu_1}. \qquad [3.36]$$

We compare this total service time with the total service time in the pre-emptive repeat identical service discipline with the same sequence X_n for the server lifetimes, and Y_n for the repair times. Denote by Z_2 the total occupation time of the server from one service. Introduce the random index

$$\nu_2 = \inf\{n;\ n \geq 1,\ X_1 + \ldots + X_n > T\}.$$

Then

$$Z_2 = T + \sum_{k=1}^{\nu_2 - 1} Y_k.. \qquad [3.37]$$

Here is our next characterization.

THEOREM 3.10.– If the coincidence in distribution $Z_1 =^d Z_2$ is valid for any lifetime of the server $X_n = x$ arbitrary close to 0, then the required service time $T \sim Exp(\mu)$ with some $\mu > 0$.

PROOF.– *Necessity.* When two distributions are identical, their transformations are also equal. We work with positive r.v., and we use LST transformations. For Z_1, we have

$$\begin{aligned}\phi_{Z_1}(s) = &\left[\int_0^\infty e^{-sx}(1 - F_X(x))dF_T(x)\right] \times \\ &\times \left[1 - \phi_Y(s) \int_0^\infty e^{-sx}(1 - F_T(x))dF_X(x)\right]^{-1}.\end{aligned} \quad [3.38]$$

This equation is proved by using that end of the first repair time $X_1 + Y_1$ is the regeneration point for the entire service process. For the LST duration in second service discipline, we obtain the equations

$$\phi_{Z_2}(s) = \int_0^\infty \beta(s,t)dF_T(t) \quad [3.39]$$

where the function $\beta(s,t)$ is determined by the equation

$$\beta(s,t) = e^{-st}(1 - F_X(t)) + \int_0^t e^{-sx}\beta(s, t - x)\phi_Y(s)dF_X(x). \quad [3.40]$$

The Laplace transform of the function $\beta(s,t)$ with respect to the argument t is determined by the equation

$$\begin{aligned}\beta^\star(s,\tau) &= \int_0^\infty e^{-t\tau}\beta(s,t)dt \\ &= [1 - \phi_X(s + \tau][(s + \tau)(1 - \phi_Y(s)\phi_X(s + \tau))]^{-1}.\end{aligned} \quad [3.41]$$

If $T \sim Exp(\mu)$, then $\phi_T(s) = \dfrac{\mu}{s + \mu}$, and from [3.38] we get

$$\phi_{Z_1}(s) = \frac{\mu}{(s + \mu)} \frac{1 - \phi_X(s + \mu)}{1 - \phi_Y(s)\phi_X(s + \mu)}.$$

On the other hand, under the same distributions of the components it is true

$$\phi_{Z_2}(s) = \int_0^\infty \beta(s,t)\mu e^{-\mu t}dt = \mu\beta^*(s,\mu).$$

When we use the same $\phi_T(s) = \frac{\mu}{s+\mu}$ in [3.41] we will be confirmed after some algebraic manipulation that

$$\phi_{Z_2}(s) = \int_0^{\infty} \beta(s,t)\mu e^{-\mu t} dt.$$

Therefore, using expression [3.39] we see that

$$\phi_{Z_1}(s) = \phi_{Z_2}(s)$$

for any distribution of the truncating sequence X_n (lifetimes of the server), and any repair time distributions of Y_n.

Sufficiency. Let

$$\phi_{Z_1}(s) = \phi_{Z_2}(s)$$

be true for any distributions of the truncating sequence X_n (lifetimes of the server), and any repair time distributions of Y_n.

Let $P\{X = x\} > 0$ for any $x > 0$. Then for this fixed x, after expanding right hand of [3.38] as a time series, and partitioning the integral on $[0, \infty)$ over intervals $[kx, (k+1)x]$ on the right of [3.38] we will have

$$\begin{aligned} \beta_1(s) &= \left[\int_0^x e^{-st} dF_T(t)\right] [1 - \phi_Y(s))e^{-sx}(1 - F_T(x))]^{-1} \\ &= \sum_{k=0}^{\infty} [\phi_Y(s)]^k e^{-ksx} [1 - F_T(x)]^k \int_0^x e^{-st} dF_T(t) \end{aligned}$$

and

$$\beta_2(s) = \sum_{k=0}^{\infty} \int_{kx}^{(k+1)x} e^{-st} dF_T(t) [\phi_Y(s)]^k.$$

Now, for non-instantaneous repairs we have $P\{Y = 0\} \neq 0$, and therefore $0 < \phi_Y(s) < 1$ for any $s > 0$ where this LST is defined, and is strictly decreasing in s. Therefore, the coefficients in front of $[\phi_Y(s) <]^k$ in the two series above coincide. Thus, we fulfill

$$[1 - F_T(x)]^k \int_0^x e^{-st} dF_T(t) = \int_{kx}^{(k+1)x} e^{-st} dF_T(t) \qquad [3.42]$$

for any $k = 0, 1, 2, \ldots$. Now, a summation over all k in these equations shows that it is true for the equation

$$\frac{\int_0^x e^{-st} dF_T(t)}{1 - e^{-sx}[1 - F_T(x)]} = \int_0^\infty e^{-st} dF_T(t) = \phi_T(s)$$

where $\phi_Y(s)$ is not presented. Hence, it is true

$$\int_0^x e^{-st} dF_T(t) = \phi_T(s)[1 - e^{-sx}(1 - F_T(x))].$$

Take the differential in x on both sides and obtain

$$e^{-sx} dF_T(x) = \phi_T(s)[1 - e^{-sx}\{s(1 - F_T(x))] + dF_T(x)\}.$$

It is equivalent to (this is true for any s arbitrary close to 0)

$$\frac{dF_T(x)}{1 - F_T(x)} = \frac{s\phi_T(s)}{1 - \phi_T(s)} = c.$$

Either of these solutions gives that $F_T(x) = 1 - e^{-cx}$, or $\phi_T(s) = c/(c + s)$, i.e. the service time is exponentially distributed.

Theorem is proved.

It is interesting when comparison between the expected total service times in the two service disciplines may imply specific class of distributions. The curious fact is that when the servers lifetimes X_n are exponentially distributed, the expected duration of the services in pre-emptive repeat different and pre-emptive repeat identical are equal if and only if the service time is exponentially distributed. A challenge is to find when in other conditions something similar is true. We can prove the following statement:

THEOREM 3.11.– If the lifetimes X_n of the server are exponentially iid distributed for a series of rates λ_j with finite point of condensation, and the iid repair times Y_n are not instant, then the equality between the expected duration of the total service times $\mathbf{E}(Z_1) = \mathbf{E}(Z_2)$ holds if and only if the service times T_n are exponentially distributed.

PROOF.– Assume that the lifetimes of the server have an exponential distribution $X \sim Exp(\lambda)$. Then using the LST of the service times

$$\phi_{Z_1}(s) = \phi_T(s + \lambda)\left[1 - \phi_Y(s + \lambda)\tfrac{s}{s+\lambda}(1 - \phi_T(s + \lambda))\right]^{-1};$$
$$\phi_{Z_2}(s) = \phi_T(s + \lambda - \lambda\phi_Y(s))$$

we obtain

$$\mathbf{E}Z_1(\lambda) = \left[\frac{1}{\lambda} + \mathbf{E}Y - \frac{1}{\lambda}\phi_T(\lambda)\right]\phi_T(\lambda) = \mathbf{E}T(\lambda\mathbf{E}Y + 1) = \mathbf{Z_2}(\lambda)$$

Simple algebraic manipulations show that then the following is fulfilled

$$\phi_I(\lambda) = \frac{1}{1 + \mathbf{E}T\lambda} \qquad [3.43]$$

Obviously, if $T \sim Exp(\mu)$ this is an identity, and sufficiency is fact. For necessity, it is sufficient to mention that when [3.43] is fulfilled for a sequence of values of parameter λ with a finite point of condensation, it is true for any values of λ, i.e. the LST of service times T, $\phi(s)$ is that of an exponential distribution.

Theorem is proved. More exotic situations discussed can be found in Dimitrov and Khalil (1994). There may be other interesting comparisons, but no more different mathematical techniques are used, and we do not point more detail here.

3.4. Probability distributions almost having lack of memory property

Here, we show that there are random variables X, for which the lack of memory property

$$P\{X - b \geq x | X \geq b\} = P\{X \geq x\} \qquad [3.44]$$

holds for any $x > 0$ and for infinitely many values of the constant b, and X do not have exponential distribution. Chukova and Dimitrov (1992) called the class of these distributions *almost-lack-of-memory (ALM) class.*

DEFINITION 3.1.– *A non-negative r.v. Z (or the distributions $F_Z(x)$ with support on $[0, \infty)$ is said to have the ALM property if the equation*

$$P\{Z - c \geq t \mid Z \geq c\} = P\{Z \geq t\} \qquad [3.45]$$

holds for all values of $t \geq 0$ and for infinitely many countable distinct values of the constant. If [3.45] is valid just for some fixed value of $c > 0$, we say that Z has lack of memory (LM) property at the moment c.

3.4.1. *Service time on an unreliable server: instantaneous repairs*

Again consider a single server queueing system with a unreliable serve where server's lifetimes are iid $\{X_n\}$ and with the pre-emptive repeat different service discipline, and iid sequence of service times $\{T_n\}$, and instantaneous repairs. With the notation

$$N = \min\{n;\, X_n < T_n, n \geq 1\} \qquad [3.46]$$

the total execution time of one service is

$$Z = \sum_{k=1}^{N} \min\{X_k,\, T_k\}. \qquad [3.47]$$

We have seen in the previous section that when the service times T_n have exponential, or geometric distribution, then it is valid for the stochastic equivalence

$$Z =^d T_n \qquad [3.48]$$

between total execution times and the required service times, no matter what is the server's lifetime distribution. And we were curious, if and when there are other distributions different from exponential and geometric, for which equations [3.48] hold. The following result is valid:

THEOREM 3.12.– If the server lifetimes are constant, i.e. if $P(X_n = c) = 1$ for some $c > 0$, then equations [3.48] hold if and only if the service times have distributions of the form

$$F_T(t) = 1 - \alpha^{\left[\frac{t}{c}\right]}\left(1 - (1-\alpha)F_0(t - \left[\frac{t}{c}\right]c)\right), \quad t \geq 0, \qquad [3.49]$$

where $\alpha \in [0, 1]$, $F_0(t)$ is a distribution with a support within the interval [0, 1], and $\left[\frac{t}{c}\right]$ is the integer part of the number t/c.

The class of distributions presented by equation [3.49] depends on three parameters, α, c and $F_0(t)$, and we denote it as $T \sim ALM(\alpha, c, F_0(t)$, where $F_0(t)$ is a distribution with support on [0, c). We call this class the *ALM class* with parameters $\alpha, c, F_0(t)$.

PROOF.– *Necessity.* Let $T \sim ALM(\alpha, c, F_0(t)$. If $\alpha = 0$, then $F_T(t) = F_o(t)$, and service is ended before the server fails, i.e. [3.48] holds. If $\alpha = 1$, then $F_T(t) = 0$, and service is never ended. Again [3.48] holds.

Let $0 < \alpha < 1$. Looking at [3.46] we understand that the r.v. N has a geometric distribution with parameter $P(T_n > X_n) = 1 - F_T(c) = \alpha$. Next we note that for any $T > 0$, there exist an integer k such that $kc \leq t < (k+ \leq t)$ and then

$$\begin{aligned} P(Z > t) &= P(N = k+1)P(T_N > (t - kc)) \\ &= 1 - (\alpha)^k[1 - (1 - \alpha)(1 - F_0(t - kc)) \\ &= P(T > t). \end{aligned}$$

Therefore, the total service time and the required service time coincide in distribution.

Sufficiency.

$$P(Z > t + c) = P(T_1 > c, T_2 > t) - (1 - P_T(c))(1 - P_T(t)),$$

i.e. the function (we drop the subscripts Z and T)

$$H(t) = 1 - P(Z > t) = 1 - P_T(t)$$

satisfies the equation

$$H(t + c) = H(c)H(t) \qquad [3.50]$$

for any $t \geq 0$). We call this property of the function $H(t)$ *LM at the point c when* $c > 0$ *is a constant.* Then using induction by letting in (3.50) $t = c, 2c, \ldots$, the following is true

$$H(kc) = [H(c)]^k \text{ for any } k = 1, 2, \ldots$$

And also

$$H(kc + t) = H(kc)H(t) = [H(c)]^k H(t)$$

for any $k - 1, 2, \ldots$ and any $t \geq 0$. Therefore the distributions of Z and T have the lack of memory property at infinitely many points of the form $kc, k = 1, 2, \ldots$ and can be named ALM distributions.

Next we consider

$$
\begin{aligned}
P(Z<t) &= P(T_1<t, T_1 \le c) + P(T_1 > c, Z < t) \\
&= \frac{P(T_1<t)}{1-P(T>c)} + \sum_{k=1}^{\infty} P(N=k)P(T_{k+1}<t-kc, T_{k+1}\le c) \\
&= \frac{P(T<t)}{1-P(T>c)} + \sum_{k=1}^{\infty} \alpha^k (1-\alpha \frac{P(T<t-kc)}{1-P(T>c)} \\
&= 1-\alpha^{\left[\frac{t}{c}\right]}\left(1-(1-\alpha)F_0(t-\left[\frac{t}{c}\right]c)\right), \quad t\ge 0,
\end{aligned}
$$

Here, we have used the notations $\alpha = P(T > c)$, and

$$
F_0(t) = P(T < t \,|\, T \le c) = \frac{F_T(t)}{1 - F_T(c)}, t \in [0,\, c]
$$

for the conditional distribution of the service time T on the interval $[0,\, c]$. Hence, $Z =^d T$ have the ALM distribution [3.49].

Theorem is proved.

Chukova and Dimitrov (1992) put more attention on the analysis of discrete case, where lifetime c of the server is an integer, and service times also are integer valued. A service of a call is considered completed even when service time and lifetime of the server end at the same instant. Since there is no principal difference in the proofs of this theorem in continuous and discrete cases, we omit these details.

The good side of the class of the ALM distribution [3.49] is that parameters α, c and the distribution $F_0(t)$ on $[0, c)$ can be any, and no special restrictions on $F_0(t)$ are required. This fact makes it flexible enough in modeling various practical situations.

In a letter to the editor Lin (1992) noted that any ALM distribution of parameter c is the distribution of the total service time Z in a queueing system where the server has constant lifetimes $X_n = c$, and Z and the required service time are statistically equivalent.

That raises an important question: How does the start of service $t_0 = 0$ affect the ALM property of the total service time? The answer is not at all.

The following statement is true:

THEOREM 3.13.– A r.v. Z has the LM property at some moment $c > 0$ if and only if Z has an ALM distribution. Actually, it is true: If $Z \sim ALM(\alpha, c, F_0(t)$, then

$$P\{Z - (nc + y) \geq t \mid Z \geq nc + y\} = P\{Z - y \geq t \mid Z \geq y\} \qquad [3.51]$$

holds for all $t \geq 0$, for arbitrary $y \in [0, c)$, and for arbitrary integer $n = 1, 2, \ldots$.

PROOF.– Denote by Z_y the total service time when the service starts at a stage $y \in [0, c[$. We will show that Z and Z_y possess the LM property at the same point c. We have

$$\begin{aligned} P(Z_y > c) &= P(Z > y + c) = P(Z > y)P(Z > c) \\ &= P(Z_y > 0)P(Z > c) = P(Z > c); \\ P(Z_y > t + c) &= P(Z < t + y + c) = P(Z > t + y)P(Z > c) \\ &= P(Z_y > t)P(Z_y > c). \end{aligned}$$

Therefore, Z_y has an ALM distribution with parameters $\alpha\, c$, and some underlying distribution parameter $F_{0,y}(t) = \frac{F_Z(y+t)}{1-F_Z(y+c)}, t \in [0,\ c]$.

The statement is proved. It says that the ALM property is invariant with respect to time shifts not exceeding the length of the period c, given that the r.v. survives the time of that shift. And surely, this result rises some interesting mathematical and statistical challenging question: What mathematical model (as a formula) provides the best fit of the origin y in regard to a real process. This question is open to researchers.

3.4.2. *Properties of ALM distributions, and equivalent presentations*

Here, we focus readers' attention on the various conventional components in presentations of probability distributions and random variables that belong to the class of ALM distributions with parameters $\alpha, c, F_0(t)$.

3.4.2.1. *Probability density*

COROLLARY 3.4.– If $Z \sim ALM(\alpha, c, F_0(t))$ and :

A) If Z is continuously distributed, then its pdf is of the form

$$f_Z(t) = \alpha^{\left[\frac{t}{c}\right]}(1 - \alpha) f_0(t - \left[\frac{t}{c}\right] c) \quad , t \geq 0,$$

where $f_0(t)$ is the pdf of the component T_0 whose cdf is $F_0(t)$ with support on $[0, c]$.

B) If Z has discrete distribution on $[0, c]$ and $c > 1$ is an integer then the pmf of Z is presented by the expression

$$P(Z = k) = \alpha^{[\frac{k}{c}]}(1 - \alpha)P(T_0 = k - \left[\frac{k}{c}\right] c) \quad , k \geq 0,$$

where $P_{T_0}(k)$ is the pmf of the component T_0 whose cumulative distribution is $F_{T_0}(k) = P(T_0 \leq k)$ defined for $k = 0, 1, \ldots, c$

The proof follows from differentiation of the cdf in equation [3.49] with respect to argument t in continuous case, or from the facts that $P(Z = k) = P(Z \geq k) - P(Z \geq k + 1)$ and

$$P(Z \geq k) = 1 - \alpha^{[\frac{k}{c}]} \left(1 - (1 - \alpha)P(T_0 \geq k - \left[\frac{k}{c}\right] c)\right), \quad k \geq 0$$

in the discrete case.

3.4.2.2. *The failure rate $\lambda(t)$ and Hazard rate $\Lambda(t)$ functions*

This observation is kind of a key to discovering relations to random processes and to many potential applications in practice and modeling. We include in our considerations only continuous-time cases. Close analogy may transfer these results to the cases of discrete-time counts.

The *failure rate function* of a continuous r.v. Z is defined as in section 3.1. Using it, we get the following.

THEOREM 3.14.– a) If the waiting time failure rate function has the form

$$\lambda_Z(t) = \frac{(1 - \alpha) f_{T_0}(t - \left[\frac{t}{c}\right] c)}{1 - (1 - \alpha) F_{T_0}(t - \left[\frac{t}{c}\right] c)}, \quad t \geq 0.$$

b) The failure rate function of the waiting time is periodic function of period c, i.e. it satisfies

$$\lambda_Z(t + kc) = \lambda_Z(t), \quad \text{for any } t \geq 0, \quad k = 1, 2, \ldots$$

PROOF.– The form of $\lambda_Z(t)$ follows from definition of failure rates, and from the expressions for the density function and the respective cdf of the r.v. Z.

The periodicity is also easy to see, from part (a). Theorem is proved.

The unit relationship between failure rate and the cdf of a r.v. allows us to identify it with either of the two functions, From definition of the failure rate (see Barlow and Proshan 1965), we have: a r.v. Z is defined by its failure rate function $\lambda_Z(t)$, since

$$F_Z(t) = 1 - exp\left\{-\int_0^t \lambda_Z(u)\,du\right\}, \quad t \geq 0.$$

Further assume $\lambda_Z(t)$ is a periodic of period $c > 0$:

$$\lambda_Z(t + c) = \lambda_Z(t).$$

THEOREM 3.15.– If for a non-negative r.v. Z the failure rate exists and is a periodic of period of duration $c > 0$, then its cdf and pdf of Z are determined by the forms given in [3.45], where

$$\alpha = exp\left\{-\int_0^c \lambda_Z(u)\,du\right\};$$

The r.v. T_0 is defined either by its cdf

$$F_{T_0}(t) = \frac{1}{1 - exp(-\int_0^t \lambda_Z(u)\,du)}\left\{1 - exp\left(-\int_0^t \lambda_Z(u)\,du\right)\right\}, \quad t \in [0,\ c).$$

or by its pdf

$$f_{T_0}(t) = \frac{\lambda_Z(t)}{1 - exp(-\int_0^t \lambda_Z(u)\,du)} exp\left(-\int_0^t \lambda_Z(u)\,du\right), \quad t \in [0,\ c).$$

PROOF.– We drop the subscript again. Use the presentation

$$\int_0^t \lambda(u)\,du = \sum_{k=0}^{[\frac{t}{c}]-1} \int_{kc}^{(k+1)c} \lambda(u)\,du + \int_{[\frac{t}{c}]c}^t \lambda(u)\,du.$$

Take into account that

$$\int_{kc}^{(k+1)c} \lambda(u)\,du = \int_0^c \lambda(u + kc)\,du = \int_0^c \lambda(u)\,du = \alpha,$$

and

$$\int_{[\frac{t}{c}]c}^{t} \lambda(u)\, du = \int_{0}^{t-[\frac{t}{c}]c} \lambda(u)\, du.$$

Finally, use that

$$exp\{\sum_{k=0}^{M-1} \int_{kc}^{(k+1)c} \lambda(u)\, du\} = \prod_{k=0}^{M-1} e^{\int_{kc}^{(k+1)c} \lambda(u)\, du} = (\alpha)^M,$$

and after some algebraic manipulations and use of the introduced notations, obtain the forms as stated in the theorem.

Another useful and interesting characteristic that also uniquely identifies the distribution of a positive r.v. is the *hazard distribution function* (HDF) of Z, defined as shown in section 3.1. The philosophic assumptions expressed there get reflection onto properties of the probability distributions, and they are expressed by the following definitions. We propose the following:

DEFINITION 3.2.– *We say that a r.v. Z has a periodic hazard distribution function (PHDF) of period c if for any t ≥ 0, it hold that*

$$\Lambda_Z(x+c,t) = \Lambda_Z(x,t) \ \textit{for all} \ x \geq 0.$$

Note that the HDF of Z is periodic only with respect to its first argument (this is the time elapsed before the first event occurs).

It is easy to show, by induction, that if the HDF of Z is a periodic function of period $c > 0$, then the following is fulfilled

$$\Lambda_Z(x+nc,t) = \Lambda_Z(x,t), \quad x,t \geq 0 \qquad [3.52]$$

for any integer $n = 0, 1, 2, ...$

We present additional properties of random variables with PHDF or PRFR. These variables are produced under the action of random environments with repeated periodic behavior.

The following statements are true:

THEOREM 3.16.– A r.v. Z has PHDF of period $c > 0$ if and only if $F_Z(t)$ has the form

$$F_Z(t) = 1 - \alpha^{\left[\frac{t}{c}\right]}\left(1 - (1-\alpha)F_{T_0}(t - \left[\frac{t}{c}\right]c)\right), \quad x \geq 0, \qquad [3.53]$$

where $\alpha \in (0,\ 1)$ and $F_{T_0}(y)$ is the c.d.f. of a r.v. T_0 with the support on the interval $[0,\ c)$.

THEOREM 3.17.– A r.v. Z has the LM property at some moment $c > 0$ if and only if Z has a PHDF of period c.

Remember, the ALM property is invariant with respect to time shifts not exceeding the length of the period c, given that the r.v. survives the time of that shift.

ALM r.v. is a sum of independent components

THEOREM 3.18.– A r.v. Z has PHDF of period $c > 0$ if and only if it can be represented by a sum

$$Z = T_0 + cN,$$

where T_0 is concentrated on the interval $[0, c)$ with probability 1, and N has the geometric distribution $P\{N = k\} = \alpha^k(1-\alpha)$, $k = 0, 1, 2, ...$ with some $\alpha \in [0, 1]$.

PROOF.– The proof of this statement relates to calculating the LST of Z and to the observation that it is a product of corresponding transforms of the r.v.'s T_0 and cN. We can see this by comparing the LSTs of the two sides in the statement. We show it in the way we got it.

Consider the LST of an ALM distribution and the relationship established in theorem 3.14. We have the relation

$$Z = (1-\alpha)T_0 + \alpha Z,$$

from which it follows

$$\phi_Z(s) = \mathbf{E}(e^{-sZ}) = \int_0^\infty e^{-st} dP(Z \leq t) = (1-\alpha)\int_0^c e^{-st} dP(T_0 < t)$$
$$+ \alpha e^{-sc}\phi_Z(s) = (1-\alpha)\varphi_{T_0}(s) + \alpha e^{-sc}\phi_Z(s).$$

Solving this equation with respect to $\phi_Z(s)$ we get

$$\phi_Z(s) = \varphi_{T_0}(s)\frac{(1-\alpha)}{1-\alpha e^{-sc}} = \varphi_{T_0}(s)\sum_{k=0}^{\infty}(1-\alpha)\left[\alpha e^{-sc}\right]^k$$

$$= \varphi_{T_0}(s)\varphi_{cN}(s) = \mathbf{E}\left(e^{-s(T_0+cN)}\right)$$

Here, we used notation $N \sim Ge(\alpha)$ for a r.v. geometrically distributed over the discrete set of points $0, c, 2c, \ldots$, and independence between T_0 and N follows from the fact that LST of Z is a product of the LSTs of T_0 and N. The statement of the theorem is proved.

In our opinion, this presentation can be useful for possible simulation of values of r.v.'s from the class $Z \sim ALM(\alpha, c, F_0(t))$ with desired parameters.

3.4.2.3. *ALM is the waiting time to the first success of extended in time Bernoulli trials*

This idea is great in modeling a variety of insurance and environmental processes, where renewal of subscriptions are used. This is also a kind of queueing service and the theory is still due to be developed.

Here, we follow a constantly evolving scheme of the classical series of Bernoulli trials. The difference with the classic model is that we include essentially some time for conducting each trial, and chances of "success" vary during each trial, but the same behavior is kept in every repetition.

The model is described below:

Bernoulli trials are repeated trials that obey the following conditions:

– each trial yields one of two outcomes, called "success" (S), and "failure" (F);

– for each trial the probability $1 - \alpha = P(S)$, is the same in each repetition, and the probability of failure F is denoted by $\alpha = P(F)$;

– the trials are independent: the probability of S in a trial does not change;

– each trial needs some time $c > 0$ to be completed. α is the probability that the event of interest will not occur during a trial, A r.v. T_0 is the time within a trial in case that the event occurs. Chances of this to occur are assessed by probability $1 - \alpha$. As soon as success is observed, it can be immediately recorded.

Our goals here are to find the distribution of the r.v. Z – the waiting time to the first occurrence of the observed event in terms of the parameters α, c, and T_0 of the model, to specify its form, and to indicate its use for various purposes in insurance, environmental modeling and related areas.

THEOREM 3.19.– A) The cdf of the waiting time Z up to the occurrence of the first success in extended in time Bernoulli trials has the form

$$F_Z(t) = 1 - \alpha^{[\frac{t}{c}]} \left(1 - (1-\alpha)F_{T_0}(t - \left[\frac{t}{c}\right] c)\right) \quad , t \geq 0;$$

B) If T_0 has a p.d.f. $f_{T_0}(y)$, then the p.d.f. $f_Z(t)$ exists and has the form shown in section 3.4.2.1.

PROOF.– In this proof, we follow the classic derivation of the geometric distribution in the conventional model of Bernoulli trials in the lights of the added conditions. We have

$$\begin{aligned}
F_Z(t) &= P\{Z \leq t\} \\
&= \sum_{n=0}^{\infty} P\{Z \leq t \mid Z = nc + Y\}\, P\{Z = nc + Y\} \\
&= \sum_{n=0}^{[\frac{t}{c}]} P\{T_0 \leq t - nc\}\alpha^n(1-\alpha) \\
&= \sum_{n=0}^{\infty} \alpha^n(1-\alpha) + P\{T_0 \leq t - \left[\frac{t}{c}\right] c\}\alpha^{[\frac{t}{c}]}(1-\alpha) \\
&= 1 - \alpha^{[\frac{t}{c}]+1} + \alpha^{[\frac{t}{c}]}(1-\alpha)F_{T_0}\left(t - \left[\frac{t}{c}\right] c\right).
\end{aligned}$$

Part (A) is proved. Part (B) follows after differentiation. Theorem is proved.

Consequently, the waiting time failure rate function has the form as shown in section 3.4.2.2, and it is periodic function of period c.

These results are consequences of our previous considerations.

REMARK.– The two last theorems justify the notation ALM(α, c, F_{T_0}) for the class of probability distributions of r.v.'s resulting from the impact of periodic random environment, the class of ALM(α, c, F_{T_0}) distributions.

We see a number of opportunities for numerous applications of this model in everyday life due to the following reasons.

3.4.3. *Periodicity in natural phenomena*

Periodicity is commonly observed in the environmental conditions surrounding real life phenomena; this is what motivated our recommendation. For instance, we can mention:

– periodic alternations between cold and warm eras on Earth;

– periodic (11 years) phases in solar activity;

– periodic tides, climatical and hydro-meteorological conditions;

– production processes with periodic fluctuation in alternating day/night, or weekly, conditions (especially important in medical research);

– car accident occurrences during the year, or yearly repeated sale rates from year to year, currency exchange rates (interesting in financial studies);

– risk-associated events, such as house fires, diseases, fertility or mortality rates, pollution intensity and consumption rates (important in insurance modeling).

We are certain that the effect of a periodic random environment (neighbored intervals of equal length) on random variables can be successfully used to model waiting times until some event of interest occurs. Later we show how the same models could be used alternatively in counting the number of events that occur within any given time interval.

3.5. Random processes with a periodic nature

We consider here non-stationary Poisson processes (NPP) with intensities $\lambda(t)$ depending on the time t. Sometimes, we call it "stress function". The integrals of the intensity are called hazard functions, namely

$$\Lambda(\tau, t) = \int_{\tau}^{t} \lambda(u)du$$

is the "integrated, accumulated stress" over the time interval $[\tau, t]$. The number of events $N[\tau, t]$ that may occur within the interval $[\tau, t]$ then has a Poisson distribution of parameter $\Lambda(\tau, t)$.

Let us introduce the r.v. X representing the *lifetime*, or *waiting time* until the first event of a NPP starting at $t_0 = 0$ occurs. Then it is true

$$F_X(t) = P(X \leq t) = 1 - P(N[0, t] = 0) = 1 - e^{-\Lambda(0, t)}.$$

Correspondingly, the failure rate function of the r.v. X at time t equals $\lambda(t)$. These considerations complete the results of section 3.4.2.2. We refer to this r.v. X as generated by respective NPP. The above functions are closely related with the concept of *minimal repairs, instant without changing the status*, used in reliability maintenance (Beichelt 1991; Block *et al.* 1993).

We noticed earlier that *the sensors of the r.v.'s recording the impact of the surrounding random environment either follow the HDF* $\Lambda_X(x,t)$, *or the failure rate function* $\lambda_X(x)$, *if it exists.*

Hence, we have: If a variable is embedded into an environment of periodic nature, these two measures would reflect these periodic properties.

Its reflection onto properties of the probability distributions is expressed by the following definitions. The following statement is true:

THEOREM 3.20.– A r.v. X has PHDF of period $c > 0$ if and only if the class of ALM(α, c, F_{T_0}) distributions, where $\alpha \in (0,\ 1)$ and $F_Y(y)$ is the c.d.f. of a r.v. Y with support on the interval $[0,\ c)$.

The proof is in section 3.4.2.2.

3.5.1. *Counting processes*

There are two counting processes that can be related with the random variable X that results from the impact of a random environment.

A renewal process generated by a sequence $\{X_n\}_{n=1}^{\infty}$ of i.i.d. r.v.'s with the same cdf as the r.v. X. It is stated by the following definition:

DEFINITION 3.3.– Given the sequence $\{X_n\}_{n\geq 1}$ of i.i.d. r.v.'s, for any $t \geq 0$ define

$$N_t^{(1)} = \max\{n; X_1 + X_2 + \ldots + X_n < t\}.$$

as the renewal process associated with the r.v. X.

Let us consider an *NPP* with an intensity rate equal to the failure rate of the r.v. X. We refer to this second process as a *counting process associated with the r.v.* X, when its *hazard function* is defined as

$$\Lambda_X(t) = \int_0^t \lambda_X(u)du,\ t > 0.$$

The number of events (points) $N_{[0,t)}$ of this NPP over the interval $[0,t)$ is a Poisson distributed r.v. with parameter $\Lambda(t)$.

Also, $N_{[\tau,\tau+t)}$, the number of events within the $[\tau, \tau+t)$, (we call τ *the initial age of the process*), is also a Poisson r.v. of parameter $\Lambda(\tau+t) - \Lambda(\tau)$, and intensity rate $\lambda(u)$, $u \in [\tau, \tau+t)$.

The above functions are closely related with the concept of *minimal repairs* used in reliability maintenance (Beichelt 1991; Block *et al.* 1993). This is a transfer of the failed item from non-working to working state, without change of its current failure rate.

We propose this NPP as the second type of counting process related with the r.v. X.

3.5.2. *Characterization of an NPP*

THEOREM 3.21.– The number of repairs of an item maintained by a minimal repair policy is an NPP with periodic failure rate of period $c > 0$, if and only if the item lifetime X belongs to the class ALM(α, c, F_Y).

PROOF.– First we show that the sequence of minimal repairs forms an NPP of HF

$$\Lambda(t) = -ln[1 - F_X(t)].$$

Let X_k be the time of kth minimal repair. The joint p.d.f. of the random vector $(X_1, X_2, ..., X_n)$ is

$$f(t_1, t_2, ..., t_n)dx_1...dx_n$$

$$= \begin{cases} dF(t_1)\frac{dF(t_2)}{1-F(t_1)}\cdots\frac{dF(t_n)}{1-F(t_{n-1})} & , if\ t_1 < ... < t_n; \\ 0 & , \quad otherwise \end{cases}$$

$$= \begin{cases} \lambda(t_1)\lambda(t_2)\lambda(t_{n-1})dF(t_n)dt_1...dt_{n-1} & if\ t_1 < ... < t_n; \\ 0 & , \quad otherwise. \end{cases}$$

The probability of having exactly n minimal repairs within the interval $[0,t)$ is then

$$P\{N_t = n\} = \underset{t_1 \leq ... \leq t_n < t; t_{n+1} \geq t}{\int\int ... \int} f(t_1, ..., t_n, t_{n+1})dt_1...dt_n dt_{n+1}$$

$$
\begin{aligned}
&= \int\int\cdots\int_{t_1\le\ldots\le t_n<t;t_{n+1}\ge t} \lambda(t_1)\ldots\lambda(t_n)dF(t_{n+1})dt_1\ldots dt_n \\
&= \left(\int_0^t \lambda(x)dx\right)^n \int_t^\infty dF(t_{n+1}) = \frac{(-ln\,(1-F(t)))^n}{n!}\,(1-F(t))\,. \\
&= \frac{(-ln\,(1-F(t)))^n}{n!}\,e^{ln(1-F(t)}.
\end{aligned}
$$

This means that the process of minimal repairs $\{N_t, t \geq 0\}$ is an NPP with HF $\mathbf{E}(N_t) = \Lambda(t) = -ln\,(1 - F(t))$.

Reversely, if X belongs to the class of $ALM(\alpha, c, F_Y)$, then Theorem 3.15 works, and we obtain the representation

$$
\begin{aligned}
&\Lambda(nc + y) = -nln\alpha - ln\,(1 - (1-\alpha)F_Y(y)) \\
&= -nln\,(1 - F_X(c)) - ln\,(1 - F_X(y)) = n\Lambda(c) + \Lambda(y), \text{ for } 0 \leq y < c.
\end{aligned}
$$

Therefore, $\Lambda(t)$ has a periodic derivative.

THEOREM 3.22.– An NPP has periodic failure rate $\lambda(t)$ of period $c > 0$, if and only if the following two properties hold:

i) For a fixed constant c and an arbitrary $t \geq 0$, it is true that

$$
P\{N_{[c,\ c+t)} = 0\} = P\{N_{[0,t)} = 0\}.
$$

ii) The r.v.'s $N_{[c,\ c+t)}$ and $N_{[0,c)}$ are independent, for any $t \geq 0$.

PROOF.– *Necessity:* Let $\lambda(u)$ be a periodic of period $c > 0$. From the periodicity of $\lambda(u)$, we conclude that

$$
\Lambda(nc + t) = \Lambda(nc) + \Lambda(t) = n\Lambda(c) + \Lambda(t) \qquad [3.54]
$$

is true for any integer $n = 0, 1, 2, \ldots$. Next we see that

$$
\begin{aligned}
&P\{N_{[nc,nc+t)} = m\} = \frac{\{\Lambda(nc+t) - \Lambda(nc)\}^m}{m!} exp\{-(\Lambda(nc+t) - \Lambda(nc))\} \\
&= \frac{\{\Lambda(t)\}^m}{m!} exp\{-\Lambda(t)\} = P\{N_{[0,c)} = m\}, \quad m = 0, 1, 2, \ldots,
\end{aligned}
$$

where n can be any integer.

Therefore, (i) is true even when replacing c with nc, with $n \geq 1$ an arbitrary but fixed integer.

Further, consider the pgf

$$P_{N_{[0,nc+t)}}(z) = \mathbf{E}(z^{N_{[0,nc+t)}}) = \sum_{m=0}^{\infty} P\{N_{[0,nc+t)} = m\} z^m$$
$$= exp\{\Lambda(nc+t)(z-1)\}.$$

Using previous equations, we observe that

$$P_{N_{[0,nc+t)}}(z) = exp\{\Lambda(nc)(z-1)\} exp\{[\Lambda(nc+t) - \Lambda(nc)](z-1)\},$$

which is the product of the p.g.f. of $N_{[0,nc)}$ and $N_{[nc,nc+t)}$.

Therefore, the two random variables $N_{[0,nc)}$ and $N_{[nc,nc+t)}$ are independent. When $n = 1$, then (ii) is also true.

Sufficiency: Let (i) and (ii) be fulfilled. Then for $m = 0$ we obtain

$$P\{N_{[0,c+t)} = 0\} = P\{N_{[0,c)} + N_{[0,nc+t)} = 0\} P\{N_{[0,c)} = 0, \ N_{[c,c+t)} = 0\}$$

$$= P\{N_{[0,c)} = 0\} P\{N_{[c,c+t)} = 0\} = P\{N_{[0,c)} = 0\} P\{N_{[0,t)} = 0\}.$$

In terms of the failure rate function, the first and last terms of this equation can be rewritten by making use of [3.54] as follows:

$$exp\{-\int_0^{c+t} \lambda(u) du\} = exp\{-\left(\int_0^c \lambda(u) du + \int_0^t \lambda(u) du\right)\}.$$

After some easy manipulations it can be shown that this equation is equivalent to the relation

$$\int_c^{c+t} \lambda(u) du\} = \int_0^t \lambda(u) du, \quad t \geq 0,$$

which is equivalent (it is seen after differentiation) to the periodicity of $\lambda(u)$ with period $c > 0$.

From the proof of this theorem, we see that (i) can be reduced to the only condition that $\{N_{[0,c)} = 0\}$ and $\{N_{[c,c+t)} = 0\}$ form a pair of mutually independent events.

We will denote $\{N_t,\ t \geq 0\}$ the counting process corresponding to an NPP, which expresses the number of events that occur within the time interval $[0, t)$. Also we say that N_t is generated by a random environment with periodic behavior of period $c > 0$ if and only if the failure rate $\lambda(t)$ of the corresponding NPP is a periodic of period c.

3.5.3. *Applications in risk modeling*

The following property of the NPP counting process N_t is of central importance for applications to risk processes embedded in a periodic environment.

THEOREM 3.23.– An NPP counting process $\{N_t; t \geq 0\}$ is generated by a random environment with periodic behavior of period $c > 0$, if and only if for any $t > 0$ it can be represented in the form

$$N_t = M_1 + M_2 + \ldots + M_{[\frac{t}{c}]} + N_{t-[\frac{t}{c}]c},$$

where $\{M_n\}_{n\geq 1}$ are i.i.d. Poisson r.v. of parameter $\Lambda(c) = \int_0^c \lambda(u)du$, independent of the component $N_{t-[\frac{t}{c}]c}$, itself a Poisson process of hazard function $\Lambda(y) = \int_0^y \lambda(u)du$, with $y = t - [\frac{t}{c}]c \in [0, c)$.

PROOF.– Let M_k be the notation for the r.v. $N_{[(k-1)c,\ kc)}$ $\quad k = 1, 2, \ldots$. Then $\{M_n\}$ is a sequence of i.i.d. Poisson r.v. of parameter $\Lambda(kc-(k-1)c) = \Lambda(c)$. The partitioning

$$N_{[0,t)} = \sum_{k=0}^{[t/c]-1} N_{[kc,\ (k-1)c)} + N_{[[\frac{t}{c}]c,\ t)}$$

is always true, and the components of this sum are independent, because of the periodicity of its intensity.

As a result, we get a presentation for the *aggregated claim process*, an important characteristic in insurance and risk studies.

Let $\{\xi_n\}_{n\geq 1}$ be the sequence of claim severity.

– the claims' arrival times of form NPP $\{N_t; t \geq 0\}$ with periodic intensity $\lambda(t)$ of periodicity $c > 0$ as the environment.

– for convenience, we re-scale time to $c = 1$.

The aggregate claims process over the interval $[0, t)$ is presented by the compound sum

$$Z_t = \sum_{n=0}^{N_t} \xi_n \quad \text{with} \quad \xi_0 = 0.$$

THEOREM 3.24.– The aggregate claim process $\{Z_t; t \geq 0\}$ driven by a periodic NPP can be decomposed into the form

$$Z_t = S_1 + S_2 + \ldots + S_{[t]} + Z_{t-[t]}. \qquad [3.55]$$

Here $\{S_n\}_{n\geq 0}$ are i.i.d. r.v.'s distributed as the compound Poisson sum

$$Z_1 = \sum_{n=0}^{N_1} \xi_n,$$

and N_1 is a Poisson r.v. of parameter $\Lambda(1) = \int_0^1 \lambda(u)du$. The term $Z_{t-[t]}$ in [3.55] is also a compound Poisson term, independent of the other S_n, with parameter $\Lambda(t) = \int_0^t \lambda(s)ds$, for $t \leq 1$.

The proof is simple consequence of the presentation of the counting process $\{N_t\}$, after regrouping the components of the compound sum Z_t within each component of the summation index $\{N_t\}$.

More about applications of NPP in risk can be found in Chukova *et al.* (1993), Dimitrov *et al.* (2000) and in works by Garrido (2004) later on.

3.6. Conclusions

Periodicity in surrounding environmental conditions is a natural phenomena. It may have a considerable impact on random events, variables and processes generated in such environment. The three parameter probability distributions we found as a results of service on unreliable server is flexible enough to describe a wide range of practical situations.

Although the choice of periodic constant c may involve some subjectivity, there should be suitable statistical approach in its determination that best fits certain statistical data. The periodic nature of the failure rate in modeling risk processes can be a powerful tool in the improved studies of ruin probabilities and other risk characteristics in this field. Moreover, it is also possible to discuss processes with double periodic rates (as did Garrido and Lu 2004).

Finally, multivariate generalizations of presented results offer some challenging new field of research in this direction. Some encouraging results have been obtained by Dimitrov *et al.* (2002).

3.7. References

Bakoucha, H. S., Christopher, C. C., Jeremias L. J. (2018). A new lifetime model with a periodic hazard rate and an application. *J. Statist. Computation and Simulation*, 88(11), 2048–2065.

Barlow R.E., Proshan F. (1965). *Mathematical Theory of Reliability*. Wiley, New York.

Barlow, R.E., Proshan, F. (1981). *Statistical Theory of Reliability and Life Testing: Probability Models (To Begin With)*. Silver Spring, MD.

Baxter, L.A. (1982). Reliability applications of the relevation transform. *Naval Res. Log. Quart.* 29, 323–330.

Beichelt, F. (1991). A unifying treatment of replacement policies with minimal repair. *Naval Research Logistics*, 39, 1221–1238.

Block, H.W., Borges, W. S., Savits, T.H. (1988). A general age replacement model with minimal repair. *Naval Research Logistics,* 35, 365–372.

Chukova, S., Dimitrov, B. (1992). On distributions having the almost lack of memory property. *J. Appl. Probab.* 29(3), 691–698.

Chukova, S., Dimitrov, B., Garrido, J. (1993). Renewal and non-homogeneous Poisson processes generated by distributions with periodic failure rate. *Statistics and Probab. Letters*, 17, 19–25.

Chukova, S., Dimitrov, B., Khalil, Z. (1993). Probability distributions similar to the exponential characterization. *Can. J. Statistics*, 21(3), 269–276.

Cinlar, E. (1974). *Introduction to Stochastic Processes*. Prentice-Hall, Englewood Clififs, NJ.

Danielyan, E., Dimitrov B. (1970). Service with priorities and preparation times. *Math. questions in production control.* Moscow State University, 165–178 (in Russian).

Danielyan, E., Dimitrov B. (1971). Service with changing priorities and preparation times. *Scientific Notices of the University of Erevan*, 1, 3–10 (in Russian).

Dimitrov, B. (1971). Service with priorities. *Mathem. Balkanica*, 1, 66–78 (in Russian).

Dimitrov, B. (1974). Asymptotic expansions of characteristics for queueing systems of the type M/G/1. *Bulletin de l'Inst. de Math., Acad. Bulg. Sci.*, XV, 237–263 (in Bulgarian).

Dimitrov, B., Chukova, S., Green, D., Jr. (1996). Probability distributions in periodic random environment and its applications. *SIAM J. Appl. Math.*, 57(2), 501–517.

Dimitrov, B., Chukova, S., Garrido, J. (2000). Compound counting processes in a periodic random environment. *J. Statistical Research*, 32(2), 99–111.

Dimitrov, B., Chukova, S., Vogt, H. (2002). A class of symmetric probability distributions similar to the double exponential. *Engineering Simulations,* 24(5), 25–37.

Dimitrov, B., Chukova, S., Khalil, Z. (1995). Definitions, characterizations and structural properties of probability distributions similar to the exponential. *J. Statistical Planning and Inference* 43, 271–287.

Dimitrov, B., Danielyan, E. (1970). Several limit theorems in reliability and queueing theory. *Mathematical Aspects of Industrial Control* [in Russian]. Moscow University, Moscow, 179–183.

Dimitrov, B., Khalil, Z. (1990). On a new characterization of the exponential distribution related to a queueing system with an unreliable server. *J. Appl. Prob.,* 27, 221–226.

Dimitrov, B., Khalil, Z. (1992). A class of new probability distributions for modeling environmental evolution with periodic behavior. *Environmetrics*, 3(4), 447–464.

Galambos, J., Hagwood, Ch. (1994). An unreliable server characterization of the exponential distribution. *J. Appl. Prob.,* 31, 274–279.

Galambos, J., Kotz, S. (1978). Characterizations of probability distributions. Lecture Notes in Mathematics, Springer-Verlag, Berlin, 675.

Garrido, J., Lu, Y. (2004). Doubly periodic non-homogeneous Poisson models for hurricanes data. *Stat. Methodol.,* 2(1), 17–35

Gnedenko, B., Danielyan, E., Klimov, G., Matveev, V., Dimitrov, B. (1973). *Prioritetnye Sistemy Obslujivania.* Moscow State University (in Russian; Engl. *Priority Queues*).

Jacobi, C.G. (1882). *Gesammelte Werke.* Bd. 2. Berlin.

Johnson, N.L., Kotz, S., Balakrishnan, N. (1994). *Continuous 120 Distributions -1, and -2*, 2nd ed. John Wiley & Sons, New York.

Kesten, H., Runnenburg, J. T. (1957). Priority in Waiting Line Problems, 60. *Koninklijke Nederlandse Akademie van Wetenschappen*, 312–336.

Khalil, Z., Dimitrov, B., Dion, J.-P. (1991). A characterization of the geometric distribution related to random sums. *Commun. Statist. Stoch. Models,* 7, 321–326.

Klimov, G.P. (1966). *Stochastic Queueing Systems*. Nauka, Moscow (in Russian).

Kotz, S., Shanbhag, D.N. (1980). Some new approaches to probability distributions. *Advances in Applied Probability,* 12, 903–921.

Krakowski, M. (1973). The relevation transform of the Gamma distribution function. *Rev. Française Automat. Recherche Operationelle*, 7(Ser. V-2), 107–120.

Lau, K.S., Prakasa Rao, B.L.S.P. (1990). Characterization of the exponential distribution by the relevation transform. *Journal of Applied Probability*, 27, 726–729.

Lin, G.D. (1993). Characterizations of the exponential distribution via the blocking time in a queueing system. *Statistica Sinica,* 3, 577–581.

Lin, G.D. (1994). A note on distributions having the almost lack of memory property. *Journal of Applied Probability*, 31, 854–856.

Omey, E., Willenkens, E. (1989). Abelian and Tauberian theorems for the Laplace transform of functions in several variables. *J. Multivariate Analysis,* 30, 292–306.

Van Harn, K., Steutel, F.W. (1991). On a characterization of the exponential distribution. *J. Appl. Prob.* 28, 947–949.

Wen-Jang, H., Jyh-Ming, S. (1993). Some characterizations of the exponential and geometric distributions arising in a queueing system with an unreliable server. *J. Appl. Prob.,* 30, 985 990.

4

The Impact of Information Structure on Strategic Behavior in Queueing Systems

Antonis ECONOMOU
Department of Mathematics, National and Kapodistrian University of Athens, Greece

The study of strategic customer behavior in service systems constitutes an important and dynamic trend in queueing theory. Indeed, the optimal design and control of service systems in real-life applications of queueing theory requires that the strategic dimension of customers is taken into account. Under this perspective, the customers are decision makers that aim to maximize their benefit, taking into account that the others have similar objectives. Therefore, the overall situation can be seen as a game among the customers and the administrator of the system. A central problem is how a social planner or a monopolist should act to incite customers to adopt a desirable behavior, one that increases the social welfare or the monopolist's revenue/profit, respectively. However, the intervention of a social planner and/or a monopolist should be indirect, as direct forcing of customers is considered unacceptable in the framework of a free market. To this end, several mechanisms have been proposed, e.g. pricing structures, priority systems and non-standard queueing disciplines. An important mechanism is the control of information that is provided to the customers. In this chapter, we will present several techniques for the control of information in a given system and their impact on strategic customer behavior, the throughput, the social welfare and a monopolist's revenue. These ideas will be explained in the simplest possible framework and then several extensions will be discussed. An overview of the corresponding literature is also included.

Queueing Theory 2,
coordinated by Vladimir ANISIMOV, Nikolaos LIMNIOS. © ISTE Editions 2020.

4.1. Introduction

Classical queueing theory assumes that customers are passive entities that do not make decisions. Under this perspective, a very large number of queueing theoretic studies have been presented in the literature that deal with the performance evaluation, optimal design and optimal control of service systems that appear in a number of situations. However, an economic evaluation of a queueing system is not credible, unless we take into account the strategic nature of its customers. This point of view was assumed in the seminal paper of (Naor 1969), who studied the join-or-balk dilemma for the customers in the M/M/1 queue, when its queue length is observable. He also considered the problem of a social planner and a monopolist who optimize the social welfare and the revenue, respectively, taking into account customer strategic behavior. Edelson and Hildebrand (1975) complemented Naor's study by considering the same problems for the unobservable version of the system. In their model, the arriving customers are not allowed to observe the number of customers in the system and make their join-or-balk decisions relying solely on the knowledge of its operational and economic parameters, assuming that the system has reached stochastic steady-state. Since then, the literature on strategic behavior in queueing systems has grown considerably. Hassin and Haviv (2003) provided a nice overview of the basic methodology and the early results in this area. The monographs by (Stidham 2009) and (Hassin 2016) contain further material and overviews about models and methodologies in this subfield of queueing theory.

One recurrent theme in the strategic queueing literature is the appropriate use of mechanisms that will incentivize customers to behave according to a social planner's or a monopolist's objective. One such mechanism is the control of information that is provided to the customers before their decisions are made. The study of the effect of the level of information on the strategic customer behavior is an important theoretical issue per se, but it is also important because it raises several issues of practical relevance. The information provision mechanism is related to the design of the system, affects the psychology of the customers and may be costly. What is interesting is that the effect of the level of information is ambiguous. More information can benefit or hurt the customers and/or the service provider. The effect of more information may be negative or positive, depending on various parameters and structural assumptions of the underlying model. It should be emphasized here that the strategic interaction of the customers is the root cause of the phenomenon that the effect of more information might be negative, unlike simple optimization. The objective of this chapter is exactly to shed some light on these issues. Because of the complexity of the problem, we will unfold the presentation in the simplest possible framework, that is of the M/M/1 queue. Indeed, this is also the framework that most of the authors have adopted in the literature. In the last part of the chapter, we also discuss more involved models.

The rest of the chapter is organized as follows. In section 4.2, we present some basic concepts from game theory that will be used in the sequel. In sections 4.3 and 4.4, we present the analysis of the two "extreme" models regarding the information that is provided to the customers: the unobservable model of (Edelson and Hildebrand 1975) and the observable model of (Naor 1969). Subsequently, in section 4.5, we present briefly the main findings from the comparison of these models. Then, we describe three significant ideas that have appeared in the literature that bridge the observable and unobservable versions of a given model. More specifically, in sections 4.6–4.8, we discuss, respectively, partially observable models, heterogeneously observable models and observable-with-delay models. Apart from the descriptions of the main ideas behind each family of models, we present the basic results, the associated methodologies and the main qualitative findings. Section 4.9 is devoted to an overview of the various ideas that appear within the chapter. Moreover, it points to several important sources in the literature, some conclusions and topics for future research.

4.2. Game-theoretical framework in queueing

In the framework of classical game theory, a game is specified by a set of players $\mathcal{N} = \{1, 2, \ldots, n\}$, sets of action plans $\mathcal{A}_i$, one for each player $i = 1, 2, \ldots, n$, and payoff (utility) functions $\mathcal{U}_i$, one for each player $i = 1, 2, \ldots, n$. The set $\mathcal{A}_i$ contains all available action plans for player i that specify what actions should be taken during the game, according to its state at every decision point. Every element of $\mathcal{A}_i$ is referred to as a pure strategy of i. A probability distribution on $\mathcal{A}_i$ is referred to as a mixed strategy of i. When player i uses a certain mixed strategy, the player chooses one of his/her pure strategies according to the probability distribution of the mixed strategy. The set of mixed strategies for of the player i is denoted by $\mathcal{S}_i$.

A strategy profile s is an ordered n-tuple of strategies, one for each player, i.e.

$$\mathbf{s} = (s_1, s_2, \ldots, s_n), \quad s_i \in \mathcal{S}_i, \ i = 1, 2, \ldots, n.$$

Given a strategy profile s, we write $\mathbf{s} = (s_i, \mathbf{s}_{-i})$, to denote that the $(n-1)$-dimensional vector $\mathbf{s}_{-i}$ contains the strategies of s except the one that corresponds to player i. The payoff function $\mathcal{U}_i$ is defined on the set of all strategy profiles and takes real values. Its value $\mathcal{U}_i(\mathbf{s}) = \mathcal{U}_i(s_i, \mathbf{s}_{-i})$ corresponds to the payoff of player i, if the strategy profile s is adopted, i.e. if player i uses his/her strategy s_i and the other players the strategies in $\mathbf{s}_{-i}$. The function $\mathcal{U}_i(\mathbf{s}) = \mathcal{U}_i(s_i, \mathbf{s}_{-i})$ is linear with respect to s_i, i.e. if the strategy s_i mixes the strategies s_i^k, $k = 1, 2, \ldots, r$ with probabilities α_k, $k = 1, 2, \ldots, r$, respectively (with $\sum_{k=1}^{r} \alpha_k = 1$), then

$$\mathcal{U}_i(s_i, \mathbf{s}_{-i}) = \sum_{k=1}^{r} \alpha_k \mathcal{U}_i(s_i^k, \mathbf{s}_{-i}).$$

If s_i^1 and s_i^2 are strategies of player i, then we say that s_i^1 weakly dominates s_i^2, if for any strategy profile for the other players, s_{-i}, then $\mathcal{U}_i(s_i^1, \mathrm{s}_{-i}) \geq \mathcal{U}_i(s_i^2, \mathrm{s}_{-i})$, with the inequality being strict for at least one strategy profile s_{-i}. We say that s_i^1 strongly dominates s_i^2, if the inequality is strict for all strategy profiles s_{-i}.

Consider, now, a player i. Given a strategy profile s_{-i} for the other players, the strategy s_i^* of i is said to be the best response against s_{-i}, if for every strategy s_i of i we have $\mathcal{U}_i(s_i^*, \mathrm{s}_{-i}) \geq \mathcal{U}_i(s_i, \mathrm{s}_{-i})$, i.e. if s_i^* maximizes $f(s_i) = \mathcal{U}_i(s_i, \mathrm{s}_{-i})$. The set of best responses to s_{-i} is denoted by $BR(\mathrm{s}_{-i})$.

A strategy s_i of i is said to be weakly (respectively, strongly) dominant, if it weakly (respectively, strongly) dominates every other strategy of i.

A strategy profile $\mathrm{s}^e = (s_1^e, s_2^e, \ldots, s_n^e)$ is said to be an equilibrium profile, if for every player i, the strategy s_i^e is the best response against s_{-i}^e. In other words, a strategy profile is an equilibrium if no player has an incentive to deviate from it unilaterally.

In the study of strategic customer behavior in queueing systems, the game-theoretical concepts that we described above are very useful. However, they cannot be applied immediately, because there are two fundamental problems. The first is the fact that the number of players is infinite, since the potential customers of the system are infinite. The second is that the customers–players do not make simultaneously their decisions since they arrive sequentially during an infinite time horizon. These problems are bypassed by defining analogous concepts and exploiting the homogeneity of the customers. For simplicity, we will assume that all customers are identical. However, the framework can be extended to allow heterogeneous customers, assuming that there are various classes with homogeneous customers within each class.

In the case of homogeneous customers, a game among them is specified by the set of their common strategies, $\mathcal{S}$, and from a payoff function $\mathcal{U}(s, s')$ that specifies the utility of a customer that uses strategy s, when all other customers follow s'. The function $\mathcal{U}(s, s')$ is linear in s, i.e. if a strategy s is the mixture of strategies s^k, $k = 1, 2, \ldots, r$ with corresponding probabilities α_k, $k = 1, 2, \ldots, r$, with $\sum_{k=1}^r \alpha_k = 1$, then

$$\mathcal{U}(s, s') = \sum_{k=1}^{r} \alpha_k \mathcal{U}(s^k, s').$$

If s^1 and s^2 are strategies of a player, then the strategy s^1 weakly dominates s^2, if for any strategy s' of the other players $\mathcal{U}(s^1, s') \geq \mathcal{U}(s^2, s')$. Moreover, we say that s^1 strongly dominates s^2, if the inequality is strict for all strategies s'.

Consider, now, a tagged customer and assume that the strategy s' is followed by the other customers. A strategy s^* of the tagged customer is said to be the best response against s', if for any strategy s of the tagged customer $\mathcal{U}(s^*, s') \geq \mathcal{U}(s, s')$, i.e. s^* maximizes $f(s) = \mathcal{U}(s, s')$. The set of best responses against s' is denoted by $BR(s')$.

A strategy s of a tagged player is said to be dominant if s is best response against any strategy of the others. A strategy s^e is said to be a (symmetric) equilibrium, if it is the best response against itself. In other words, the strategy s^e is equilibrium if

$$\mathcal{U}(s^e, s^e) \geq \mathcal{U}(s, s^e), \quad s \in \mathcal{S},$$

or equivalently if $s^e \in BR(s^e)$. It should be noted here that because of the linearity of the payoff function with respect to its first argument, the indifference principle of a mixed equilibrium strategy holds: If an equilibrium mixed strategy assigns positive probability to some pure strategies, then all these pure strategies ensure the same payoff to a tagged customer given that the population of customers follow the equilibrium strategy.

A basic step for the study of strategic customer behavior concerns the computation of the payoff function $\mathcal{U}(s, s')$. The fundamental assumption for this computation is that if we consider a tagged customer who follows a strategy s, when all others follow a strategy s', then the tagged customer's strategy does not influence the performance measures of the system. The general behavior of the system and the corresponding performance measures are determined by the strategy s' that the other customers follow, since the impact of the tagged customer is negligible. Moreover, it is assumed that the system has reached stochastic steady-state. To determine the dominant and equilibrium customer strategies in a queueing system, a general methodology is applied using the following steps:

Step 1: the steady-state behavior of the system under an arbitrary strategy s' of the population of the customers is studied.

Step 2: the utility function $\mathcal{U}(s, s')$ of a tagged customer that follows strategy s, when all other customers follow strategy s', is computed.

Step 3: the best response of the tagged customer against an arbitrary strategy, s', of the population of the customers is computed:

$$BR(s') = \{s \in \mathcal{S} : \mathcal{U}(s, s') \geq \mathcal{U}(\hat{s}, s'), \hat{s} \in \mathcal{S}\}.$$

Step 4: all strategies with the property $s^e \in BR(s^e)$ are identified. These are exactly the equilibrium strategies. If an equilibrium strategy s^e satisfies the stronger condition $s^e \in BR(s)$, for all $s \in \mathcal{S}$, then it is a dominant strategy.

A related problem from a social planner's point of view is the maximization of the social welfare per customer given that a symmetric strategy s is followed by the population of customers. This is defined to be the quantity $\mathcal{U}(s,s)$. Then, a socially optimal symmetric strategy s_{soc} is such that $\mathcal{U}(s_{soc}, s_{soc}) \geq \mathcal{U}(s,s)$, for $s \in \mathcal{S}$. To determine such strategies we must solve the optimization problem $\max_{s\in\mathcal{S}} \mathcal{U}(s,s)$.

4.3. The unobservable model

We consider an M/M/1 queue, where customers arrive according to a Poisson process at rate λ and the service times are exponentially distributed with rate μ. Each customer receives mean reward R for his/her service completion, whereas he/she accumulates waiting costs at rate C, during his/her sojourn time in the system (assuming that the cost is accumulated with the same rate whether the customer stays in the waiting space or receives service). The dilemma of the customers is whether to join or balk. In this section, we consider the unobservable version of the model, i.e. we assume that the customers make their join-or-balk decisions, without observing the queue length in the system. However, the various operational and economic parameters of the system, λ, μ, R and C are common knowledge to the customers.

In this case, the pure strategies of a customer are two: Join (1) or balk (0). A mixed strategy is specified by a join probability $q \in [0,1]$. Edelson and Hildebrand (1975) studied the equilibrium customer strategies for the unobservable M/M/1 queue regarding the join-or-balk dilemma and the associated social optimization and revenue maximization problems of the administrator of the system. We summarize their main findings, using the 4-step methodology that applies as follows: Suppose that the population of the customers follow a join probability q. Then, because of the thinning property of the Poisson process, the system becomes an M/M/1 queue with arrival rate λq and service rate μ; hence the mean sojourn time of a customer who decides to join is $\frac{1}{\mu-\lambda q}$ as long as $\lambda q < \mu$ and 0 otherwise (see, e.g. (Hassin and Haviv 2003), section 1.4). Consider, now, a tagged customer who joins with probability q', when the others join with probability q. Then, his/her expected utility is

$$\mathcal{U}(q',q) = (1-q')\cdot 0 + q'\left(R - \frac{C}{\mu-\lambda q}\right).$$

Therefore, to find his/her best response against q, the tagged customer has to solve the problem $\max_{q'\in[0,1]} \mathcal{U}(q',q)$. However, the function $\mathcal{U}(q',q)$ is linear with respect to q', so the tagged customer bases his/her decision on the sign of the quantity

$$S_{ind}^{(un)}(q) = R - \frac{C}{\mu-\lambda q}.$$

The superscript "(un)" in the notation of the function $S_{ind}^{(un)}(q)$ indicates that we refer to the unobservable model, whereas the subscript "ind" refers to the individualistic point of view that we consider here. Similar notational conventions will be used in the sequel. Let

$$\bar{q}_e = \frac{1}{\lambda}\left(\mu - \frac{C}{R}\right).$$

be the root of $S_{ind}^{(un)}(q)$. Then, the set of best responses against q, $BR(q)$, is (as far as $q \in [0,1]$ and $\lambda q < \mu$)

$$BR(q) = \begin{cases} \{0\}, & \text{if } q > \bar{q}_e, \\ [0,1], & \text{if } q = \bar{q}_e, \\ \{1\}, & \text{if } q < \bar{q}_e. \end{cases}$$

We can now proceed to the computation of the equilibrium strategies:

The strategy of "always balk" ($q_e = 0$) is equilibrium strategy, if and only if $0 \in BR(0)$, i.e. $0 \geq \bar{q}_e$, which reduces to $R \leq \frac{C}{\mu}$.

A strategy $q_e \in (0,1)$ is equilibrium strategy, if and only if $q_e \in BR(q_e)$, i.e. $q_e = \bar{q}_e$, which reduces to $q_e = \frac{1}{\lambda}\left(\mu - \frac{C}{R}\right)$. This is valid as far as $\bar{q}_e \in (0,1)$, which occurs if and only if $\frac{C}{\mu} < R < \frac{C}{\mu-\lambda}$.

Finally, the "always join" ($q_e = 1$) is equilibrium strategy, if and only if $1 \in BR(1)$, i.e. $1 \leq \bar{q}_e$, which reduces to $R \geq \frac{C}{\mu-\lambda}$.

In summary, we have the following result:

THEOREM 4.1.– For the join-or-balk customer dilemma in the unobservable M/M/1 queue, a unique equilibrium strategy q_e exists, given by the formula

$$q_e = \begin{cases} 0, & R \leq \frac{C}{\mu}, \\ \frac{1}{\lambda}\left(\mu - \frac{C}{R}\right), & \frac{C}{\mu} < R < \frac{C}{\mu-\lambda}, \\ 1, & R \geq \frac{C}{\mu-\lambda}. \end{cases}$$

Therefore, unless $R > \frac{C}{\mu-\lambda}$, the social welfare under the equilibrium strategy is 0. We now consider the problem of the system administrator that can act as a social planner who wants to induce a socially optimal strategy, q_{soc}, to maximize the social welfare per time unit. To this end, the administrator of the system can impose an admission fee (entrance or service price) p that will change the service reward from

R to $R-p$. Note that p may be even negative, corresponding to a subsidy for service. Using this price mechanism, the administrator may induce whatever join probability, q, he desires. Then, the system behaves as an M/M/1 queue with arrival rate λq and service rate μ. The revenue of the administrator per time unit will be λqp, whereas the total customer utility will be $\lambda q\left(R-p-\frac{C}{\mu-\lambda q}\right)$. Therefore, the social welfare per time unit is given by

$$S_{soc}^{(un)}(q)=\lambda q\left(R-p-\frac{C}{\mu-\lambda q}\right)+\lambda qp=\lambda q\left(R-\frac{C}{\mu-\lambda q}\right),$$

which is independent of p (since the transfer payments do not appear in the social welfare function). The subscript "soc" in the notation of the function $S_{soc}^{(un)}(q)$ shows that we adopt here the point of view of a social planner.

Note, now, that

$$\frac{d}{dq}S_{soc}^{(un)}(q)=\lambda\left(R-\frac{C\mu}{(\mu-\lambda q)^2}\right),$$
$$\frac{d^2}{dq^2}S_{soc}^{(un)}(q)=-\frac{2C\mu\lambda^2}{(\mu-\lambda q)^3}<0.$$

Therefore, the function $S_{soc}^{(un)}(q)$ is concave for $q\in[0,\frac{\mu}{\lambda})$ and attains its maximum at the root $\bar{q}_{soc}$ of $\frac{d}{dq}S_{soc}^{(un)}(q)$, which is given by the formula

$$\bar{q}_{soc}=\frac{1}{\lambda}\left(\mu-\sqrt{\frac{C\mu}{R}}\right). \qquad [4.1]$$

When $\bar{q}_{soc}\in[0,1]$, we deduce that the socially optimal strategy (i.e. admission probability) is given by the formula [4.1], otherwise the maximum of the social welfare is attained at 0 or 1. More concretely, we have the following result:

THEOREM 4.2.– For the social planner's admission problem in the unobservable M/M/1 queue, a unique socially optimal strategy exists, given by the formula

$$q_{soc}=\begin{cases}0, & R\le\frac{C}{\mu},\\ \frac{1}{\lambda}\left(\mu-\sqrt{\frac{C\mu}{R}}\right), & \frac{C}{\mu}<R<\frac{C\mu}{(\mu-\lambda)^2},\\ 1, & R\ge\frac{C\mu}{(\mu-\lambda)^2}.\end{cases}$$

We can easily see that $q_{soc} \leq q_e$, for all parameter values λ, μ, R and C. Indeed, for a mixed strategy $q \in (0,1)$, the individual condition of optimality is $R = \frac{C}{\mu - \lambda q}$, whereas the first-order condition for social optimality is $R = \frac{C\mu}{(\mu-\lambda q)^2}$. The quadratic term in the latter appears because of the negative external effects of joining (for details, see the excellent paper of (Haviv and Oz 2018)). Therefore, we observe that, without an admission fee, the customers tend to use the system more than what is socially desirable.

We now consider the monopolist's pricing problem, when the objective of the system administrator is the maximization of the revenue per time unit. By imposing an admission fee, p, the customers adopt the corresponding equilibrium strategy q_e that we described above that corresponds to the service reward $R - p$. For inducing join probability $q \in (0,1)$, the administrator should impose an admission fee $p = R - \frac{C}{\mu - \lambda q}$. For inducing $q = 1$, he should impose the maximum possible price that allow all customers to join, which is $p = R - \frac{C}{\mu - \lambda}$. When he induces join probability $q = 0$, by imposing a large admission fee, his profit will be 0. Therefore, we see that in any case, the function of the revenue per time unit is

$$S_{prof}^{(un)}(q) = \lambda q p = \lambda q \left(R - \frac{C}{\mu - \lambda q} \right),$$

i.e. it is identical to the function of the social welfare per time unit.

We conclude that the optimal join probability for the system administrator when he acts as a monopolist is the socially optimal join probability. This occurs because of symmetric information (i.e. all information known to the customers is also known to the monopolist) and customer homogeneity that allow the monopolist to extract all utility surplus from the customers. Then, the objective functions of the social welfare and the monopolist's revenue coincide. Therefore, the administrator of the system imposes an admission fee

$$p_{prof} = R - \frac{C}{\mu - \lambda q_{soc}}$$

which maximizes the social welfare, and collects all of this revenue for himself. Substituting the socially optimal join probability into the formula for p_{prof} yields

$$p_{prof} = \begin{cases} R - \frac{C}{\mu - \lambda}, & \text{if } \lambda < \mu - \sqrt{\frac{C\mu}{R}} \\ R - \sqrt{\frac{RC}{\mu}}, & \text{if } \lambda \geq \mu - \sqrt{\frac{C\mu}{R}}, \end{cases}$$

which is a decreasing and ultimately constant function of λ. This may seem a paradox at first glance, since λ can be interpreted as the demand for service and therefore we expect that an increase in the demand will make the monopolist raise the price. However, an increase in the demand induces a significant decrease in the "quality" of the service, because of increasing delays. Therefore, the customers become more reluctant to buy the service and the monopolist cannot increase the price.

We further notice that pricing is not the only way for the regulation of the unobservable M/M/1 queue. Indeed, (Haviv and Oz 2018) describe eight different regulation schemes for this model.

4.4. The observable model

Consider, now, the join-or-balk customer dilemma in the M/M/1 queue with the same operational and economic parameters as in section 4.3, assuming that each customer makes his/her decision after observing the number of present customers upon arrival. In this case, a mixed customer strategy is specified by a sequence $\mathbf{q} = (q_0, q_1, q_2, \ldots)$, where $q_n \in [0, 1]$, $n = 0, 1, 2, \ldots$, is the join probability when an arriving customer finds n customers in the system (without counting herself/himself). Naor (1969) studied the equilibrium customer strategies for this case and the associated social optimization and revenue maximization problems. We will present the corresponding findings in this section.

First, we focus on the equilibrium strategies of the customers. When the population of the customers follow a strategy $\mathbf{q}$, the system behaves as an M/M/1 queue with state-dependent arrival rates $\lambda_n = \lambda q_n$, and any performance measure can be easily computed using standard results from birth–death processes. More importantly, the conditional mean sojourn time of a tagged customer, given that he/she finds n customers in the system, does not depend on the strategy $\mathbf{q}$ employed by the population of customers. Because of the FCFS queueing discipline and the memoryless property of the exponential service times, we can easily argue that his/her conditional mean sojourn time is $\frac{n+1}{\mu}$.

Suppose that the tagged customer follows strategy $\mathbf{q}' = (q'_0, q'_1, q'_2, \ldots)$, when the population of the others follows strategy $\mathbf{q} = (q_0, q_1, q_2, \ldots)$. If the tagged customer finds n customers upon arrival, then his/her expected utility is

$$\mathcal{U}(\mathbf{q}', \mathbf{q}|n) = (1 - q'_n) \cdot 0 + q'_n \left(R - \frac{C(n+1)}{\mu} \right).$$

Therefore, his/her best response depends on the sign of the quantity

$$S_{ind}^{(obs)}(n) = R - \frac{C(n+1)}{\mu}.$$

The set of best responses $BR(\mathbf{q}|n)$ of the tagged customer against a strategy $\mathbf{q}$ of the others, when he/she finds n customers in the system is

$$BR(\mathbf{q}|n) = \begin{cases} \{0\}, & \text{if } \frac{R\mu}{C} - 1 < n, \\ [0,1], & \text{if } \frac{R\mu}{C} - 1 = n, \\ \{1\}, & \text{if } \frac{R\mu}{C} - 1 > n, \end{cases}$$

which clearly does not depend on $\mathbf{q}$. Therefore, we have the following result:

THEOREM 4.3.– For the join-or-balk customer dilemma in the observable M/M/1 queue, the n_e-threshold strategy with

$$n_e = \left\lfloor \frac{R\mu}{C} \right\rfloor, \qquad [4.2]$$

that prescribes a customer to join the system as long as the number of customers in the system including his/her is at most n_e is the dominant strategy (and therefore the equilibrium strategy).

We now turn to the problem of the administrator of the system, when he acts as a social planner, aiming to induce a socially optimal strategy $\mathbf{q}_{soc}$,to maximize the social welfare per time unit, i.e. the quantity

$$S_{soc}^{(obs)}(\mathbf{q}) = S_{soc}^{(obs)}(q_0, q_1, \ldots) = \lambda_e^{(obs)}(\mathbf{q})R - CE_{\mathbf{q}}[Q], \qquad [4.3]$$

where $\lambda_e^{(obs)}(\mathbf{q})$ is the steady-state throughput, under strategy $\mathbf{q}$, and $E_{\mathbf{q}}[Q]$ the corresponding expected steady-state number of customers in the system. The classical approach for solving this problem is the use of stochastic dynamic programming (Stidham 1985), where it is shown that the optimal strategy is of threshold type. Indeed, the social optimization problem can be solved by considering an appropriate Markov decision process. Since the state of this process is fully observable, there always exists a non-randomized optimal policy that can be seen to be a pure threshold optimal policy. However, in what follows, we will focus on the version of this problem where the social planner optimizes the social welfare per time unit by imposing a common admission fee (price) to all arriving customers. To this end, we suppose that the administrator charges an admission fee p. Then, the customers follow a threshold strategy $n = \left\lfloor \frac{(R-p)\mu}{C} \right\rfloor$ and the system reduces to an M/M/1/n queue. We limit the study, for simplicity, to the case $\rho < 1$ (but the case $\rho \geq 1$ can be treated similarly). Then, using standard formulas for the M/M/1/n queue regarding its throughput and mean queue length (see, e.g., section 7.3.2 in

(Kulkarni 2010)), and substituting them into [4.3], we conclude that the social welfare, when the n-threshold strategy has been imposed is given by

$$S_{soc}^{(obs)}(n) = \lambda R \frac{1-\rho^n}{1-\rho^{n+1}} - C\left[\frac{\rho}{1-\rho} - \frac{(n+1)\rho^{n+1}}{1-\rho^{n+1}}\right].$$

After a bit of algebraic manipulation, we see that

$$S_{soc}^{(obs)}(n) - S_{soc}^{(obs)}(n-1) = \frac{\lambda R(1-\rho)^2\rho^{n-1}}{(1-\rho^{n+1})(1-\rho^n)}$$

$$+ \frac{C((n+1)\rho - \rho^{n+1} - n)\rho^n}{(1-\rho^{n+1})(1-\rho^n)}.$$

For $\rho < 1$,

$$\begin{aligned} S_{soc}^{(obs)}(n) - S_{soc}^{(obs)}(n-1) \geq 0 \Leftrightarrow \lambda R(1-\rho)^2 \\ \geq C\rho(n + \rho^{n+1} - (n+1)\rho) \\ \Leftrightarrow \frac{R\mu}{C} \geq \frac{n + \rho^{n+1} - (n+1)\rho}{(1-\rho)^2}. \end{aligned}$$

Let $g(n)$ be the quantity in the right-hand side of the last inequality. Then,

$$\begin{aligned} g(n) &= \frac{n + \rho^{n+1} - (n+1)\rho}{(1-\rho)^2} \\ &= \frac{1}{(1-\rho)^2}(n(1-\rho) - \rho(1-\rho^n)) \\ &= \frac{1}{1-\rho}\left(n - \sum_{k=1}^{n} \rho^k\right) \\ &= \frac{1}{1-\rho}\sum_{k=1}^{n}(1-\rho^k), \end{aligned} \qquad [4.4]$$

which is obviously strictly increasing in n. Moreover, $g(0) = 0$ and $\lim_{n\to\infty} g(n) = \infty$. Therefore, a unique number n_{soc} exists such that $g(n) \leq \frac{R\mu}{C}$, for $n \leq n_{soc}$, whereas $g(n) > \frac{R\mu}{C}$, for $n > n_{soc}$. Hence, $S_{soc}^{(obs)}(n) - S_{soc}^{(obs)}(n-1) \geq 0$ for $n \leq n_{soc}$, whereas $S_{soc}^{(obs)}(n) - S_{soc}^{(obs)}(n-1) < 0$ for $n > n_{soc}$. We conclude that $S_{soc}^{(obs)}(n)$ is unimodal with a maximum at n_{soc}. In a nutshell, we have the following result:

THEOREM 4.4.– For the social planner's admission problem in the observable M/M/1 queue, the n_{soc}-threshold strategy with

$$n_{soc} = \max\left\{n : g(n) \le \frac{R\mu}{C}\right\},$$

where the function $g(n)$ is given by [4.4], is the socially optimal strategy. The threshold n_{soc} is induced by the administrator of the system, by imposing an admission fee p_{soc}, such that $n_{soc} = \left\lfloor \frac{(R-p_{soc})\mu}{C} \right\rfloor$, i.e. by imposing any $p_{soc} \in \left(R - \frac{C(n_{soc}-1)}{\mu}, R - \frac{Cn_{soc}}{\mu}\right]$.

In addition, it is easy to see that $n_{soc} \le n_e$, i.e. in the absence of an admission fee, the individually optimal threshold is greater than or equal to the socially optimal threshold. Indeed, we have

$$\begin{aligned} g(n) - n &= \frac{1}{1-\rho}\sum_{k=1}^{n}(1-\rho^k) - \frac{1}{1-\rho}n(1-\rho) \\ &= \frac{\rho}{1-\rho}\sum_{k=1}^{n}(1-\rho^{k-1}) \ge 0, \end{aligned}$$

whence $g(n_{soc}) \ge n_{soc}$. But $g(n_{soc}) \le \frac{R\mu}{C}$, by the very definition of n_{soc}, so $n_{soc} \le \frac{R\mu}{C}$ that yields $n_{soc} \le \left\lfloor \frac{R\mu}{C} \right\rfloor = n_e$.

Therefore, we see that, without imposing an admission fee, the customers overuse the system. This occurs because they neglect the negative externalities of their joining decisions on future customers. The same phenomenon was also observed in the unobservable counterpart of the model.

We now consider the monopolist's problem. In this case, the administrator of the system imposes an admission fee, aiming to the maximization of his own revenue. If he imposes admission fee p, then the customers adopt the corresponding threshold strategy and his revenue becomes $\lambda_e^{(obs)}p$, where $\lambda_e^{(obs)}$ stands for the corresponding equilibrium throughput. To determine the threshold n_{prof} that maximizes the revenue, we express the revenue as a function of the imposed threshold n. For inducing a threshold n to the customers, the administrator should impose an admission fee p such that $\left\lfloor \frac{(R-p)\mu}{C} \right\rfloor = n$. In the monopolist's problem, the administrator benefits from imposing the maximum price that induces the threshold, i.e. he should impose $p = R - \frac{Cn}{\mu}$. Then, the monopolist's revenue is

$$S_{prof}^{(obs)}(n) = \lambda\frac{1-\rho^n}{1-\rho^{n+1}}\left(R - \frac{Cn}{\mu}\right) = \lambda R\frac{1-\rho^n}{1-\rho^{n+1}}\left(1 - \frac{n}{\nu_e}\right),$$

where $\nu_e = \frac{R\mu}{C}$. This equation shows that for $n > n_e = \lfloor \frac{R\mu}{C} \rfloor$ we have $S_{prof}(n) < 0$, since the administrator should set a negative admission fee (i.e. he should subsidized the entrance) to induce a threshold, which is greater than n_e. Therefore, we conclude that $n_{prof} \le n_e$.

To study the monotonicity behavior of the function $S^{(obs)}_{prof}(n)$, we consider the ratio $S^{(obs)}_{prof}(n)/S^{(obs)}_{prof}(n-1)$. We present again only the case where $\rho < 1$.

$$\frac{S^{(obs)}_{prof}(n)}{S^{(obs)}_{prof}(n-1)} = \frac{(1-\rho^n)^2(\nu_e - n)}{(1-\rho^{n+1})(1-\rho^{n-1})(\nu_e - n + 1)}.$$

Then,

$$\begin{aligned}
\frac{S^{(obs)}_{prof}(n)}{S^{(obs)}_{prof}(n-1)} \ge 1 &\Leftrightarrow \frac{1-\rho^n}{1-\rho^{n+1}}(\nu_e - n) \ge \frac{1-\rho^{n-1}}{1-\rho^n}(\nu_e - n + 1)\\
&\Leftrightarrow \frac{(1-\rho^n)^2 - (1-\rho^{n-1})(1-\rho^{n+1})}{(1-\rho^{n+1})(1-\rho^n)}(\nu_e - n) \ge \frac{1-\rho^{n-1}}{1-\rho^n}\\
&\Leftrightarrow \nu_e - n \ge \frac{(1-\rho^{n-1})(1-\rho^{n+1})}{\rho^{n-1}(1-\rho)^2}\\
&\Leftrightarrow \frac{R\mu}{C} \ge n + \frac{(1-\rho^{n-1})(1-\rho^{n+1})}{\rho^{n-1}(1-\rho)^2}.
\end{aligned}$$

It can be seen that the function $h(n)$ with

$$h(n) = n + \frac{(1-\rho^{n-1})(1-\rho^{n+1})}{\rho^{n-1}(1-\rho)^2} \qquad [4.5]$$

is increasing in n, so a unique number n_{prof} exists such that $h(n) \le \frac{R\mu}{C}$, for $n \le n_{prof}$, whereas $h(n) > \frac{R\mu}{C}$, for $n > n_{prof}$. Therefore, the function $S^{(obs)}_{prof}(n)$ is unimodal with the maximum at n_{prof}. Therefore, we have the following result:

THEOREM 4.5.– For the monopolist's admission problem in the observable M/M/1 queue, the n_{prof}-threshold strategy with

$$n_{prof} = \max\left\{n : h(n) \le \frac{R\mu}{C}\right\},$$

where the function $h(n)$ is given by [4.5], is the revenue-maximizing strategy. The threshold is induced by the administrator of the system by imposing an admission fee $p_{prof} = R - \frac{Cn_{prof}}{\mu}$.

Using some quite involved algebraic manipulations, it can be seen that the inequality $n_{prof} \leq n_{soc} \leq n_e$ holds. This is known as *Naor's inequality* and is valid in a number of situations. Recently, (Hassin and Snitkovsky 2018) provided general conditions for the validity of this inequality in a general framework. If a state-dependent admission fee is permitted, then due to information symmetry a monopolist can extract all customer surplus and hence the monopolist's optimal threshold strategy and the socially optimal threshold strategy coincide, as in the unobservable model (for details, see Chen and Frank (2001)).

4.5. Comparison of the unobservable and the observable models

Hassin (1986) compared the observable and the unobservable versions of the M/M/1 queue with strategic customers regarding their join-or-balk dilemma, by focusing on the social welfare and a monopolist's revenue under a revenue-maximizing admission fee. Let λ, μ, R and C be the parameters of a model, defined as in section 4.3. Hassin showed that if $R\mu \leq 2C$, then the revenue under a revenue-maximizing admission fee is larger for the observable model, for all $\lambda > 0$. Hence, a monopolist prefers to reveal the queue length to the customers. If, however, $R\mu > 2C$, then a unique potential arrival rate λ^Z exists such that the revenue under a revenue-maximizing admission fee is larger for the observable model, if and only if $\lambda \geq \lambda^Z$. Thus, in this case, a monopolist prefers to reveal the queue length only when $\lambda \geq \lambda^Z$. Thus, there is a range of the parameters ($R\mu > 2C$ and $\lambda < \lambda^Z$), where the provision of more information to the customers hurts the service provider. The same properties hold also for the social welfare under a revenue-maximizing fee, but with a different critical value λ^S in place of λ^Z. Therefore, in this case, for $\lambda < \lambda^S$, it is socially preferable for the monopolist to be unable to inform customers of the queue length. Thus, there is a range of the parameters where the provision of more information hurts the society as a whole. Note also that $\lambda^S < \lambda^Z$, so for arrival rates λ with $\lambda^S < \lambda < \lambda^Z$, the profit maximizer prefers to conceal the queue length, although it is socially preferable to induce him to reveal it. However, when the profit maximizer prefers to reveal the queue length (i.e. when $\lambda \geq \lambda^Z$), this is certainly socially preferable as well.

Chen and Frank (2004) compared the observable and the unobservable versions of the M/M/1 queue by focusing on the equilibrium effective arrival rate (which is the same as the throughput since there are no abandonments), under an arbitrary fixed admission fee. For a given potential arrival rate λ, let $\lambda_e^{(obs)}(\lambda)$ and $\lambda_e^{(un)}(\lambda)$ denote the corresponding equilibrium effective arrival rates in the observable and unobservable versions, respectively. Chen and Frank proved that $\lambda_e^{(obs)}(\lambda) - \lambda_e^{(un)}(\lambda)$ monotonically increases in λ and there exists a critical value λ^* such that $\lambda_e^{(obs)}(\lambda^*) - \lambda_e^{(un)}(\lambda^*) = 0$. Therefore, to attract more customers to the system, it is advisable to conceal the queue length for potential arrival rates λ with $\lambda < \lambda^*$, and to reveal it when $\lambda > \lambda^*$. This is intuitively plausible. Indeed, in an unobservable M/M/1 queue with low arrival rates all customers join, whereas in the corresponding observable queue there will be always some customers (those that arrive during high congestion periods) who balk. Hence, for sufficiently low arrival rates the equilibrium effective arrival rate is higher for the unobservable model. For high arrival rates, we have the opposite situation, i.e. all customers balk in the unobservable model, whereas a positive fraction of customers who find low congestion do enter the observable model.

Shone *et al.* (2013) considered the same problem of the comparison of the equilibrium throughputs $\lambda_e^{(obs)}$ and $\lambda_e^{(un)}$, between the observable and unobservable versions of the M/M/1 queue. They provided necessary and sufficient conditions on the system parameters under which the equilibrium throughputs are equal in the two versions. Moreover, they investigated the behavior of the equilibrium throughputs in the two informational cases as functions of the normalized service value $\frac{R\mu}{C}$. In particular, they showed that the number of distinct values of the normalized service value for which $\lambda_e^{(obs)} = \lambda_e^{(un)}$ is monotonically increasing with respect to the utilization rate $\rho = \frac{\lambda}{\mu}$ and tends to infinity as $\rho \to 1$.

Guo and Zipkin (2007) compared the observable, the unobservable and the workload-observable versions of the M/M/1 queue, under a general reward–cost structure that generalizes the standard Naor's linear reward–cost structure. Under this framework, the service value is R, but a customer's waiting cost is $\theta E[c(W)]$, where W stands for the steady-state waiting time, $c(w)$ is a common basic cost function for all customers and θ is a customer-specific parameter that represents the sensitivity to delay. In other words, a customer with delay sensitivity θ has expected utility $R - \theta E[c(W)]$, if he/she decides to join. The authors showed that the maximum equilibrium throughput of the system may correspond to different information levels according to the values of the underlying parameters. The main conclusion is that the primary factor that determines whether information is good or bad for the service provider and the customers is the distribution function of the customer delay sensitivity and not the common basic cost function.

The aforementioned papers show that neither the observable nor the unobservable versions of the M/M/1 queue are preferable for the whole range of the underlying operational and economic parameters. Therefore, a number of authors studied the M/M/1 queue with strategic customers under information structures that lie between the observable and unobservable versions. To the best of our knowledge, there are three main ideas that have appeared in the literature that bridge the observable and unobservable versions of the M/M/1 queue: partially observable models, heterogeneously observable models and observable-with-delay models. We present these ideas in sections 4.6–4.8.

4.6. Partially observable models

In partially observable models, the state-space of the queue length of a given service system is partitioned into subsets and the arriving customers are not informed about the exact queue length, but rather about the subset it belongs to. If the state-space is partitioned into subsets of consecutive integers, the waiting space can be considered to be "compartmented" and the customers are informed only about the compartment in which they are going to be placed. Economou and Kanta (2008) considered the case of regular compartmentalization (all compartments being of the same size) in the M/M/1 queue and studied the customer strategic behavior and the associated social optimization and revenue maximization administrator's problems. In what follows, we present the corresponding main results.

The model of interest is an M/M/1 queue with the same operational and economic parameters λ, μ, R and C that were introduced in section 4.3. However, we assume that the space of the system is partitioned in compartments of fixed capacity of a customers and we consider two information cases for the customers:

– N: Known compartment number: Customers observe the number of the compartment in which they are going to enter but not the position within it. More specifically, if there exist n customers in the system just before the arrival of a tagged customer, his/her information will be the compartment number $i = \lfloor n/a \rfloor + 1$ in which he/she enters if he/she decides to join the system.

– P: Known compartment position: Customers observe the position of the compartment in which they are going to enter but not the number of the compartment. The information of an arriving customer is the position $i = (n \mod a) + 1$ in which he/she enters if he/she decides to join the system.

The decisions of the customers are irrevocable, i.e. retrials of balking customers and reneging of entering customers are not allowed.

We now limit our exposition within the framework of the known compartment number (N) case. Then, a pure strategy is specified by a set $A \subseteq \{1, 2, \ldots\}$, which shows the "favorable" compartment numbers for a customer, i.e. a customer decides to join the system if he/she knows that he/she will enter a compartment with a number belonging to A. We consider a tagged customer and assume that all other customers follow a strategy A. Then the Markov chain describing the number of customers in the system will be eventually absorbed in the set $\{0, 1, 2, \ldots, i^*a\}$, where i^* is the maximum integer such that $\{0, 1, 2, \ldots, i^*\} \subseteq A$. Indeed, a moment of reflection shows that under strategy A all other states become transient and the system behaves as an M/M/1/i^*a queue. Consider now the tagged customer who is to decide whether to join the system or not, given the information that he/she can enter the compartment number i. So he/she knows that the number of customers at his/her arrival is $n \in$

$\{(i-1)a, (i-1)a+1, \dots, ia-1\}$. If he/she decides to enter, then his/her expected net individual benefit is

$$S_{ind}^{(po-N)}(i) = R - \frac{C(E_A[Q^- | Q^- \in \{(i-1)a, \dots, ia-1\}] + 1)}{\mu}, \qquad [4.6]$$

where Q^- is a random variable with steady-state distribution of the number of customers at an arrival instant in the M/M/1/i^*a queue and the subscript "A" in the expectation signifies that the population of the customers follow the pure strategy A. Because of the PASTA property, the distribution of Q^- coincides with the distribution of the number of customers in continuous-time, Q. Using standard formulas for the M/M/1/n queue (see, e.g. section 7.3.2 in Kulkarni (2010)), we obtain that for $\rho \neq 1$

$$S_{ind}^{(po-N)}(i) = R - \frac{C}{\mu}\left(ia - \frac{a}{1-\rho^a} + \frac{1}{1-\rho}\right). \qquad [4.7]$$

For $\rho = 1$, we can see that $S_{ind}^{(po-N)}(i) = R - \frac{C}{\mu}\{ia - \frac{a-1}{2}\}$, which is the limiting case of [4.7] for $\rho \to 1$. Therefore, in the rest of this section, we present the formulas under the assumption that $\rho \neq 1$, with the understanding that the results are also valid for $\rho = 1$ with the appropriate limits in the formulas (as $\rho \to 1$).

The tagged customer will decide to enter if and only if $S_{ind}^{(po-N)}(i) \geq 0$, that is when $i \leq \lfloor \frac{R\mu}{aC} + \frac{1}{1-\rho^a} - \frac{1}{a(1-\rho)} \rfloor$. Therefore, we conclude with the following result:

THEOREM 4.6.– For the join-or-balk customer dilemma in the partially N-observable a-compartmented M/M/1 queue, the i_e-threshold strategy with

$$i_e = \lfloor x_e \rfloor, \qquad [4.8]$$

$$x_e = \frac{R\mu}{aC} + \frac{1}{1-\rho^a} - \frac{1}{a(1-\rho)}, \qquad [4.9]$$

that prescribes a customer to join the system as long as the compartment number that will be assigned is at most i_e is the dominant strategy (and therefore the equilibrium strategy).

The problem of social optimization can be also solved along similar lines with the observable M/M/1 queue. Indeed, using an appropriate admission fee, the administrator of the system can induce any desired i-threshold strategy. Then, the corresponding total social welfare per time unit is $S_{soc}^{(po-N)}(i) = \lambda^{(po-N)}(i)R - CE_i[Q]$, where $\lambda^{(po-N)}(i)$ is the throughput and $E_i[Q]$ the expected steady-state number of customers in the system, given that the customers follow the i-threshold strategy. In this case, the system behaves as an M/M/1/ia queue and using the corresponding steady-state distribution yields

$$S_{soc}^{(po-N)}(i) = \lambda R \frac{1-\rho^{ia}}{1-\rho^{ia+1}} - C\left(\frac{\rho}{1-\rho} - \frac{(ia+1)\rho^{ia+1}}{1-\rho^{ia+1}}\right). \quad [4.10]$$

The unimodality of $S_{soc}^{(po-N)}(i)$ is a result of the unimodality of $S_{soc}^{(obs)}(n)$ (since $S_{soc}^{(po-N)}(i) = S_{soc}^{(obs)}(ai)$) and we obtain the next result that characterizes the socially optimal policy:

THEOREM 4.7.– For the social planner's admission problem in the partially N-observable a-compartmented M/M/1 queue, the i_{soc}-threshold strategy with

$$i_{soc} = \lfloor x_{soc} \rfloor \quad [4.11]$$

with x_{soc} being the unique solution of the equation $g(x) = i_c$ in $[1, \infty]$ with

$$g(x) = \frac{(xa+1)(1-\rho^a) - a(1-\rho^{xa+1})}{a(1-\rho)(1-\rho^a)} + \frac{1}{1-\rho^a} - \frac{1}{a(1-\rho)}, \quad [4.12]$$

is the socially optimal strategy.

The monopolist's revenue-maximization problem is also solved similarly to the corresponding problem in the framework of the observable M/M/1 queue. Indeed, in light of the equations [4.8]–[4.9] the maximum entrance fee that can be imposed by the administrator of the system in order to force the customers to adopt a given threshold i is

$$p(i) = R - \frac{aC}{\mu}\left(i - \frac{1}{1-\rho^a} + \frac{1}{a(1-\rho)}\right). \quad [4.13]$$

Then, his expected revenue per time unit is $S_{prof}^{(po-N)}(i) = \lambda^{(po-N)}(i)p(i)$ where $\lambda^{(po-N)}(i)$ is the throughput given that the customers follow the threshold i. The system behaves as an M/M/1/ia queue and therefore using the corresponding steady-state distribution (see, e.g. section 7.3.2 in Kulkarni (2010)), we obtain after some algebra that

$$S_{prof}^{(po-N)}(i) = \frac{\lambda a C(1-\rho^{ia})}{\mu(1-\rho^{ia+1})}(x_e - i), \quad [4.14]$$

where x_e is given by [4.9]. This function is unimodal and its maximum point is characterized in the following result (for more detail, see Economou and Kanta (2008)):

THEOREM 4.8.– For the monopolist's admission problem in the partially N-observable a-compartmented M/M/1 queue, the i_{prof}-threshold strategy with

$$i_{prof} = \lfloor x_{prof} \rfloor \qquad [4.15]$$

with x_{prof} being the unique solution of the equation $h(x) = x_e$ in $[1, \infty]$ with

$$h(x) = x + \frac{(1 - \rho^{xa-a})(1 - \rho^{xa+1})}{\rho^{xa-a}(1 - \rho)(1 - \rho^a)}, \qquad [4.16]$$

is the revenue-maximizing strategy.

Using these theoretical results, a number of numerical experiments were carried out by (Economou and Kanta 2008) with various interesting findings. Moreover, a number of secondary theoretical results were proved. The P case has been also studied in detail by (Economou and Kanta 2008) and the equilibrium, socially optimal and revenue-maximizing strategies have been characterized in almost-explicit forms. We do not report the results in detail here, since this case is more difficult to implement. We now summarize the main findings of this study:

In both information cases, equilibrium threshold balking strategies exist. The corresponding equilibrium, social and revenue-maximizing thresholds can be explicitly computed.

In the N case, the equilibrium threshold strategy is the dominant strategy. This is quite exceptional behavior since dominant strategies are reported in the literature only for fully observable models. In addition, a kind of Naor's inequality holds: $i_{prof} \leq i_{soc} \leq i_e$, i.e. the individual optimization leads to longer queues that it is socially desirable (while the revenue maximization induces even shorter queues). The equilibrium threshold i_e is a decreasing function of λ. This is in contrast to (Naor 1969) model where the equilibrium threshold n_e does not depend on the arrival rate λ.

In the N case, the maximum number of customers in the system is $i_e a$, when the customers follow an equilibrium strategy. When keeping all parameters fixed and letting a varies, this number is minimized for compartment sizes a that are near to the integer divisors of Naor's equilibrium threshold, n_e, of the observable model. This suggests that it is desirable for the designer of the system to construct the compartments according to one of these a, if he wants to minimize the waiting space required for the proper function of the system.

And finally, perhaps the most interesting finding in the N case is that the administrator's revenue is a unimodal function of the compartment size a, when keeping the other parameters fixed. Therefore, there exists an optimal a that maximizes the administrator's revenue. This suggests that apart from other mechanisms suggested in the literature (information pricing, state-dependent pricing, service discipline, etc.), the compartmentalization and an adequate selection of the compartment size can be used for increasing the administrator's revenue.

In the P case, the equilibrium threshold is a decreasing function of λ while the social and profit maximizing thresholds exhibit unimodal behavior. This is in agreement with the corresponding findings for (Edelson and Hildebrand 1975) model. However, a crucial difference between the P case and the unobservable model is the fact that the objectives of a revenue-maximizer monopolist and a social planner do not coincide, except for the case $a = 1$. This happens because of the partial information of the customers that enable them to secure a positive surplus.

When $a = 1$, the N case reduces to (Naor 1969) model, while the P case reduces to (Edelson and Hildebrand 1975) model. For $a \to \infty$, we have the opposite situation, i.e. the N case behaves as (Edelson and Hildebrand 1975) model and the P case as (Naor 1969) model. In general, we can say that for small values of a, the N case is more informative because then the knowledge of the compartment number determines the exact position of the customer with error of at most a. For large values of a, we know that almost all the customers enter the first compartment so the N case is less informative. Then, knowing the position in the compartment is much more important for the customer to assess his/her gains.

There are many other papers dealing with partially observable variants of the M/M/1 queue. Guo and Zipkin (2009) considered the general case of compartments with possibly different sizes and proved several interesting results about the comparison of two partitions of the state space, one a refinement of the other. More recently, (Simhon *et al.* 2016) considered the M/M/1 queue with strategic customers that face the dilemma of joining/balking, when the administrator informs the customers about the current queue length only when it is short, i.e. when it does not exceed a certain threshold D. This corresponds to the partition of the state-space to the subsets $\{0\}, \{1\}, \{2\}, \ldots, \{D\}$ and $\{D+1, D+2, \ldots\}$. The authors proved that the equilibrium throughput is a monotone function of D and therefore if the administrator's goal is to maximize throughput, then the optimal policy is one of the extremes, either the observable or the unobservable queue. Kim and Kim (2017) considered the generalization of the last model by assuming that the customers are informed about the current queue length only when it belongs to a subset O. This corresponds to the partition of the state-space comprising the singletons of the elements of O and the complement of O (which contains the unobservable states). The authors proved the counterintuitive result that the optimal partition for the maximization of the throughput of the system corresponds to a set O that contains all the states above a threshold, i.e. it is preferable to allow the customers to observe the queue length only when it is large. Finally, (Hassin and Koshman 2017) considered a model where the arriving customers are informed whether the queue length is less than an exogenously given threshold N or not. They focused on the monopolist's problem for the dynamic pricing version of this model (i.e. different prices are offered to the customers according to whether the queue length is below N or not) and proved the interesting result that the choice of $N = 1$ guarantees at least half of the maximum value that can be generated by the system.

4.7. Heterogeneously observable models

In heterogeneously observable models, the population of customers is divided into observing and uninformed (non-observing) customers. A simple model that encompasses this characteristic has been studied by (Economou and Grigoriou 2015) and (Hu *et al.* 2018). In what follows, we present some of the reported results for this model.

The system of interest is an M/M/1 queue with the same operational and economic parameters with the main model that we introduced in section 4.3. Every arriving customer is observing or uninformed, with probabilities p_o and $p_u = 1 - p_o$, respectively, independently of the other customers. Observing customers are informed about the number of customers in the system before making their decisions whether to join or balk, whereas uninformed customers do not. The decisions of the customers are irrevocable (i.e. neither retrials nor reneging are permitted). All the parameters of the model, including the proportion of observing customers, p_o, are assumed common knowledge for the customers.

An observing customer faces a situation like the one that we described for the observable M/M/1 queue in section 4.4. Thus, his/her dominant strategy is to join according to Naor's threshold n_e given from [4.2]. On the other hand, a general mixed strategy for an uninformed customer is specified by a single number, the join probability q.

We are now focusing on the best response of a tagged uninformed customer. To this end, suppose that the observing customers follow the n_e-threshold strategy, i.e. they join according to Naor's threshold, and that the uninformed customers follow a mixed strategy q. Then, the tagged customer bases his/her decision on the sign of his/her expected net benefit if he/she decides to join. This is given as

$$S_{ind}^{(ho)}(q) = R - \frac{C(E_q[Q^-] + 1)}{\mu}, \qquad [4.17]$$

where $E_q[Q^-]$ is the mean number of customers in system found by an arriving uninformed customer, given that the others follow the strategy (n_e, q) (i.e. the observing customers follow the n_e-threshold strategy and the uninformed customers follow the q-mixed strategy). However, the uninformed customers arrive according to a Poisson process with rate λp_u and therefore, because of the PASTA property, $E_q[Q^-]$ coincides with the mean number of customers in the system $E_q[Q]$, when the strategy (n_e, q) is employed. Therefore, for the study of strategic customer behavior, we need to compute $E_q[Q]$, for any strategy (n_e, q). Under such a strategy the number of customers in the system is a continuous-time Markov chain of birth–death type with transition rates

$$q_{i,j} = \begin{cases} \lambda_1 \text{ if } 0 \leq i \leq n_e - 1, \ j = i + 1 \\ \lambda_2 \text{ if } i \geq n_e, \ j = i + 1 \\ \mu \ \text{ if } i \geq 1, \ j = i - 1 \\ 0 \ \text{ otherwise,} \end{cases} \qquad [4.18]$$

where

$$\lambda_1 = \lambda p_o + \lambda(1 - p_o)q, \ \lambda_2 = \lambda(1 - p_o)q. \qquad [4.19]$$

Using the well-known formula for the steady-state distribution of birth–death processes and standard summation techniques, we can easily derive the steady-state distribution in closed-form and its mean. More concretely, we have the following result:

PROPOSITION 4.1.– The steady-state distribution of the number of customers in the heterogeneously observable M/M/1 queue, when the customers follow an (n_e, q) strategy, with $n_e, q > 0$, is given by

$$\pi_n = \begin{cases} B\rho_1^n & \text{if } 0 \leq n \leq n_e - 1, \\ B\rho_1^{n_e}\rho_2^{n-n_e} & \text{if } n \geq n_e, \end{cases} \qquad [4.20]$$

where

$$\rho_1 = \frac{\lambda_1}{\mu}, \ \rho_2 = \frac{\lambda_2}{\mu} \qquad [4.21]$$

and

$$B = \frac{(1 - \rho_1)(1 - \rho_2)}{1 - \rho_2 - \rho_1^{n_e+1} + \rho_1^{n_e}\rho_2}. \qquad [4.22]$$

The corresponding mean steady-state number of customers is

$$E_q[Q] = \frac{(1-\rho_2)[(n_e-1)\rho_1^{n_e+1} - n_e\rho_1^{n_e} + \rho_1]}{(1-\rho_1)[1-\rho_2-\rho_1^{n_e+1}+\rho_1^{n_e}\rho_2]} + \frac{(1-\rho_1)[n_e\rho_1^{n_e} - (n_e-1)\rho_1^{n_e}\rho_2]}{(1-\rho_2)[1-\rho_2-\rho_1^{n_e+1}+\rho_1^{n_e}\rho_2]}. \qquad [4.23]$$

The cases where one or both of n_e and q are 0 can be easily derived from the above formulas by taking the appropriate limits. We do not report the formulas for brevity, but we discuss them briefly: For $n_e = 0$ and $q = 0$, the system is continuously empty. For $n_e = 0$ and $q > 0$, the system behaves as an M/M/1 queue with arrival rate $\lambda_2 = \lambda(1-p_o)q$ and service rate μ. Similarly, for $n_e > 0$ and $q = 0$, the system behaves as an M/M/1/n_e queue with arrival rate $\lambda_1 = \lambda p_o$ and service rate μ. Note also that the quantity $E_q[Q]$ is increasing in q. This is intuitively clear, but it can be also formally proven by using (Kirstein 1976) rate sufficient conditions for the strong comparability of Markov chains.

We can now return to the study of a tagged uninformed customer's best response when a strategy (n_e, q) is employed by the other customers. Let $\hat{q}_e$ be the root of $S_{ind}^{(ho)}(q)$. Then, the analysis proceeds along the same lines with the unobservable model of section 4.3. The set of best responses against (n_e, q), $BR((n_e, q))$, is (as far as $q \in [0,1]$ and $\lambda(1-p_o)q < 1$)

$$BR((n_e,q)) = \begin{cases} \{(n_e,0)\}, & \text{if } q > \hat{q}_e, \\ \{n_e\} \times [0,1], & \text{if } q = \hat{q}_e, \\ \{(n_e,1)\}, & \text{if } q < \hat{q}_e. \end{cases}$$

We can now proceed to the computation of the equilibrium strategies:

The strategy $(n_e, 0)$ is equilibrium strategy, if and only if $(n_e, 0) \in BR((n_e, 0))$, i.e. $0 \geq \hat{q}_e$, which reduces to $R \leq \frac{C(E_0[Q]+1)}{\mu}$.

A strategy (n_e, q_e) with $q_e \in (0,1)$ is equilibrium strategy, if and only if $(n_e, q_e) \in BR((n_e, q_e))$, i.e. $q_e = \hat{q}_e$. This is valid as far as $\hat{q}_e \in (0,1)$, which occurs if and only if $\frac{C(E_0[Q]+1)}{\mu} < R < \frac{C(E_1[Q]+1)}{\mu}$.

Finally, the strategy $(n_e, 1)$ is the equilibrium strategy, if and only if $(n_e, 1) \in BR((n_e, 1))$, i.e. $1 \leq \hat{q}_e$, which reduces to $R \geq \frac{C(E_1[Q]+1)}{\mu}$.

In summary, we have the following result:

THEOREM 4.9.– For the join-or-balk customer dilemma in the heterogeneously observable M/M/1 queue, where a fraction of customers are observing and the rest customers are uninformed, a unique equilibrium strategy (n_e, q_e) exists. The threshold n_e for the entrance of the observing customers is Naor's threshold given from [4.2]. The join probability q_e for the uninformed customers is given by the formula

$$q_e = \begin{cases} 0, & R \leq \frac{C(E_0[Q]+1)}{\mu}, \\ \hat{q}_e, & \frac{C(E_0[Q]+1)}{\mu} < R < \frac{C(E_1[Q]+1)}{\mu}, \\ 1, & R \geq \frac{C(E_1[Q]+1)}{\mu}, \end{cases}$$

where $\hat{q}_e$ is the unique root of $S_{ind}^{(ho)}(q)$ given from [4.17] and $E_q[Q]$ is computed from [4.23].

Economou and Grigoriou (2015) determined the equilibrium strategies in the slightly more general case where the service value R and the waiting cost rate C are different for informed and uninformed customers and provided a preliminary analysis. Hu *et al.* (2018) considered the homogeneous reward–cost framework of this section and studied in depth the effect of the fraction p_o of observing customers on the equilibrium threshold and social welfare. They showed that different behaviors emerge, according to the arrival rate of the customers. If the throughput is the focal performance measure, then $p_o = 1$ maximizes throughput if the arrival rate is high enough, whereas $p_o = 0$ maximizes throughput if the arrival rate is low enough. If the arrival rate is in an intermediate range, the maximum throughput is achieved at a p_o strictly between 0 and 1. Therefore, in this range, it is optimal to have a segment of uninformed customers or reveal the queue-length only to a fraction of customers.

If social welfare is the focal performance measure, the service provider should reveal the queue length information and encourage its dissemination when the arrival rate is relatively small. In other situations, it is optimal to have a segment of uninformed customers or, equivalently, to hide the queue-length from a certain fraction of customers.

In a nutshell, the results of (Hu *et al.* 2018) showed that throughput and social welfare are in general unimodal and not monotonous in the fraction of observing customers. In other words, information heterogeneity in a population can lead to more efficient outcomes, in terms of the system throughput or social welfare, than information homogeneity. Moreover, it was shown that for an overloaded system (with utilization factor sufficiently higher than 1), social welfare always attains its maximum when some fraction of customers is uninformed.

4.8. Observable-with-delay models

In models with delayed observations, the customers decide whether to join or balk without knowing the state of the system, but later on they are informed about their current position and may renege. Burnetas *et al.* (2017) considered a simple model of this kind, which boils down to an M/M/1 queue where the administrator of the system makes periodic announcements to the customers about their current positions. The model was motivated by a situation that occurs when people submit petitions through certain web-based systems. Then, upon submission, the customers receive a confirmation message with the registration number of their petition. Later on they learn the number of pending petitions in front of them. This is done either by periodic refreshments of a web page that indicates the registration number of the currently processed petition or by periodic bulk emails that announce the status of pending petitions. In what follows, we describe the model and the corresponding findings in some detail.

We consider an M/M/1 queue with the same operational and economic parameters with the main model that we introduced in section 4.3. Each customer, upon arrival, decides whether to join or balk, without observing the number of customers in the system. However, the administrator of the system announces to the customers their positions in the system, at the points of a Poisson process at rate θ. The customers, after an announcement, reevaluate their expected benefit of staying in the system, and will renege if it is negative.

Because of the exponentiality assumptions, it makes sense that a customer makes decisions only at the instants of system state transitions, i.e. at his/her arrival instant and at the times of the subsequent announcements of the administrator of the system. At his/her arrival instant, the system is unobservable to the customer. Therefore, his/her join-or-balk strategy is specified by a join probability, say q.

The analysis of the reneging behavior of a customer is trivial. At the epoch of the first announcement following his/her entrance, the system becomes observable to his/her. Therefore, the customer is willing to stay if his/her position n in the system is such that $R - C\frac{n}{\mu} \geq 0$, i.e. he/she decides to stay in the system, only if $n \leq n_e$, where n_e is Naor's threshold given by [4.2]. If a customer does not renege after the first announcement, he/she will not renege later. This fact follows from the Markovian (memoryless) nature of the model.

Therefore, the behavior of the system is specified by two parameters: the join probability q of the customers and their renege threshold n_e at the time of the first announcement after their arrival. Therefore, we need to analyze the behavior of the system when the customers follow a strategy (n_e, q). Under such a strategy, the number of customers in the system is represented by a continuous-time Markov chain with transition rates

$$q_{i,j} = \begin{cases} \lambda q, & \text{if } i \geq 0, \;\; j = i+1, \\ \mu, & \text{if } i \geq 1, i \neq n_e + 1, \;\; j = i-1, \\ \theta, & \text{if } i \geq n_e + 2, \;\; j = n_e, \\ \mu + \theta, & \text{if } i = n_e + 1, \; j = n_e, \\ 0, & \text{otherwise.} \end{cases} \quad [4.24]$$

The steady-state distribution of the number of customers in the system, Q, and its mean can be computed in closed-form, by solving the system of the corresponding balance equations.

PROPOSITION 4.2.– The steady-state distribution of the number of customers in the observable-with-delay M/M/1 queue, when the customers follow an (n_e, q) strategy, with $n_e, q > 0$ and $\lambda q \neq \mu$, is given by [4.20] where

$$\rho_1 = \frac{\lambda q}{\mu}, \; \rho_2 = \frac{\lambda q + \mu + \theta - \sqrt{(\lambda q + \mu + \theta)^2 - 4\lambda q \mu}}{2\mu}, \quad [4.25]$$

and B is given by [4.22]. The corresponding mean steady-state number of customers is given by [4.23].

The singular cases where $n_e = 0$ or $q = 0$ or $\lambda q = \mu$ can be easily derived from the formulas of Proposition 4.2 by taking appropriate limits.

We now define $S^{(owd)}_{ind-cond}(n|n_e)$ to be the conditional expected net benefit of a customer who joins the system, given that he/she finds n customers, when all customers (including herself/himself) use the same renege threshold n_e. To compute it, for a certain n, consider a tagged customer who arrives and decides to join, when the system has n other customers. We consider two cases according to whether $n \leq n_e - 1$ or not.

In the first case, where $n \leq n_e - 1$, the customer will certainly receive the service reward R, since he/she has no incentive to renege later. Moreover, his/her sojourn time in the system will be the sum of $n+1$ service times and we conclude that

$$S^{(owd)}_{ind-cond}(n|n_e) = R - \frac{C(n+1)}{\mu}. \quad [4.26]$$

In the second case, where $n \geq n_e$, the net benefit of the tagged customer, S_n, has the representation

$$S_n = (R - C(Y_n + Z))1_{\{X \geq Y_n\}} - CX1_{\{X < Y_n\}}, \quad [4.27]$$

where X, Y_n and Z are independent random variables with X being an exponential distribution with rate θ, Y_n being an Erlang distribution with $n+1-n_e$ phases and rate μ and Z being an Erlang distribution with n_e phases and rate μ. This representation is deduced by interpreting X as the time until the first announcement after the arrival of the tagged customer, Y_n the time of $n+1-n_e$ service times until the tagged customer has no incentive to renege and Z the time after Y_n until the departure of the tagged customer (if he/she does not renege). Moreover, note that the customer reneges if $X < Y_n$, in which case his/her sojourn time is X. On the other hand, when $X \geq Y_n$, the customer stays in the system for $Y_n + Z$ time units and receives the reward for service. Taking expected values in [4.27] yields

$$S^{(owd)}_{ind-cond}(n|n_e) = (R - CE[Z])\mathrm{P}[X \geq Y_n] - CE[\min(X, Y_n)]. \qquad [4.28]$$

We can easily deduce that

$$\mathrm{P}[X \geq Y_n] = \left(\frac{\mu}{\mu+\theta}\right)^{n-n_e+1}, \quad E[\min(X, Y_n)] = \frac{1}{\theta}\left(1 - \left(\frac{\mu}{\mu+\theta}\right)^{n-n_e+1}\right),$$

and $E[Z] = \frac{n_e}{\mu}$. Combining and simplifying [4.26] and [4.28] (for details, see Burnetas *et al.* (2017)) yields the following result:

LEMMA 4.1.– Consider the observable-with-delay M/M/1 queue, where the customers follow an (n_e, q) strategy. We have

$$S^{(owd)}_{ind-cond}(n|n_e) = \begin{cases} R - \frac{C(n+1)}{\mu}, & \text{if } 0 \leq n \leq n_e - 1, \\ \left(R - \frac{Cn_e}{\mu} + \frac{C}{\theta}\right)\left(\frac{\mu}{\mu+\theta}\right)^{n-n_e+1} - \frac{C}{\theta}, & \text{if } n \geq n_e. \end{cases} \qquad [4.29]$$

We can now compute the (unconditional) expected net benefit for a customer who decides to join, when the others follow a strategy (n_e, q). We denote this quantity by $S^{(owd)}_{ind}(q)$. Then,

$$S^{(owd)}_{ind}(q) = \sum_{n=0}^{\infty} \pi_n S^{(owd)}_{ind-cond}(n|n_e),$$

where (π_n) is the steady-state distribution given in Proposition 4.2 and the quantity $S^{(owd)}_{ind-cond}(n|n_e)$ is computed in Lemma 4.1. Evaluating the corresponding geometric sums (for details, see Burnetas *et al.* (2017)), we deduce the following result:

PROPOSITION 4.3.– Consider the observable-with-delay M/M/1 queue, where the customers follow an (n_e, q) strategy, with $n_e, q > 0$ and $\lambda q \neq \mu$. The expected net benefit of a customer who decides to join is given by

$$\begin{aligned} S_{ind}^{(owd)}(q) &= B\left(R - \frac{C}{\mu}\right)\frac{1-\rho_1^{n_e}}{1-\rho_1} - B\frac{C}{\mu}\frac{(n_e-1)\rho_1^{n_e+1} - n_e\rho_1^{n_e} + \rho_1}{(1-\rho_1)^2} \\ &+B\left(R - \frac{Cn_e}{\mu} + \frac{C}{\theta}\right)\frac{\mu\rho_1^{n_e}}{\mu+\theta-\mu\rho_2} - B\frac{C}{\theta}\frac{\rho_1^{n_e}}{1-\rho_2}, \qquad [4.30] \end{aligned}$$

where B, ρ_1 and ρ_2 are given in proposition 4.2.

Furthermore, using coupling arguments and basic properties of stochastic orders, it has been shown that the expected net benefit $S_{ind}^{(owd)}(q)$ is a strictly decreasing function of q, which shows an avoid-the-crowd behavior in the model. Using this fact, it is easy to show the existence and uniqueness of the equilibrium strategy.

More concretely, we consider a tagged arriving customer and assume that the strategy (n_e, q) is employed by the other customers. Let $\tilde{q}_e$ be the root of $S_{ind}^{(owd)}(q)$. Then, the analysis proceeds along the same lines with the heterogeneously observable model of section 4.7. The set of best responses against (n_e, q), $BR((n_e, q))$, is (as far as $q \in [0, 1]$)

$$BR((n_e,q)) = \begin{cases} \{(n_e,0)\}, & \text{if } q > \tilde{q}_e, \\ \{n_e\} \times [0,1], & \text{if } q = \tilde{q}_e, \\ \{(n_e,1)\}, & \text{if } q < \tilde{q}_e. \end{cases}$$

We can now proceed to the computation of the equilibrium strategies:

The strategy $(n_e, 0)$ is equilibrium strategy, if and only if $(n_e, 0) \in BR((n_e, 0))$, i.e. $0 \geq \tilde{q}_e$, which reduces to $S_{ind}^{(owd)}(0) \leq 0$.

A strategy (n_e, q_e) with $q_e \in (0, 1)$ is equilibrium strategy, if and only if $(n_e, q_e) \in BR((n_e, q_e))$, i.e. $q_e = \tilde{q}_e$. This is valid as far as $\tilde{q}_e \in (0, 1)$, which occurs if and only if $S_{ind}^{(owd)}(1) < 0 < S_{ind}^{(owd)}(0)$.

Finally, the strategy $(n_e, 1)$ is the equilibrium strategy, if and only if $(n_e, 1) \in BR((n_e, 1))$, i.e. $1 \leq \tilde{q}_e$, which reduces to $S_{ind}^{(owd)}(1) \geq 0$.

In summary, we have the following result:

THEOREM 4.10.– For the join-or-balk and stay-or-renege customer dilemmas in the observable-with-delay M/M/1 queue, a unique equilibrium strategy (n_e, q_e) exists.

The threshold n_e for reneging at the first announcement instant following customers' arrivals is Naor's threshold given from [4.2]. The join probability q_e is given by the formula

$$q_e = \begin{cases} 0, & S_{ind}^{(owd)}(0) \leq 0, \\ \tilde{q}_e, & S_{ind}^{(owd)}(1) < 0 < S_{ind}^{(owd)}(0), \\ 1, & S_{ind}^{(owd)}(1) \geq 0, \end{cases}$$

where $\tilde{q}_e$ is the unique root of $S_{ind}^{(owd)}(q)$ given from [4.30].

For the social optimization and the profit maximization in this model, the administrator may impose two kinds of fees: an admission fee paid by all customers that decide to join the system and a service fee paid by those customers who do receive service (i.e. those who do not abandon before being served). Burnetas *et al.* (2017) studied these problems using a combination of theoretical results and numerical experiments. The more important managerial take-away messages that have been deduced from the analytical results and the numerical experimentation are discussed in the sequel.

An important finding is that the equilibrium throughput of the system is a unimodal function of the announcement rate θ. Thus, if the administrator of the system is interested in maximizing the throughput (for example if he receives an exogenous payment per served customer), then there is an ideal announcement rate to achieve this objective. The optimal announcement rate lies strictly between 0 and ∞. In other words, some delay in providing information to the customers is beneficial in terms of throughput.

In the unobservable model, (Edelson and Hildebrand 1975) showed that the socially optimal join probability is always smaller than or equal to the equilibrium join probability (because of the negative externalities – see the comment just after theorem 4.2). However, in the observable-with-delay model this is no longer valid: The socially optimal join probability may be greater than the equilibrium join probability in some cases. It is important to stress here that this not due to positive externalities, but due to n_{soc} being less than or equal to n_e.

Finally, the possibility of using two prices for the entrance and the service of the customers may be efficient for coordinating a given system. Even if this is not possible, this pricing mechanism can induce an almost efficient system, where the equilibrium behavior of the customers is very near to the socially desirable.

Another model with delayed observation characteristics is the so-called "armchair decision" problem introduced by (Hassin and Roet-Green 2014) (see also Roet-Green (2013)). In this model, the customers observe the queue length before reaching it, using probably some web-based application. Then, they decide whether to leave their armchair and go to the service facility or not, but when they arrive at the system they are informed about the current queue length and should make their second decision to join or balk.

4.9. Conclusions and literature review for further study

The advancement of technology and its incorporation in the contemporary service systems have provided a great flexibility to the administrators of the systems to share real-time information about congestion with the customers. Yet, given the strategic and usually selfish nature of the customers, the provision of more information is not always beneficial for the administrator or for them. In this work, we presented a number of models that lie between full information and no information for the customers and showed that the optimal level of information is strictly between these two extremes, in most cases. However, much more work remains to be done. In particular, it seems important to extend the studies for the comparison of information levels in more general models. This will enable to drive robust conclusions about the effect of information provision in service systems. Another important issue is the study of pricing for information acquisition. The pricing can be used as an additional mechanism for differentiating customers and may be particularly useful for social optimization and profit maximization.

Apart from the three basic ideas for bridging observable and unobservable models that have been presented in this chapter, a recent study of (Dimitrakopoulos *et al.* 2019) shows that alternating a system between observable and unobservable periods can be also advantageous for increasing the equilibrium throughput and/or the equilibrium social welfare.

After reading the present short introduction, readers who wish to deepen their understanding of the subject should read the papers that are mentioned in the references, in particular the recent and excellent survey of (Ibrahim 2018) on sharing delay information in service systems and the recent book of (Hassin 2016).

4.10. Acknowledgments

I would like to express my gratitude to my friend and colleague Professor Apostolos Burnetas who introduced me in the field of Strategic Queueing, during a series of seminars, some years ago. I also owe many thanks to Professors Refael Hassin and Moshe Haviv who shaped this field with their monograph (Hassin and Haviv 2003). It is certainly the book that has influenced most my academic career. In addition, various inspiring discussions with Professor Hassin sowed the seed for the present work. Finally, I would like to thank the editors of this volume, Professors

Nikolaos Limnios and Vladimir Anisimov for their kind invitation to write this chapter and an anonymous referee who helped a lot with feedback to improve the content and the presentation of the chapter.

4.11. References

Burnetas, A., Economou, A., Vasiliadis, G. (2017). Strategic customer behavior in a queueing system with delayed observations. *Queueing Syst.*, 86, 389–418.

Chen, H., Frank, M. (2001). State dependent pricing with a queue. *IIE Trans.*, 33, 847–860.

Chen, H., Frank, M. (2004). Monopoly pricing when customers queue. *IIE Trans.*, 36, 569–581.

Dimitrakopoulos, Y., Economou, A., Leonardos, S. (2019). Strategic customer behavior in a queueing system with alternating information structure. Working Paper.

Economou, A., Grigoriou, M. (2015). Strategic balking behavior in a queueing system with a mixed observation structure. *Proceedings of the 10th Conference on Stochastic Models of Manufacturing and Service Operations (SMMSO 2015)*. University of Thessaly Press, Volos, 51–58.

Economou, A., Kanta, S. (2008). Optimal balking strategies and pricing for the single server Markovian queue with compartmented waiting space. *Queueing Syst.*, 59, 237–269.

Edelson, N.M., Hildebrand, K. (1975). Congestion tolls for Poisson queueing processes. *Econometrica*, 43, 81–92.

Guo, P., Zipkin, P. (2007). Analysis and comparison of queues with different levels of delay information. *Manage. Sci.*, 53, 962–970.

Guo, P., Zipkin, P. (2009). The effects of the availability of waiting-time information on a balking queue. *Eur. J. Oper. Res.*, 198, 199–209.

Hassin, R. (1986). Consumer information in markets with random products quality: The case of queues and balking. *Econometrica*, 54, 1185–1195.

Hassin, R. (2016). *Rational Queueing*. CRC Press, Taylor and Francis Group, Boca Raton, FL.

Hassin, R., Haviv, M. (2003). *To Queue or Not to Queue: Equilibrium Behavior in Queueing Systems*. Kluwer Academic Publishers, Boston.

Hassin, R., Koshman, A. (2017). Profit maximization in the M/M/1 queue. *Oper. Res. Lett.*, 45, 436–441.

Hassin, R., Roet-Green, R. (2014). The armchair decision: Depart or stay home. Working Paper.

Hassin, R., Snitkovsky, R. (2018). Social and monopoly optimization in observable queues. Working Paper.

Haviv, M., Oz, B. (2018). Self-regulation of an unobservable queue. *Manage. Sci.*, 64, 2380–2389.

Hu, M., Li, Y., Wang, J. (2018). Efficient ignorance: Information heterogeneity in a queue. *Manage. Sci.*, 64, 2650–2671.

Ibrahim, R. (2018). Sharing delay information in service systems: A literature survey. *Queueing Syst.*, 89, 49–79.

Kim, B., Kim, J. (2017). Optimal disclosure policies in a strategic queueing model. *Oper. Res. Lett.*, 45, 181–186.

Kirstein, B.M. (1976). Monotonicity and comparability of time-homogeneous Markov processes with discrete state space. *Math. Operationsforsch. Statist.*, 7, 151–168.

Kulkarni, V.G. (2010). *Modeling and Analysis of Stochastic System*, 2nd ed. CRC Press, Taylor and Francis Group, Boca Raton, FL.

Naor, P. (1969). The regulation of queue size by levying tolls. *Econometrica*, 37, 15–24.

Roet-Green, R. (2013). Information in queueing systems with strategic customers. PhD Thesis, School of Mathematical Sciences, Tel-Aviv University.

Shone, R., Knight, V.A., Williams, J.E. (2013). Comparisons between observable and unobservable M/M/1 queues with respect to optimal customer behavior. *Eur. J. Oper. Res.*, 227, 133–141.

Simhon, E., Hayel, Y., Starobinski, D., Zhu, Q. (2016). Optimal information disclosure policies in strategic queueing games. *Oper. Res. Lett.*, 44, 109–113.

Stidham, S. Jr. (1985). Optimal control of admission to a queueing system. *IEEE T. Automat. Contr.*, 30, 705–713.

Stidham, S. Jr. (2009). *Optimal Design of Queueing Systems*. CRC Press, Taylor and Francis Group, Boca Raton, FL.

5

Non-extensive Maximum Entropy Formalisms and Inductive Inference of a Stable M/G/1 Queue with Heavy Tails

Demetres D. KOUVATSOS[1] and Ismail A. MAGEED[1]
[1]*NetPEn – Networks and Performance Engineering Research Group, School of Electrical Engineering and Computer Science, University of Bradford, UK*

An exposition is undertaken of Rényi's and Tsallis's non-extensive maximum entropy (NME) formalisms as methods of inductive inference for systems with long-range interactions of order, q $(0.5 < q < 1)$. The analytic forms of the respective NME state probabilities $p_{q,R}(S_n)$ and $p_{q,T}(S_n)$ of a general physical system Q with a countably infinite set S of states $\{S_n, n = 1, 2, \cdots\}$ are characterized, based on the maximization of the respective entropy functionals, subject to normalization and a finite number of suitable mean value constraints. The credibility of the NME formalisms is justified in terms of four consistency axioms associated with the classical Shannon's extensive maximum entropy (EME) formalism of systems with short-range interactions of order q equal to 1. Consequently, focusing on an inductive application of Rényi's and Tsallis's formalisms on the analysis of a stable M/G/1 queue, novel state probabilities $p_{q,R}(n)$ and $p_{q,T}(n), n = 1, 2, \cdots$ are explicitly determined, subject to normalization, server utilization (SU) and Pollaczeck Khinchin (P K) mean queue length (MQL) constraints. It is shown that via the embedded Markov chain (EMC) these NME state probabilities are exact when the underlying service times of the stable M/G/1 queue are distributed according to two new generalized exponential (GE) type distributions,

Queueing Theory 2,
coordinated by Vladimir ANISIMOV, Nikolaos LIMNIOS.

namely $GE_{q,R}$ and $GE_{q,T}$, respectively. Two case studies, based on typical numerical experiments, illustrate the impact of Rényi's and Tsallis's NME state probabilities $p_{q,R}(n)$ and $p_{q,T}(n), n = 1, 2, \cdots$ of order q $(0.5 < q < 1)$ on the stable $M/GE_{q,R}/1$ and $M/GE_{q,T}/1$ queues with heavy tails. Concluding remarks and future research directions are included.

5.1. Introduction

Non-extensive maximum entropy (NME) formalisms, as analytic methods of inductive inference for physical systems with long-range interactions of order, q $(0.5 < q < 1)$, were advanced by (Rényi 1961) and (Tsallis 1988) in Information Theory and Statistical Physics, respectively. These formalisms are based on respective generalizations of the extensive maximum entropy (EME) formalism with short-range interactions proposed by (Shannon 1948) in Information Theory.

Based on two earlier applications of the EME formalism on the stable M/G/1 queue, the state probabilities $\{p(n), n = 0, 1, 2, \cdots\}$ of a stable M/G/1 queue were derived by maximizing the EME functional, subject to two respective sets of mean value constraints, namely

1) Set of constraints: {normalization, $\sum_n p(n) = 1$ and Pollaczeck–Khinchin (P-K) mean queue length (MQL), $< n > = \sum_n np(n)\}$ by (Shore 1982);

2) $\{Set_1$ constraints and server utilization (SU), $1 - p(0)\}$ by (El-Affendi and Kouvatsos 1983).

Note that the EME state probability $\{p(n), n = 1, 2, \cdots\}$ of a stable M/G/1 queue devised in Shore (1982) turned out to be exact for a stable M/M/1 queue. Moreover, the state probability proposed in El-Affendi and Kouvatsos (1983) was shown to be exact for a stable M/G/1 queue when the underlying service (S) times are distributed according to the generalized exponential (GE) distribution of the form (see Figure 5.1)

$$F_s(t) = P(S \leq t) = 1 - \tau_s\, e^{-\mu t \tau_s} \quad [5.1]$$

$$\tau_s = \frac{2}{1 + C_s^2} \quad [5.2]$$

where μ is the mean service rate and C_s^2 is the squared coefficient of variation (SCV) of the service times. The GE distribution turned out to be versatile, possessing pseudo-memoryless properties, which made the derivation of exact and approximate solutions of GE-type queueing systems analytically tractable (e.g. (El-Affendi and Kouvatsos 1983; Kouvatsos 1986, 1988)).

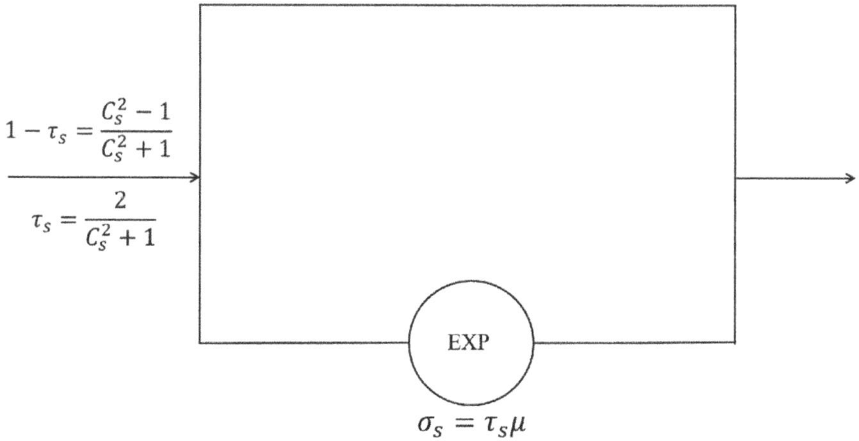

Figure 5.1. *The GE-type distribution with parameters* $\{1/\mu, C_s^2 > 1\}$

Specifically, the GE-type distribution has as a counting process a compound Poisson process (CPP) with parameter $2\mu/(C_s^2+1)$ and geometrically distributed bulk sizes with mean $1/\tau_s = (1+C^2)/2$ and SCV, $1-\tau_s = \frac{(C_S^2-1)}{(C_S^2+1)}$, namely

$$P(N_{cp}=n) = \begin{cases} \sum_{i=1}^{n} \frac{\sigma_s^i}{i!} e^{-\sigma_s} \binom{n-1}{i-1} \tau_s^i (1-\tau_s)^{n-i} & if\ n \geq 1 \\ e^{-\sigma} & if\ n=0 \end{cases} \quad [5.3]$$

where Ncp is a CPP random variable of the number of service completions per unit time corresponding to a stationary GE-type service time random variable with $\sigma_s = \tau_s \mu$ and $\tau_s = 2/(C_s^2+1)$.

In an extended context, an approximate analytic study was conducted in Kouvatsos (2010) based on (Tsallis 1988) NME state probability, $p_{q,T}(S_n)$ of a general system Q with a finite or countably infinite set S of states $\{S_n, n = 1, 2, \cdots\}$, subject to normalization and a finite number of suitable mean value constraints. In particular, the credibility of Tsallis' NME formalism, as a method of inductive inference for non-extensive systems with long-range interactions, was assessed (Kouvatsos 2010) in terms of four consistency axioms proposed by (Shore 1980) for Shannon's EME formalism with short-range interactions, namely uniqueness, invariance, system independence and subset independence. Based on Tsallis' NME formalism, approximate analyses of single server gS-S/GE/1 and gS-S/GE/1/N queues with respective infinite and finite N (>1) capacities, general self-similar (gS-S) arrival process and GE-type service times were carried out. In this context, the NME state probabilities of these queues were characterized as generalizations of the Zipf-Mandelbrot (Z-M) type distribution (Adamic 2002) depicting heavy queue tails and asymptotic power law behavior. Moreover, an extensive review by (Kouvatsos 2011) reported on the applications of Shannon's and Tsallis' ME entropy formalisms for the analysis of the stable GE/GE/1 and gS-S/GE/1 queues with bursty and heavy tails, respectively.

This work reports a comparative exposition into (Rényi 1961) and (Tsallis 1988) NME formalisms as methods of inductive inference for systems with long-range interactions of order, q $(0.5 < q < 1)$. In this context, the NME state probabilities, namely $p_{q,R}(S_n)$ and $p_{q,T}(S_n)$, of a general physical system Q with a finite or countably infinite set S of states $\{S_n, n = 1, 2, \cdots\}$ are explored based on the maximization of the respective functionals, subject to normalization and a finite number of suitable mean value constraints $\{F_{k,q}, 1 \leq k \leq m < +\infty\}$. Moreover, a credibility study on Rényi's NME formalism with long-range interactions is carried out, as the one contacted for Tsallis's formalism in Kouvatsos (2010), in terms of four consistency axioms associated with the classical Shannon's EME formalism proposed in Shore (1980). Consequently, analytic expressions of the state probabilities $p_{q,R}(n)$ and $p_{q,T}(n), n, 1, 2, \cdots$, are derived for a stable M/G/1 queue by maximizing the respective Rényi's and Tsallis' NME functionals, subject to normalization, SU and P-K MQL constraints. Moreover, based on the embedded Markov chain and P-K transform equation (Kleinrock 1976), further investigation is carried out in order to establish and suitably interpret the underlying service time distributions of a stable M/G/1 queue that make both Rényi's and Tsallis' NME state probabilities $p_{q,R}(n)$ and $p_{q,T}(n), n, 1, 2, \cdots$ exact.

This paper is organized as follows: Shannon's EME formalism and Rényi's and Tsallis's NME formalisms for a general physical system Q are reviewed in section 5.2. The credibility of these NME formalisms is explored in section 5.3 in terms of four consistency axioms associated with Shannon's EME formalism (Shore 1980), namely uniqueness, invariance, system independence and subset independence. Novel analytic expressions of Rényi's and Tsallis' NME state probabilities for a stable M/G/1 queueing system, subject to normalization, P-K MQL and SU are devised in section 5.4. Moreover, new GE-type underlying service time distributions that correspond to exact Rényi's and Tsallis' NME state probabilities, $p_{q,R}(n)$ and $p_{q,T}(n), n, 1, 2, \cdots$, respectively, are determined in section 5.5. Two case studies with typical numerical experiments and related interpretations on the impact of different degrees of long-range interactions of order q on Rényi's and Tsallis' NME state probabilities of a stable M/G/1 queue are presented in section 5.6. Concluding remarks and future research directions follow in section 8.7. The credibility of Rényi's NEM formalism is quantitatively verified in terms of Shannon's EME consistency axioms in the Appendix.

5.2. General systems and inductive ME formalisms

5.2.1. *"Classical" Shannon's EME formalism with short-range interactions*

For a general physical system Q with countably infinite set S of configurations or states $\{S_n, n = 1, 2, \cdots\}$ and short-range interactions with information parameter q equal to 1, the classical Shannon's EME entropy functional $H_{1,S}(p)$ (Shannon 1948) is defined by

$$H_{1,S}(p) = -c \sum_{S_n \in S} p_{1,S}(S_n) \, log \, p_{1,S}(S_n) \qquad [5.4]$$

where c (c>0) is a positive constant and $\{p_{1,S}(S_n), S_n \in S = \{S_n, n = 0, 1, 2, \cdots\}\}$ are the EME state probabilities.

The entropy function $H_{1,S}(p)$ can be interpreted as a measure of information that is implied by $p_{1,S}(S_n)$ about a physical system Q with short-range interactions such as "holding matter together" in statistical physics (Statistics 2007). In this context, quantities such as "energy and entropy" are considered as extensive variables in the sense that the overall energy of the system Q with short-range interactions is "proportional to the system size" (Statistics 2007). By analogy, traffic flows in queues with short-range dependence (SRD), such as those with Poisson (regular) and compound Poisson (bursty) (Kleinrock 1976) traffic processes, influence the creation of single queues and queueing networks, where the state and entropy variables are extensive.

In this context, the EME formalism (Shore 1980), as a method of inductive inference for determining the form of the "least biased" EME state probability distribution is based on the maximization of the extensive entropy functional,$H_{1,S}(p)$, subject to normalization and say $(m + 1)$ suitable mean value constraints

$$\sum_{S_n \in S} p_{1,S}(S_n) = 1 \qquad [5.5]$$

$$\sum_{S_n \in S} g_{k,1,S}(S_n) \, p_{1,S}(S_n) = G_{k,1,S}, 1 \leq k \leq m < \infty \qquad [5.6]$$

where $c(c > 0)$ is a positive constant, and$\{G_{k,1,S}, k = 1, 2, \cdots m\}$ are the prescribed mean values defined on the set of functions $\{g_{k,1,S}(S_n), k = 1, 2, \cdots m\}$.

The implementation of the EME formalism is achieved by applying the method of Lagrange's undetermined multipliers leading to the characterization of the state probability distribution, $p_{1,S}(S_n), S_n \in S$, namely

$$p_{1,S}(S_n) = \frac{1}{Z_{1,S}} exp\Big\{ - \sum_{k=1}^{m} \beta_{k,1,S}\, g_{k,1,S}(S_n)\Big\} \quad [5.7]$$

$$Z_{1,S} = exp\{\beta_0\} = \sum_{S_n \in S} exp\Big\{ - \sum_{k=1}^{m} \beta_{k,1,S}\, g_{k,1,S}(S_n)\Big\} \quad [5.8]$$

where $\{\beta_{k,1,S}, k = 1, 2, \cdots m\}$ are the Lagrangian multipliers determined from the set of constraints [5.6] and $Z_{1,S}$, known in statistical physics as the "partition" function, is given by $Z_{1,S}$ = $\exp\{\beta_0\}$, where β_0 is the Lagrangian coefficient determined by the normalization constraint [5.5]. Note that maximizing the entropy $H_{1,S}(p)$ for an extensive system Q with a finite number of states, subject only to the normalization constraint, the EME solution of the state probability distribution is characterized, as expected, by the uniform distribution as the least biased distribution estimate.

5.2.2. *Rényi's and Tsallis's NME formalisms with long-range interactions*

For a general system Q with an integer number of possible (microscopic) configurations or states $N(> 0)$ and long-range interactions, such as gravity in statistical physics, "energy and entropy are no longer extensive quantities" (Statistics 2007). This increases the complexity of the physical system Q for which the state probability distribution associated, say with energy, has heavy tails and power law behavior and therefore, it can no longer be determined by maximizing Shannon's extensive entropy, $H_{1,S}(p)$.

To this end, for the aforementioned physical system Q with a finite or countably infinite set S of configurations or states $\{S_n, n = 0, 1, 2, \cdots\}$ and long-range interactions, the NME functionals proposed by Tsallis's (Tsallis 1988) and Rényi's (Rényi 1961) are defined by

$$H_{q,T}(p) = \frac{c}{q-1}\Big\{1 - \sum_{S_n \in S} p_{q,T}(S_n)^q\Big\} \quad [5.9]$$

$$H_{q,R}(p) = \frac{c}{1-q} ln\Big\{ \sum_{S_n \in S} p_{q,R}(S_n)^q\Big\} \quad [5.10]$$

where c is a positive constant and q $(0.5 < q < 1)$ is the order of long-range interactions.

Using the Lagrange's undetermined multipliers to maximize the NME functionals [5.9] and [5.10], we subject to mean value constraints

$$\sum_{S_n \in S} p_{q,X}(S_n) = 1 \qquad [5.11]$$

$$\sum_{S_n \in S} g_{k,q,X}(S_n)\, p_{q,X}(S_n) = G_{k,q,X}, 1 \leq k \leq m < \infty \qquad [5.12]$$

where X stands for either T (Tsallis) or R (Rényi) and $\{F_{k,q,X}, k = 1, 2, \cdots m\}$ are the prescribed mean values defined on the set of functions $\{f_{k,q,X}(S_n), k = 1, 2, \cdots m\}$, subject to normalization and the m+1 mean value constraints [5.11] and [5.12], the following minimally biased NME probabilities are obtained:

$$p_{q,T}(S_n) = \frac{1}{Z_{q,T}}\left(1 + q(1-q)\sum_{k=1}^{m} \beta_{k,q,T}\, g_{k,q,T}(S_n)\right)^{\frac{1}{q-1}} \qquad [5.13]$$

$$p_{q,R}(S_n) = \frac{1}{Z_{q,R}}\left(1 + \frac{1-q}{q}\sum_{k=1}^{m} \beta_{k,q,R}\, g_{k,q,R}(S_n)\right)^{\frac{1}{q-1}} \qquad [5.14]$$

where $\{\beta_{k,q,T}, \beta_{k,q,R}, k = 1, 2, \cdots m\}$ are the Lagrangian multipliers corresponding to the constraints [5.6] and $Z_{q,X}$ is the normalization constant given by

$$Z_{q,T} = \sum_{S_n \in S}\left(\left(1 + q(1-q)\sum_{k=1}^{m} \beta_{k,T}\, g_{k,q,T}(S_n)\right)^{\frac{1}{q-1}}\right) \qquad [5.15]$$

$$Z_{q,R} = \sum_{S_n \in S}\left(\left(1 + \frac{1-q}{q}\sum_{k=1}^{m} \beta_{k,R}\, g_{k,q,R}(S_n)\right)^{\frac{1}{q-1}}\right) \qquad [5.16]$$

5.3. NME formalisms and EME consistency axioms

The EME formalism proposed by Shannon (Shannon 1948) with short-range interactions was shown by Shore and Johnson (Shore 1980) to be a uniquely correct method of inductive inference for extensive systems, subject to new information given in the form of suitable mean value constraints. The analytic approach was based on the major assumption that the "Application of the EME principle, as a credible method of inductive inference, should lead to consistent state probabilities for a physical system Q when there are different ways to analyzing it, which account with the same prior information constraints" (Shore 1980). This fundamental requirement for EME systems was formalized by Shore and Johnsonin terms of four consistency axioms, namely uniqueness, invariance, system independence and subset independence (Shore 1980).

Specifically, as it was highlighted in Shore (1980), the uniqueness axiom advocates that "if the same problem is solved twice in exactly the same way, the same answer is expected in both cases, i.e. the solution should be unique". The invariance axiom states that "the same solution should be obtained if the same inference problem is solved twice in two different coordinate systems" (Shore 1980). Moreover, the axiom of system independence, stipulates that "it should not matter whether one accounts for independent information about independent systems separately in terms of different marginal probabilities or together in terms of the joint probability" (Shore 1980). Finally, the axiom of subset independence implies that "it does not matter whether one treats an independent subset of system states in terms of a separate conditional density or in terms of the full system density" (Statistics 2007). Thus, given new prior information in terms of mean value constraints, there is a unique EME state probability distribution satisfying these constraints, which can be devised by a procedure based on EME formalism based on the proposed four consistency axioms (Shore 1980).

Note that the credibility of Tsallis' NME formalism was assessed by Kouvatsos and Assi (Kouvatsos 2010, 2011) in terms of the aforementioned four consistency axioms for extensive systems. It was shown that Tsallis' formalism, even though satisfied the consistency axioms of uniqueness, invariance and subset independence, nevertheless defied, as it should, the axiom of system independence due to the existence of long-range interactions.

In this context, the credibility of Rényi's NME formalism, as a method of inductive inference, is also explored in terms of the four EME consistency axioms in the Appendix. It transpires that the joint NME state probability distribution of two independent non-extensive systems Q and V defies, due to the presence of long-range interactions (as in the Tsalis' case), the axiom of system independence (Shore 1980).

Thus, the Rényi's and Tsallis's NME formalisms, as a methods of inductive inference, are clearly suitable for the quantitative studies of non-extensive dynamic systems with heavy queue tails.

5.4. A stable M/G/1 queue with long-range interactions

In this section, novel analytic solutions are derived for the steady-state probabilities of a stable M/G/1 queueing system based on Rényi's and Tsallis' generalized NME formalisms, subject to the normalization, SU and MQL constraints.

Throughout the analysis, the following notation is adopted:

– $X = S$ (Shannon) with $q = 1$, T (Tsallis) and R (Rényi) with $0.5 < q < 1$;

– $p_{q,X}(n)$, steady-state probability of having n jobs in the M/G/1 queue, $n = 0, 1, 2, \cdots$;

– λ, mean arrival rate; μ, mean service rate; $\rho = \lambda/\mu, SU$; $< n >= \sum_{n=0}^{\infty} n p_{q,X}(n)$, MQL;

– $C^2_{s,q,X}$, SCV of the GE-type service (s) time distribution, $GE_{q,X}$;

– $f_{s,q,X}(t)$ and $F'_{s,q,X}(t)$, probability density and cumulative functions, respectively, of the service time distribution;

– $F^*_{s,q,X}(\theta)$, Laplace transform of $f_{o,q,X}(t)$.

5.4.1. *Background: Shannon's EME state probability of a stable M/G/1 queue*

It has been shown (El-Affendi and Kouvatsos 1983) that the EME steady-state probability of a stable M/G/1 queue that maximizes Shannon's entropy function (Shannon 1948),

$$H(p_{1,S}) = -\sum_{n=0}^{\infty} p_{1,S}\,(n)\; ln(p_{1,S}) \qquad [5.17]$$

subject to prior information in terms of the following mean value constraints,

– Normalization,

$$\sum_{n=0}^{\infty} p_{1,S}\,(n) \;= 1 \qquad [5.18]$$

– SU,

$$p_{1,S}(0) = \sum_{n=0}^{\infty} h(n)\, p_{1,S}\,(n) \;= 1 - \rho = \lambda/\mu \qquad [5.19]$$

where

$$h(n) = \begin{cases} 1 & \text{for } n = 0 \\ 0 & \text{for } n = 1, 2, \cdots \end{cases} \qquad [5.20]$$

– P-K MQL,

$$<n>=\sum_{n=0}^{\infty} n\, p_{1,S}\,(n) \;=\frac{\rho}{2}\Big(1+\frac{1+\rho C^2_{s,1,S}}{1-\rho}\Big) \qquad [5.21]$$

is given by

$$p_{1,S}(n)=\begin{cases} p_{1,S}(0) & n=0 \\ p_{1,S}(0)\tau_s x^n & n>0 \end{cases} \qquad [5.22]$$

where $p_{1,S}(0)=1-\rho, \tau_s=\frac{2}{(1+C^2_{s,1,S})}$ and $x=\frac{<n>-\rho}{<n>}$.

5.4.2. *Tsallis' and Rényi's NME state probabilities of a stable M/G/1 queue*

THEOREM 8.1.– The Rényi's and Tsallis' NME solutions for the steady-state probability, $p_{q,X}(n), n \;=\; 0,1,2,\cdots$ of the stable M/G/1 queue, subject to normalization [5.18] and SU [5.19] and P/K MQL [5.21] constraints are expressed by

$$p_{q,R}(n)=\begin{cases} p_{q,R}(0) & n=0 \\ p_{q,R}(0)\tau_s^{\frac{1}{q}} x^n & n>0 \end{cases} \qquad [5.23]$$

$$p_{q,T}(n)=\begin{cases} p_{q,T}(0) & n=0 \\ p_{q,T}(0)\tau_s^{q} x^n & n>0 \end{cases} \qquad [5.24]$$

with the idle state probabilities $p_{q,R}(0)$ and $p_{q,T}(0)$ satisfying the SU constraint [5.19], namely

$$p_{q,R}(0)=p_{q,T}(0)=1-\rho \qquad [5.25]$$

while τ_s and x are the Lagrangian multipliers corresponding to SU [5.19] and P-K MQL [5.21] constraints, respectively, are determined by

$$\tau_s=\frac{2}{(1+C^2_{s,1,S})} \qquad [5.26]$$

$$x=\frac{<n>-\rho}{<n>}=\begin{cases} \dfrac{\rho}{\left(\rho+(1-\rho)\left(\frac{2}{1+C^2_{s,1,S}}\right)^{\frac{1}{q}}\right)}, & \text{Rényi} \\[2ex] \dfrac{\rho}{\left(\rho+(1-\rho)\left(\frac{2}{1+C^2_{s,1,S}}\right)^{q}\right)}, & \text{Tsallis} \end{cases} \qquad [5.27]$$

with

$$\frac{\rho(1-x)}{(1-\rho)x} = \begin{cases} \tau_s^{\frac{1}{q}}, & \text{Rényi} \\ \\ \tau_s^{q}, & \text{Tsallis} \end{cases} \quad [5.28]$$

PROOF.– Maximizing Rényi's entropy functional subject to constraints [5.18], [5.19] and [5.21] yields the Lagrange's equation

$$\begin{aligned} &\Bigg[\frac{q}{(1-q)\Big(\sum_{n=0}^{\infty}\big(p_{q,R}(n)\big)^{q}\Big)}\sum_{n=0}^{\infty}\Big(p_{q,R}(n)\Big)^{q-1} \\ &-\alpha\Big(\sum_{n=0}^{\infty}h(n)\Big)-\beta\Big(\sum_{n=0}^{\infty}1\Big)-\gamma\Big(\sum_{n=0}^{\infty}n\Big)\Bigg]=0 \end{aligned} \quad [5.29]$$

Hence,

$$\Big(p_{q,R}(n)\Big)^{q-1} = \left(\frac{(1-q)\Big(\sum_{n=0}^{\infty}\big(p_{q,R}(n)\big)^{q}\Big)}{q}\right)\Big(\alpha h(n)+\beta\Big)\Big(1+\frac{\gamma}{\alpha h(n)}\frac{(1-q)}{q}n\Big)$$

Therefore, it is implied that $p_{q,R}(n)$ is in the form

$$p_{q,R}(n) = a\Big(1+b\frac{(1-q)}{q}n\Big)^{\frac{1}{q-1}} \quad [5.30]$$

where

$$\begin{aligned} a &= \left(\frac{(1-q)\Big(\sum_{n=0}^{\infty}\big(p_{q,R}(n)\big)^{q}\Big)}{q}\right)\Big(\alpha h(n)+\beta\Big)^{\frac{1}{q-1}}, \\ b &= \frac{\gamma}{(\alpha h(n)+\beta)}\frac{(1-q)}{q} \end{aligned} \quad [5.31]$$

From [5.30], it clearly follows that

$$a = p_{q,R}(0) \quad [5.32]$$

Hence, the $p_{q,R}(n)$ is expressed by

$$p_{q,R}(n) = p(0)\Big(1+b\frac{(1-q)}{q}n\Big)^{\frac{1}{q-1}} \quad [5.33]$$

where $b > 0$ and $q > 0$ and different than 1. The analytic formula [5.33] may be rewritten by replacing at the expression $\left(1 + b\frac{(1-q)}{q}n\right)^{\frac{1}{q-1}}$ with its equivalent expression, $\tau_S^{\frac{1}{q}} x^n$ get to the form, namely

$$p_{q,R}(n) = p_{q,R}(0)\tau_S^{\frac{1}{q}} x^n \qquad [5.34]$$

where

$$\tau_S^{\frac{1}{q}} x^n = \left(1 + b\frac{(1-q)}{q}n\right)^{\frac{1}{q-1}} \qquad [5.35]$$

This proof is concluded by using the constraint [5.18], together with the derived formula [5.31], clearly holds that

$$1 = p(0) + p(0)\tau_S^{\frac{1}{q}}\frac{x}{(1-x)},\ implying\ \tau_S^{\frac{1}{q}} = \frac{\rho(1-x)}{(1-\rho)x} \qquad [5.36]$$

From the analysis of M/G/1 queue, it clearly follows that τ_s is expressed by $\tau_s = \frac{2}{1+y_s}$, where $y_s = \frac{(C^2_{s,q,R}-1)}{2}$, yielding $\tau_s = \frac{2}{(1+C^2_{s,q,R})}$ and $\tau_S^{\frac{1}{q}} = \frac{\rho(1-x)}{(1-\rho)x}$, which clearly implies $x = \frac{\rho}{\left(\rho+(1-\rho)\left(\frac{2}{1+C^2_{s,1,S}}\right)^{\frac{1}{q}}\right)}$. Focusing on the maximization of Tsallis' NME functional, subject to constraints [5.18], [5.19] and [5.21], it follows that the Lagrange's equation is expressed by

$$\left[q\sum_{n=0}^{\infty}\left(p_{q,T}(n)\right)^{q-1} - \alpha'\left(\sum_{n=0}^{\infty}h(n)\right) - \beta'\left(\sum_{n=0}^{\infty}1\right) - \gamma'\left(\sum_{n=0}^{\infty}n\right)\right] = 0 \qquad [5.37]$$

where $p_{q,T}(n)$ is in the form

$$p_{q,T}(n) = a'(1 + w(1-q)n)^{\frac{1}{q-1}}, \qquad [5.38]$$

and

$$a' = \left[\left(\frac{1}{q}\right)(\alpha' h(n) + \beta')\right]^{\frac{1}{q-1}}, w = \frac{\gamma'}{(\alpha' h(n) + \beta')}(1-q) \qquad [5.39]$$

From [5.38] to [5.39], it clearly follows that

$$a' = p_{q,T}(0) \tag{5.40}$$

Hence, it is implied that

$$p_{q,T}(n) = p_{q,T}(0)(1 + wq(1-q)n)^{\frac{1}{q-1}}\ b > 0 \tag{5.41}$$

where $q > 0$ and different from 1. Therefore, the formula [5.41] can be rewritten as

$$p_{q,T}(n) = p_{q,T}(0)\tau_s^q x^n \tag{5.42}$$

where

$$\tau_s^q x^n = (1 + wq(1-q)n)^{\frac{1}{q-1}} \tag{5.43}$$

To conclude the proof, using the normalization constraint [5.18], together with the derived formula [5.31], it clearly holds that $1 = p_{q,T}(0)+p_{q,T}(0)\ \tau_s^q \frac{x}{(1-x)}$, which implies that

$$\tau_s^q = \frac{\rho(1-x)}{(1-\rho)x}, \tag{5.44}$$

Moreover, it clearly follows that

$$x = \frac{\rho}{\left(\rho + (1-\rho)\left(\frac{2}{1+C_{s,1,S}^2}\right)^q\right)} \tag{5.45}$$

5.4.3. *Exact Rényi's and Tsallis' NME state probabilities with distinct GEq-type service time distributions*

This section investigates the characterization of the underlying service time probability density functions (pdfs) and cumulative distribution functions (CDFs) of a stable M/G/1 queue that makes exact the Rényi's and Tsallis' NME $p_{q,R}(n)$ and $p_{q,T}(n)$ state probabilities, respectively. Particular attention is devoted on the influence of the information parameter q, which appeared explicitly in the expressions for the NME state probabilities $\{p_{q,R}(n),\ p_{q,T}(n), n = 1, 2, \cdots$ (see [5.23] and [5.24]) on the pdfs and CDFs of a stable M/G/1 queue in section 5.4.2.

5.4.3.1. *The underlying service time probability density functions*

In this context, the following theorem 8.2 is presented.

THEOREM 8.2.– The Rényi's and Tsallis' NME steady-state state probabilities, $\{p_{q,R}(n), p_{q,T}(n), n = 1, 2, \cdots$ of a stable M/G/1 queue are exact when the underlying service time pdfs are of the $GE_{q,X} - type$, where $X = R, T$ and are given by

$$f_{s,q,R}(t) = \left(1 - \tau_s^{\frac{1}{q}}\right) u_0(t) + \mu\tau_s^{\frac{2}{q}} e^{-\mu t \tau_s^{\frac{1}{q}}} \qquad [5.46]$$

$$f_{s,q,T}(t) = \left(1 - \tau_s^{q}\right) u_0(t) + \mu\tau_s^{2q} e^{-\mu t \tau_s^{q}} \qquad [5.47]$$

where $u_0(t)$ is the unit impulse function

$$u_0(t) = \begin{cases} \infty, & t = 0 \\ 0, & t \neq 0 \end{cases} \qquad [5.48]$$

such that $\int_{-\infty}^{\infty} u_0(t) = 1$ and $\tau_s = \frac{2}{(1+C_{s,1,S}^2)}$.

PROOF.– Focusing on Rényi's case, the z-transform $Q_{q,R}(z)$ of the NME solution $p_{q,R}(n), n = 0, 1, 2, \cdots$ is given by Kleinrock (1976)

$$Q_{q,R}(z) = \sum_{n=0}^{\infty} p_{q,R}(n) z^n, |z| < 1 \qquad [5.49]$$

Hence, by replacing $p_{q,R}(n)$ of [5.23] into [5.49], it follows that

$$Q_{q,R}(z) = \sum_{n=0}^{\infty} p_{q,R}(n) z^n = p_{q,R}(0) + \frac{p_{q,R}(0)\tau_s^{\frac{1}{q}} xz}{(1-xz)} = \frac{p_{q,R}(0)\left(1 - xz\left(1 - \tau_s^{\frac{1}{q}}\right)\right)}{(1-xz)} \qquad [5.50]$$

Using the P/K transform equation (Kleinrock 1976), namely

$$Q_{q,R}(z) = \frac{p_{q,R}(0)(1-z)\left(F_{s,q,R}^*(\lambda - \lambda z)\right)}{F_{s,q,R}^*(\lambda - \lambda z) - z} \qquad [5.51]$$

where

$$F^*_{s,q,R}(\theta) = E\left[e^{-\theta s}\right] = \int_0^\infty e^{-\theta t} f_{s,q,R}(t)\, dt \qquad [5.52]$$

is the Laplace–Stieltjes transform of the service time pdf. Clearly, $Q_{q,R}(0) = p_{q,R}(0)$ and $Q_{q,R}(1) = 1$. Equating the RHS of [5.50] and [5.51], it is implied that

$$F^*_{s,q,R}(\lambda - \lambda z) = \frac{\mu\tau_s^{\frac{1}{q}} + (\lambda - \lambda z)(1 - \tau_s^{\frac{1}{q}})}{\mu\tau_s^{\frac{1}{q}} + (\lambda - \lambda z)} \qquad [5.53]$$

Substituting θ for $(\lambda - \lambda z)$, the expression [5.53] becomes

$$F^*_{s,q,R}(\theta) = \frac{\mu\tau_s^{\frac{1}{q}} + \theta(1 - \tau_s^{\frac{1}{q}})}{\mu\tau_s^{\frac{1}{q}} + \theta} \qquad [5.54]$$

By inverting Laplace–Stieltjes transform, $F^*_{s,q,R}(\theta)$ the $GE_{q,R} - type$ pdf $f_{s,q,R}(t)$ (see [5.46]) of Theorem 3 is obtained. For the Tsallis' case, following similar steps as for the Rényi's case, the z-transform Q(z) of the NME solution $p_{q,T}(n), n = 0, 1, 2, \cdots$ is given by

$$Q_{q,T}(z) = \sum_{n=0}^{\infty} p_{q,T}(n) z^n = \frac{p_{q,T}(0)(1 - xz(1 - \tau_s^q))}{(1 - xz)} \qquad [5.55]$$

Using the P/K transform equation

$$Q_{q,T}(z) = \frac{p_{q,T}(0)(1 - z)\big(F^*_{s,q,T}(\lambda - \lambda z)\big)}{F^*_{s,q,T}(\lambda - \lambda z) - z} \qquad [5.56]$$

where

$$F^*_{s,q,T}(\theta) = E\left[e^{-\theta s}\right] = \int_0^\infty e^{-\theta t} f_{s,q,T}(t)\, dt \qquad [5.57]$$

is the Laplace–Stieltjes transform of the service time distribution. It can be verified that

$Q_{q,T}(0) = p_{q,T}(0)$ and $Q_{q,T}(1) = 1$

Equating the RHS of [5.55] and [5.56], it is implied that

$$F^*_{s,q,T}(\lambda - \lambda z) = \frac{\mu\tau_s^q + (\lambda - \lambda z)(1 - \tau_s^q)}{\mu\tau_s^q + (\lambda - \lambda z)} \quad [5.58]$$

Substituting θ for $(\lambda - \lambda z)$, the expression [5.58] becomes

$$F^*_{s,q,R}(\theta) = \frac{\mu\tau_s^q + \theta(1 - \tau_s^q)}{\mu\tau_s^q + \theta} \quad [5.59]$$

By inverting the Laplace–Stieltjes transform, the $GE_{q,T} - type$ pdf $f_{s,q,T}(t)$ (see [28]) of Theorem 3 is devised. Corollary 5.1 addresses the characterization of service time cumulative distribution functions, $\{F_{s,q,R}(t)\}$ and $\{F_{s,q,T}(t)\}$.

COROLLARY 5.1.– The CDFs $\{F_{s,q,R}(t), F_{s,q,T}(t)\}$ of the $GE_{q,R} - type$ and $GE_{q,T} - type$ types of service times with the pdfs $\{f_{s,q,R}(t), f_{s,q,T}(t)\}$ of theorem 8.3 are characterized by the cumulative distribution functions $\{F_{s,q,R}(t), F_{s,q,T}(t)\}$ of, $S_{q,R}$ and $S_{q,T}$, respectively, and are determined by

$$F_{s,q,R}(t) = 1 - \tau_s^{\frac{1}{q}} e^{-\mu\tau_s^{\frac{1}{q}} t} \quad [5.60]$$

$$F_{s,q,T}(t) = 1 - \tau_s^q e^{-\mu\tau_s^q t} \quad [5.61]$$

where $\tau_s = \frac{2}{(1+C^2_{s,1,S})}$.

PROOF.– Focusing on Rényi's NME formalism, the service time CDF $F_{s,q,R}(t) = P(S_{q,R} \leq t)$ is determined as follows:

$$\begin{aligned} F_{s,q,R}(t) &= \int_0^t f_{s,q,R}(x)\,dx \\ &= \int_0^t \left(1 - \tau_s^{\frac{2}{q}}\right) u_0(x)dx + \mu\tau_s^{\frac{2}{q}} \int_0^t e^{-\mu\tau_s^{\frac{1}{q}} x} dx \\ &= \left(1 - \tau_s^{\frac{1}{q}}\right) + \frac{\mu\tau_s^{\frac{2}{q}}}{\mu\tau_s^{\frac{1}{q}}}\left(1 - e^{-\mu\tau_s^{\frac{1}{q}} t}\right) \\ &= 1 - \tau_s^{\frac{1}{q}} e^{-\mu\tau_s^{\frac{1}{q}} t} \end{aligned}$$

Moreover, the Tsallis's service time CDF,$F_{s,q,T}(t)$ is characterized as follows:

$$
\begin{aligned}
F_{s,q,T}(t) &= \int_0^t f_{s,q,T}(x)\, dx \\
&= \int_0^t \left(1-\tau_s^q\right) u_0(x) dx + \mu\tau_s^{2q} \int_0^t e^{-\mu\tau_s^q x} dx \\
&= \left(1-\tau_s^q\right) + \frac{\mu\tau_s^{2q}}{\mu\tau_s^q}(1-e^{-\mu\tau_s^q t}) \\
&= 1-\tau_s^q e^{-\mu\tau_s^q t}
\end{aligned}
$$

Note that in the limiting case $q \longrightarrow 1$, which corresponds to the classical EME Shannon's entropy formalism, both expressions [5.60] and [5.61] of the service time CDF reduce to the original service time CDF obtained in El-Affendi and Kouvatsos (1983), namely $F_{s,q,T}(t) = 1-\tau_s e^{-\mu t\tau_s}$ with $\tau_s = \frac{2}{(1+C^2_{s,1,S})}$.

COROLLARY 5.2.– For the service time pdfs of [5.46] and [5.47], the following mean values and SCVs of the service time, S, are determined by

$$E\left(s_{q,R}\right) = \frac{1}{\mu} \tag{5.62}$$

$$E\left(s^2_{q,R}\right) = \frac{2}{\mu^2\tau_s^{\frac{1}{q}}} \tag{5.63}$$

$$C^2_{s,q,R} = \frac{E\left(s^2_{q,R}\right)}{\left(E\left(s_{q,R}\right)\right)^2} - 1 = \frac{(2-\tau_s^{\frac{1}{q}})}{\tau_s^{\frac{1}{q}}} \tag{5.64}$$

$$E\left(s_{q,T}\right) = \frac{1}{\mu} \tag{5.65}$$

$$E\left(s^2_{q,T}\right) = \frac{2}{\mu^2\tau_s^q} \tag{5.66}$$

$$C^2_{s,q,T} = \frac{E\left(s^2_{q,T}\right)}{\left(E\left(s_{q,T}\right)\right)^2} - 1 = \frac{(2-\tau_s^q)}{\tau_s^q} \tag{5.67}$$

where $\tau_s = \frac{2}{(1+C^2_{s,1,S})}$.

PROOF.–

The mean of $S_{q,R}$ is given by

$$E\Big(s_{q,R}\Big) = \int_0^\infty t f_{s,q,R}(t)\,dt \qquad [5.68]$$

$$= \int_0^\infty t\mu\tau_s^{\frac{2}{q}} e^{-\mu\tau_s^{\frac{1}{q}}t} dt$$

$$= \mu\tau_s^{\frac{2}{q}} \int_0^\infty t\, e^{-\mu\tau_s^{\frac{1}{q}}t} dt$$

By employing the gamma function, namely

$$\Gamma(m) = \int_0^\infty w^{m-1}\, e^{-w} dw \qquad [5.69]$$

and substituting $w = \mu\tau_s^{\frac{1}{q}} t$ in [5.69] and since $\Gamma(2) = 1$, it is implied that

$$E\Big(s_{q,R}\Big) = \frac{\mu\tau_s^{\frac{2}{q}}}{\mu^2\tau_s^{\frac{2}{q}}}\Gamma(2) = \frac{1}{\mu}$$

Moreover, it can be seen that

$$E\Big(s^2_{q,R}\Big) = \int_0^\infty t^2 f_{s,q,R}(t)\,dt$$

$$= \int_0^\infty t^2\mu\tau_s^{\frac{2}{q}} e^{-\mu\tau_s^{\frac{1}{q}}t} dt$$

$$= \mu\tau_s^{\frac{2}{q}} \int_0^\infty t^2\, e^{-\mu\tau_s^{\frac{1}{q}}t} dt \qquad [5.70]$$

and setting $w = \mu\tau_s^{\frac{1}{q}} t$, $E(s^2_{q,R})$ is given by [5.63] and subsequently

$C^2_{s,q,R}$ by [5.64]. Following the same methodology, the required formulae [5.65]–[5.67] for Tsallis' case can be obtained.

5.5. Numerical experiments and interpretations

This section presents for illustration purposes two case studies featuring two numerical applications of Rényi's and Tsallis's NME formalisms as methods of inductive inference of stable M/G/1 queues.

Case study 1

Case study 1 focuses on the assessment of the impact of long-range interactions of order q on the queue tails of Rényi's and Tsallis's state probabilities $p_{q,T}(n)$ and $p_{q,R}(n), n = 1, 2, \cdots$ of a stable M/G/1 queue.

Numerical experiments are carried out in Figures 5.2–5.11 for the computation and plotting of the state probabilities $p_{q,T}(n)$ and $p_{q,R}(n)$ versus $n = 1, 2, \cdots$ for each information parameter $q, q = 0.55, 0.70, 0.85, 1.00$ and typical input data: (i) mean arrival rate, $\lambda = 0.75$, (ii) mean service rate, $\mu = 1.00$ (i.e. SU, $\rho = 75\%$) and (iii) selected SCVs of service times, $C^2_{s,1,S} = 2, 5, 10, 20$ and 50 of the P-K MQL constraint.

It can be observed that Rényi's and Tsallis' state probabilities $p_{q,T}(n)$ and $p_{q,R}(n), n = 1, 2, \cdots$ of the stable M/G/1 queue capture heavy queue tails due to long-range interactions of order q (see Figures 5.2–5.6). Rényi's state probabilities $p_{q,R}(n), n = 1, 2, \cdots$ display heavier queue tails than the ones of Tsallis' state probabilities $p_{q,T}(n), n = 1, 2, \cdots$ for all combined values of SCVs, $C^2_{s,1,S} = 2, 5, 10, 20,$ and 50 and information order $q, q = 0.55$, 0.70, 0.85, 1.00, as appropriate. As $C^2_{s,1,S}$ is increasing, the plots of the state probabilities are progressively "flatter" with respect to n and the associated curves eventually becoming straight lines (see $C^2_{s,1,S} = 50$). Moreover, due to the different impact of "$1/q$" and "q" as exponents of the analytic Rényi's and Tsallis' state probabilities, the $p_{q,R}(n), n = 1, 2, \cdots$ display heavier queue tails for decreasing values of q as $q \longrightarrow 0.55$ while $p_{q,T}(n), n = 1, 2, \cdots$ show heavier queue tails for increasing values of q, as $q \longrightarrow 1.00$.

Case study 2

Case study 2 investigates the limiting cases of the q-dependent family of SCVs $C^2_{s,q,R}$ and $C^2_{s,q,T}$ of the CDFs $\{F_{s,q,R}(t), F_{s,q,T}(t)\}$ of the $GE_{q,R}$ and $GE_{q,T}$ types of service times, respectively, for each initially selected value of SCV, $C^2_{s,1,S}$ (of the P-K MQL formula) and increasing values of the long-range interactions order q.

In this context, numerical experiments are performed in Figures 5.7–5.11 in order to draw the curves of $C^2_{s,q,R}$ and $C^2_{s,q,T}$ versus $q, q = 0.55, 0.60, 0.70, 0.8, 0.9, 1.00$ for each SCV, $C^2_{s,1,S} = 2, 5, 10, 20,$ and 50. It is clearly validated that the limiting case of each of the two plots for the SCVs $C^2_{s,q,R}$ and $C^2_{s,q,T}$ versus q of Figures 5.7–5.11, respectively, converges, as expected, at $C^2_{s,1,S}$ (with $q = 1$) for all increasing values of information parameter q and $C^2_{s,1,S}$. Moreover, it can be seen from Figures 5.7–5.11 that $C^2_{s,q,T} < C^2_{s,q,R}$ for all increasing values of q and $C^2_{s,1,R}$, i.e. the $GE_{q,T}$-type CDF $F_{s,q,T}(t)$ has lower variability than the $GE_{q,R}$-type CDF $F_{s,q,R}(t)$

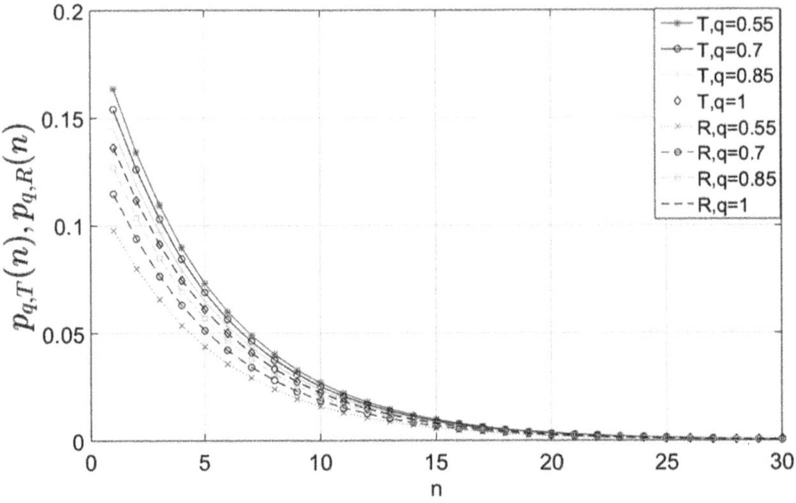

Figure 5.2. $p_{q,T}(n), p_{q,R}(n)$ *versus n for* $C^2_{s,1,S} = 2$*. For a color version of this figure, see www.iste.co.uk/anisimov/queueing2.zip*

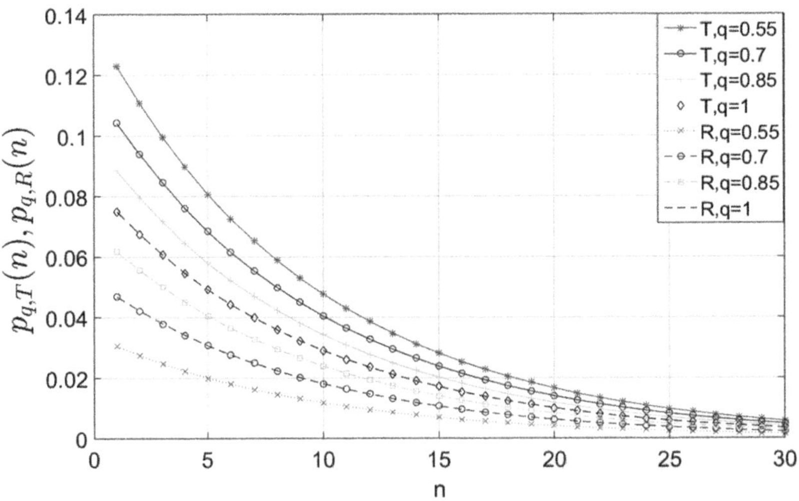

Figure 5.3. $p_{q,T}(n), p_{q,R}(n)$ *versus n for* $C^2_{s,1,S} = 5$*. For a color version of this figure, see www.iste.co.uk/anisimov/queueing2.zip*

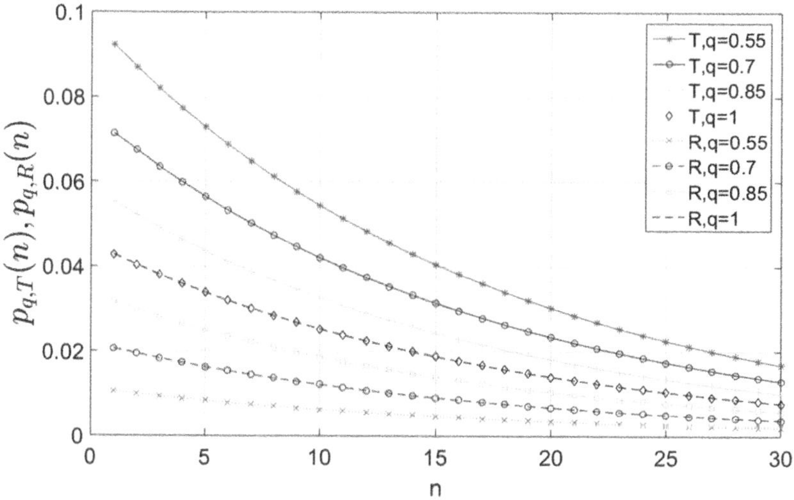

Figure 5.4. $p_{q,T}(n), p_{q,R}(n)$ *versus n for* $C^2_{s,1,S} = 10$*. For a color version of this figure, see www.iste.co.uk/anisimov/queueing2.zip*

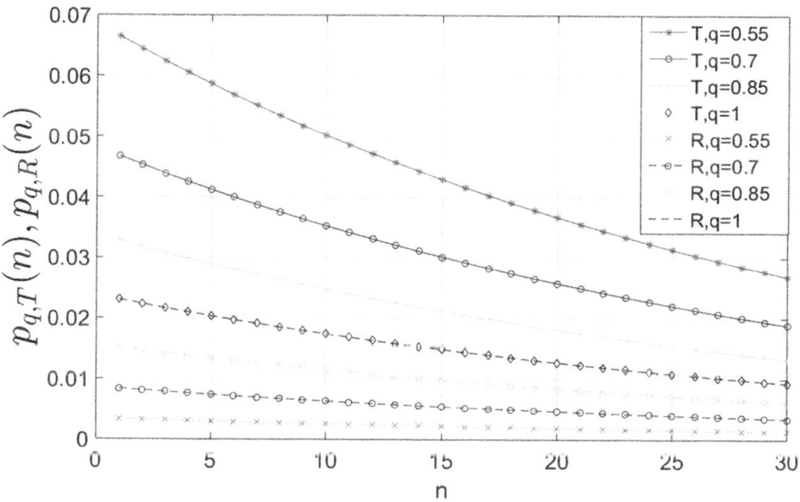

Figure 5.5. $p_{q,T}(n), p_{q,R}(n)$ *versus n for* $C^2_{s,1,S} = 20$*. For a color version of this figure, see www.iste.co.uk/anisimov/queueing2.zip*

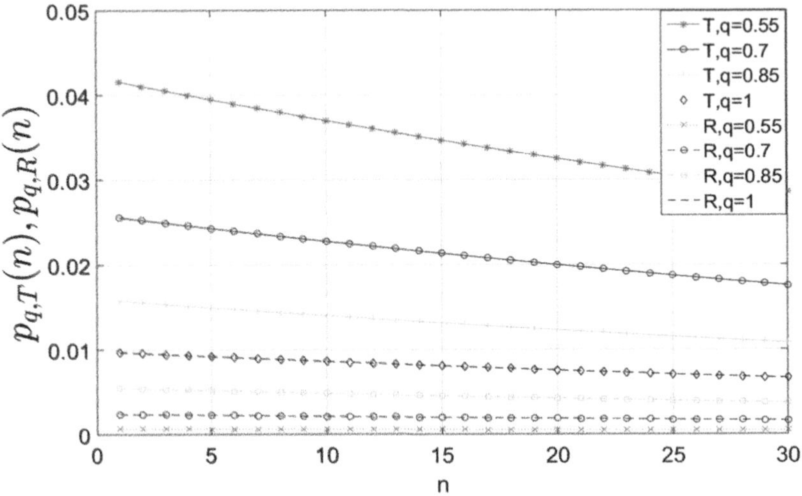

Figure 5.6. *$p_{q,T}(n), p_{q,R}(n)$ versus n for $C^2_{s,1,S} = 50$. For a color version of this figure, see www.iste.co.uk/anisimov/queueing2.zip*

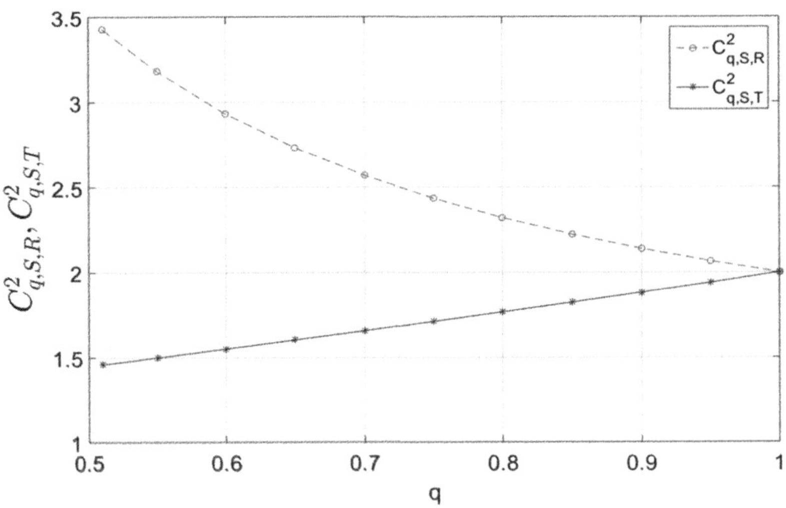

Figure 5.7. *$C^2_{q,s,R}$ and $C^2_{q,s,T}$ versus q for original $C^2_{s,1,S} = 2$. For a color version of this figure, see www.iste.co.uk/anisimov/queueing2.zip*

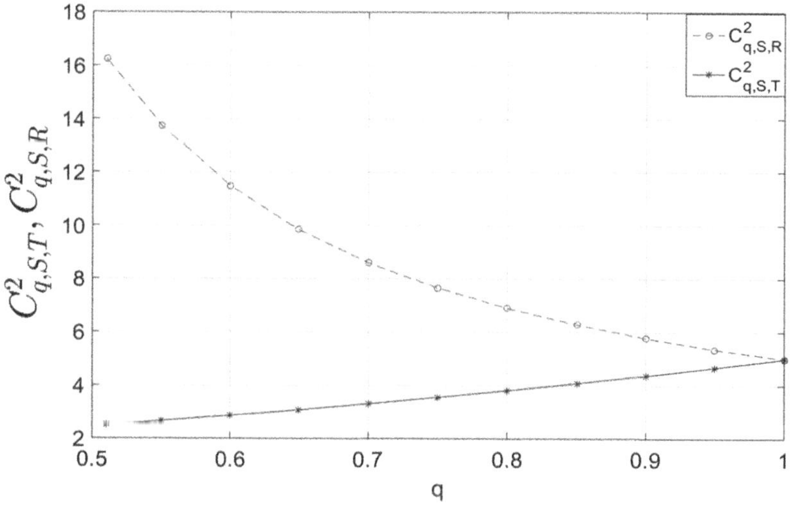

Figure 5.8. $C^2_{q,s,R}$ and $C^2_{q,s,T}$ versus q for original $C^2_{s,1,S} = 5$. *For a color version of this figure, see www.iste.co.uk/anisimov/queueing2.zip*

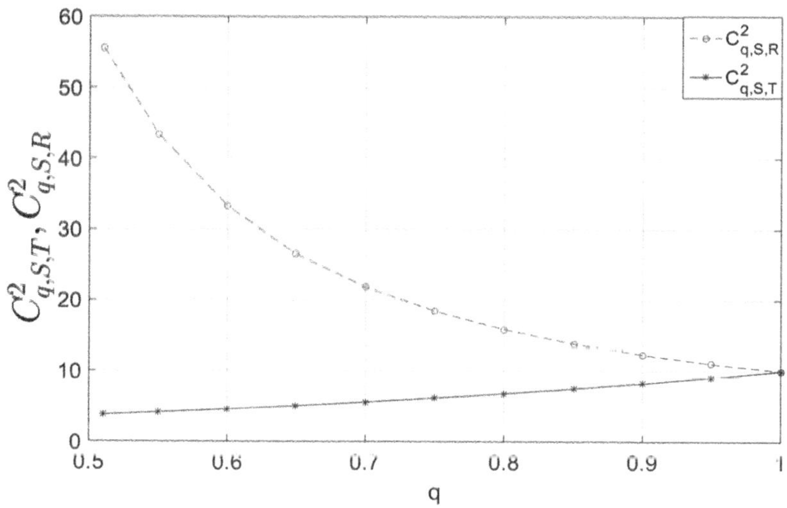

Figure 5.9. $C^2_{q,s,R}$ and $C^2_{q,s,T}$ versus q for original $C^2_{s,1,S} = 10$. *For a color version of this figure, see www.iste.co.uk/anisimov/queueing2.zip*

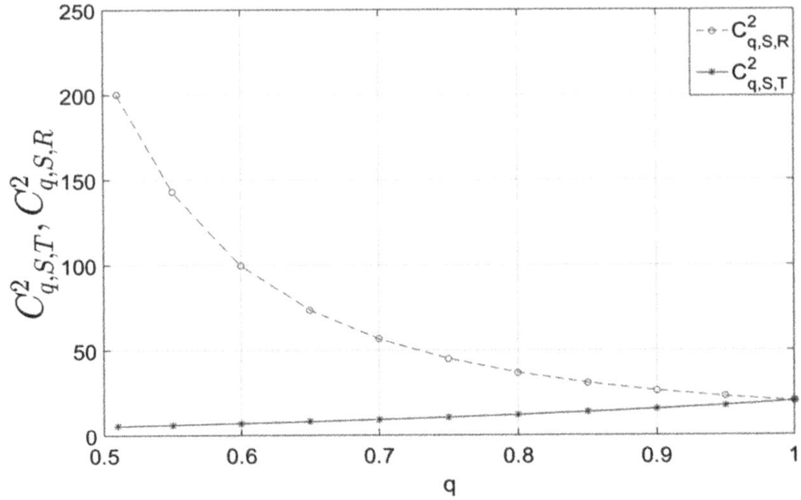

Figure 5.10. *$C^2_{q,s,R}$ and $C^2_{q,s,T}$ versus q for original $C^2_{s,1,S} = 20$. For a color version of this figure, see www.iste.co.uk/anisimov/queueing2.zip*

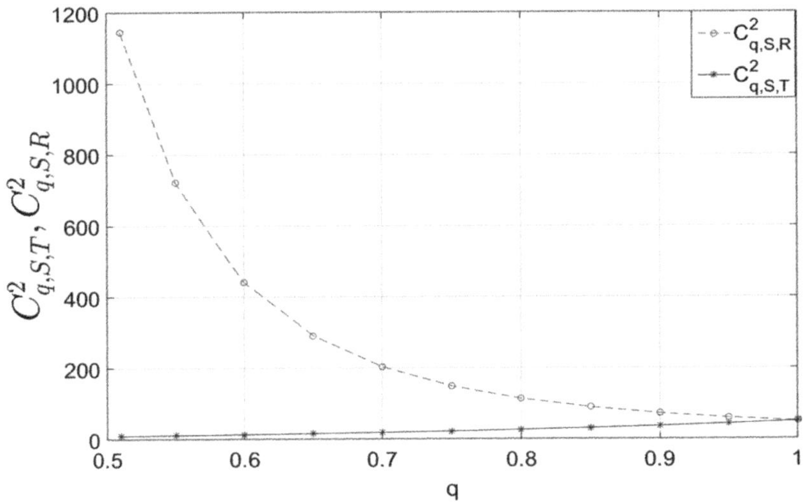

Figure 5.11. *$C^2_{q,s,R}$ and $C^2_{q,s,T}$ versus q for original $C^2_{s,1,S} = 50$. For a color version of this figure, see www.iste.co.uk/anisimov/queueing2.zip*

5.6. Conclusions

An exploration was carried out of Rényi's and Tsallis' NME formalisms as methods of inductive inference for a physical system Q with a countably infinite set S of states $\{S_n, n = 0, 1, 2, \cdots\}$ and long-range interactions of order, q $(0.5 < q < 1)$. In this context, the respective NME state probabilities $p_{q,R}(S_n)$ and $p_{q,T}(S_n)$ were analytically derived, based on the maximization of the respective entropy functionals, subject to normalization and a finite number of suitable mean value constraints. Moreover, the credibility of these NME formalisms, as methods of inductive inference, was justified in terms of four consistency axioms, namely uniqueness, invariance, system independence and subset independence, which were associated with Shannon's EME formalism of systems with short-range interactions. Specifically, it was verified that both Rényi's and Tsallis's formalisms, satisfied the consistency axioms of uniqueness, invariance and subset independence but, nevertheless, defied as expected, the axiom of system independence due to the existence of long-range interactions.

Consequently, focusing on an inductive application of Rényi's and Tsallis' formalisms on the analysis of a stable M/G/1 queue, novel expressions of the state probabilities $p_{q,R}(n)$ and $p_{q,T}(n), n = 0, 1, 2, \cdots$ were explicitly determined as functions of information parameter q, subject to normalization, SU and P-K MQL constraints. By employing the EMC, it was shown that these NME state probabilities are exact when the underlying service times of the stable M/G/1 queue are distributed according to two new distinct families of underlining CDFs of service times $\{F_{s,q,R}(t), F_{s,q,T}(t)\}$ of the $GE_{q,R}$ and $GE_{q,T}$ types. Moreover, for illustration purposes, two case studies were presented based on numerical experiments on (i) the impact of Rényi's and Tsallis' NME state probabilities $p_{q,R}(n)$ and $p_{q,T}(n), n = 0, 1, 2, \cdots$ with long-range interactions of order, q $(0.5 < q < 1)$, on the stable M/$GE_{q,R}$/1 and M/$GE_{q,T}$/1 queues with heavy tails and (ii) the convergence of q-dependent families of SCVs $C^2_{s,q,R}$ and $C^2_{s,q,T}$ of the CDFs of service times at $C^2_{s,1,S}$ (with $q = 1$) for all increasing values of information parameter q and $C^2_{s,1,S}$.

NME formalisms with long-range interactions, such as those of Rényi's and Tsallis', open new avenues of investigation for the inductive inference of queueing systems with heavy queue tails. It is expected that additional higher moment constraints will continue to reveal novel state probabilities for single and multiple server queues with new $GE_{q,R}$ and $GE_{q,T}$ information theoretic distributional structures for both interarrival and service time distributions. Moreover, approximate queue by queue decomposition algorithms for queueing network models with long-range interactions of order q $(0.5 < q < 1)$ could be established based on the merging, splitting and departing $GE_{q,R}$ and $GE_{q,T}$ type traffic flows.

5.7. Acknowledgments

The authors wish to express their sincerest thanks to Mr. Kabiru K. M. Maiyama, PhD research student, Networks and Performance Engineering Research Group, Department of Computer Science, University of Bradford, UK for the most professional drawing of the curves of Figures 8.2–8.11 and the presentational conversion of our paper from Word into Latex format.

5.8. Appendix: Rényi's NME formalisms versus EME consistency axioms

5.8.1. *Uniqueness*

According to this axiom, "If the same problem is solved twice in exactly the same way, the same answer is expected in both cases, i.e. the solution should be unique" (Shore 1980). Focusing on the Rényi's generalized extensive entropy function (see [5.4]), let Q be a general non-extensive system having, without loss of generality, a finite set S of $N, N > 0$ possible discrete states $\{S_n, n = 1, 2, \cdots N\}$ and Ω be a closed convex set of all probability distributions $\{p_{q,R}(S_n), S_n \in S\}$ such that $p_{q,R}(S_n) > 0$ for $S_n \in S, n = 1, 2, \cdots, N$ and $\sum_{n=1}^{N} p_{q,R}(S_n) = 1$. Let $f_{q,R}$ and $h_{q,R}$ be two probability distributions defined on S with the same extended entropy functions, namely

$$H_q^*\left(f_{q,N}\right) = H_q^*\left(h_{q,N}\right) \qquad [5.71]$$

Hence, $\sum_{n=1}^{N} f_{q,N}^q = \sum_{n=1}^{N} h_{q,N}^q$, which clearly implies $\sum_{n=1}^{N}\left(f_{q,N}^q - h_{q,N}^q\right) = 0$. Moreover, by assuming the converse statement, i.e. $f_{q,N}^q \neq h_{q,N}^q$, it follows that there exists a constant $\gamma > 1$ satisfying

$$f_{q,N}^q = \gamma h_{q,N}^q \qquad [5.72]$$

Hence,

$$\frac{1}{1-q}\left(ln\left(\sum_{n=1}^{N} f_{q,N}^q\right)\right) = \frac{1}{1-q}(ln(\gamma)) + \frac{1}{1-q}\left(ln\left(\sum_{n=1}^{N} h_{q,N}^q\right)\right) \qquad [5.73]$$

Therefore, if $f_{q,N}^q = h_{q,N}^q$, it is implied $\frac{1}{1-q}(ln(\gamma)) = 0$, i.e. $\gamma = 1$, which is a contradiction. Therefore, there cannot be two distinct probability distributions $f_{q,N}, h_{q,N} \in \Omega$ with the same maximum non-extensive entropy in Ω.Thus, Rényi's NME formalism satisfies the axiom of uniqueness (Shore 1980).

5.8.2. *Invariance*

The invariance axiom states that "The same solution should be obtained if the same inference problem is solved twice in two different coordinate systems" (Shore 1980). Following the analytic methodology proposed in Shore (1980) and adopting the notation of section 5.8.1, let Ξ be a coordinate transformation from state $\{S_n, n = 1, 2, \cdots, N\}$ to state $\{M_n, n = 1, 2, \cdots, N\}$, where M be a transformed set of N possible discrete states, namely $\{M = M_n, n = 1, 2, \cdots, N\}$ with $\Gamma(p_{q,N}(M_n)) = \Xi^{-1}(p_{q,N}(S_n))$, where J is the Jacobian $J = \frac{\partial(M_n)}{\partial(S_n)}$. Moreover, let $\Gamma\Omega$ be the closed convex set of all probability distributions Γ defined on M such that $\Xi(p_{q,N}(M_n)) > 0$ for all $M_n \in M, n = 1, 2, \cdots, N$ and $\sum_{n=1}^{N} \Xi(p_{q,N}(M_n)) = 1$. It can be clearly seen that, transforming variables from $S_n \in S$ into $R_n \in R$, the extended Rényi's entropy function (see [5.4]) is transformation invariant (Kayal 2017; Steinbrecher *et al.* 2016) namely

$$H_q^*\left(p_{q,N}\right) = H_q^*\left(\Xi\left(p_{q,N}\right)\right) \tag{5.74}$$

Thus, the EME formalism satisfies the axiom of invariance since the minimum in $\Xi\Omega$ corresponds to the minimum in "Ω" (Shore 1980).

5.8.3. *System independence*

This axiom of system independence, which is also referred as the "additivity property", states that "It should not matter whether one accounts for independent information about independent systems separately in terms of different probabilities or together in terms of the joint probability" (Shore 1980). In this context, consider two general non-extensive systems Q and V, each having a finite set of $N, N > 0$ possible discrete states $\{x_n, n = 1, 2, \cdots, N\}$ and $\{y_n, n = 1, 2, \cdots, N\}$, respectively. Moreover, let X and Y be the random variables describing the state of the systems Q and V with corresponding state probabilities $\{f_{q,N}(x_n) = Pr\{X = x_n\}\}$ and $g_{q,N}(y_n) = Pr\{Y = y_n\}$, respectively. Assuming that Q and M are independent systems, then the joint probability is expressed by

$$h_{q,N}\left(x_k, y_n\right) = Pr\left(X = x_k, Y = y_n\right) = f_{q,N}\left(x_k\right) g_{q,N}\left(y_n\right) \tag{5.75}$$

For the joint probability $h_{q,N}(x_n, y_n)$, the Rényi's generalized entropy function (see [5.4]) can be written using [5.10],

$$H_q^*\left[h_{q,N}\right] = \frac{1}{1-q} ln\left(\sum_k \sum_n h_{q,N}^q\right) \tag{5.76}$$

From the definition of [5.10], it clearly follows that

$$H_q^*\left[h_{q,N}\right] \neq H_q^*\left[f_{q,N}\right] + H_q^*\left[g(Y_{q,N})\right] \quad [5.77]$$

The inequality [5.77] implies that, in information theoretic terms, the joint EME state probability distribution of two independent non-extensive systems Q and V defies, due to the presence of long-range interactions, the axiom of system independence (Shore 1980). Thus, this attribute of the ME formalism, as a method of inductive inference, is clearly most suitable for the quantitative studies of non-extensive dynamic systems with heavy queue tails and asymptotic power law behavior.

Note that in the limiting case $q \longrightarrow 1$, equation [5.77] becomes equality, namely

$$H_1^*\left[h_{1,N}\right] = H_1^*\left[f_{1,N}\right] + H_1^*\left[g(Y_{1,N})\right] \quad [5.78]$$

Expression [5.78] verifies that the joint EME state probability distribution, as expected, satisfies the axiom of system independence (Shore 1980). This is "an appropriate property of EME formalism, as a method of inductive inference, for the study of extensive systems with short-range interactions" (Kouvatsos 2010).

5.8.4. *Subset independence*

The axiom of subset independence states that "It does not matter whether one treats an independent subset of system states in terms of a separate conditional density or in terms of the full system density" (Statistics 2007).

Consider a general non-extensive system Q that has a finite number, $L(L > 0)$ of disjoint sets of discrete states $\{S_i^*, i = 1, 2, \cdots, L\}$, whose union is S. Let $\{x_{ij}, i = 1, 2, \cdots, L, j = 1, 2, \cdots, L\}$ be a conditional state in S_i^*. Let ξ_i be the probability that a state of the system Q is in the set $\{S_i^*, i = 1, 2, \cdots, L\}$ such that $\sum_i \xi_i = 1$. Moreover, let probability $f_{q,i}(x_{ij}) \in \Omega_i$, where Ω_i is the closed convex set of all probability distributions on S_i^*, i.e. $\{f_{q,i}(x_{ij}) = Pr\{X_i = x_{ij}\}\}$, where X_i is the state conditional random variable of the system $S_i^*, i = 1, 2, \cdots, L$. Moreover, let x be an aggregate state of system Q and probability $f_q(x) \in \Omega$, where Ω is the closed convex set of all probability distributions on S, i.e. $f_q(x) = Pr\{X = x\}$, where X is the random variable describing the aggregate state of the system S. Clearly, $\xi_i, i = 1, 2, \cdots, L$ can be expressed by

$$\sum_{S_i^*} f_{q,i}(x_{ij}) = \xi_i \quad [5.79]$$

The overall non-extensive entropy function of system $Q, H_q^*(f_q)$ is defined on the total number of states in the union S of states $\{S_i^*, i = 1, 2, \cdots, L\}$. Using the extended entropy function (see [5.10]), it can be shown that

$$H_q^*\left(f_q\right) = \frac{1}{1-q} ln\left(\sum_i \sum_{S_i} \xi_i f_{q,i}^q\left(x_{ij}\right)\right) \qquad [5.80]$$

where $f_q(x) \in \Omega$. Equation [5.29] can be rewritten in the form

$$H_q^*\left(f_q\right) = \frac{1}{1-q} ln\left[\sum_i \xi_i \sum_{S_i} f_{q,i}^q\left(x_{ij}\right)\right] \qquad [5.81]$$

However, the conditional extensive entropy, $H_i^*(f_{q,i})$, defined on the set of states $\{S_i, i = 1, 2, \cdots, L\}$ is expressed by

$$H_{q,i}^*\left(f_{q,i}\right) = \frac{1}{1-q} ln\left(\sum_{S_i^*} f_{q,i}^q\left(x_{ij}\right)\right) \qquad [5.82]$$

Clearly, it follows that

$$\sum_{S_i^*} f_{q,i}^q\left(x_{ij}\right) = e^{(1-q)H_q^*(f_{q,i})} \qquad [5.83]$$

Hence,

$$H_q^*\left(f_q\right) = \frac{1}{1-q} ln\left(\sum_i \xi_i e^{(1-q)H_q^*(f_{q,i})}\right) \qquad [5.84]$$

Therefore, maximizing the generalized aggregate entropy function, $H_q^*(f_q)$, subject to an aggregate set of available constraints, it is equivalent to maximizing each generalized conditional entropy function, $H_{q,i}^*\ (f_{q,i})$ individually, subject to a conditional set of available constraints. Thus, Rényi's ME formalism satisfies the axiom of subset independence (c.f., Shore and Johnson (1980)).

5.9. References

Adamic, L.A., Huberman, B.A. (2002). Zipf's law and the internet. *Glottometrics*, 3(1), 143–150.

El-Affendi, M.D., Kouvatsos, D. (1983). A maximum entropy analysis of the m/g/1 and g/m/1 queueing systems at equilibrium. *Acta Informatica*, 19(4), 339–355.

Kayal, S., Kumar, S. (2017). Estimating renyi entropy of several exponential distributions under an asymmetric loss function. *REVSTAT-Statistical Journal*, 15(4), 501–522.

Kleinrock, L. (1976). *Queueing systems, volume 2: Computer applications*. Wiley, New York.

Kouvatsos, D.D. (1986). A maximum entropy queue length distribution for the g/g/1 finite capacity queue. In *ACM SIGMETRICS Performance Evaluation Review, ACM*, 224–236.

Kouvatsos, D.D. (1988). A maximum entropy analysis of the g/g/1 queue at equilibrium. *Journal of the Operational Research Society*, 39(2), 183–200.

Kouvatsos, D.D., Assi, S.A. (2010). On the analysis of queues with heavy tails: a non-extensive maximum entropy formalism and a generalisation of the zipf-mandelbrot distribution. In *International Workshop on Performance Evaluation of Computer and Communication Systems*, Springer, 99–111.

Kouvatsos, D.D., Assi, S.A. (2011). Generalised entropy maximisation and queues with bursty and/or heavy tails. In *Network Performance Engineering*, Springer, 357–392.

Rényi, A. (1961). On measures of information and entropy. In *Proceedings of the 4th Berkeley Symposium on Mathematics, Statistics and Probability*, 547–561.

Shannon, C.E. (1948). A mathematical theory of communication. *Bell System Technical Journal*, 27(3), 379–423.

Shore, J.E. (1982). Information theoretic approximations for m/g/1 and g/g/1 queuing systems. *Acta Informatica*, 17(1), 43–61.

Shore, J., Johnson, R. (1980). Axiomatic derivation of the principle of maximum entropy and the principle of minimum cross-entropy. *IEEE Transactions on Information Theory*, 26(1), 26–37.

Statistics, T. (2007). Statistical mechanics for non-extensive systems and long-range interactions. Notebooks. Available: http://www.cscs.umich.edu/.

Steinbrecher, G., Sonnino, A., Sonnino, G. (2016). Category theoretic properties of the a. rényi and c. tsallis entropies. *Journal of Modern Physics*, 7, 251–266.

Tsallis, C. (1988). Possible generalization of boltzmann-gibbs statistics. *Journal of Statistical Physics*, 52 (1-2), 479–487.

6

Inventory with Positive Service Time: a Survey

Achyutha KRISHNAMOORTHY[1], Dhanya SHAJIN[2] and Viswanath C. NARAYANAN[3]

[1] *Department of Mathematics, CMS College, Kottayam, India.*

[2] *Department of Mathematics, S. N. College, Chempazhanthy, Thiruvananthapuram, India.*

[3] *Department of Mathematics, Government Engineering College, Thrissur, India*

In classical inventory, a customer is served the demanded item immediately on his arrival, provided at least one item is available. However, in inventory with positive service time, it takes a certain amount of time to serve the item. As a result, a queue of demands builds up. We intent to overview the work so far done in mathematical inventory models involving positive service time. These include perishable items, stock supplied through production or through various control policies such as $(s, S), (r, Q)$, random reorder quantity, etc. Stochastic decomposition results are also discussed. A brief account of the work done in queues with requirement of additional items for service is also provided.

6.1. Introduction

In classical inventory model, it is assumed that the item demanded is served instantly, provided it is available. This is the case of negligible service time. If the item is not available, the following possibilities arise: (i) no backlog of demands, (ii) finite backlog and (iii) infinite backlog. In case (i), demands arising are lost to the system forever when inventory level is zero (unless such customers keep retrying). In

Queueing Theory 2,
coordinated by Vladimir ANISIMOV, Nikolaos LIMNIOS.

case (ii), a finite backlog of demands is entertained, whereas in case (iii) all demands that take place during inventory dry period are met after adequate replenishment of items, depending on the type of replenishment policy adopted. The lead time (time duration required for replenishment of inventory, starting from the epoch the replenishment order is placed) may be zero or positive. In the latter case, it can be a constant or a random variable. Various control policies are employed for replenishment. These are discussed in section 5.2. In cases (ii) and (iii), the queue formed vanishes the moment at least as many items as the number of customers in the system arrives. However, in inventory with positive service time (several authors call it queueing-inventory system (see, for example: Babak and Alireza 2011; Cui and Wang 2013; Koroliuk *et al.* 2018), a queue of customers is formed even if inventory is available, because each customer requires a "positive service time" to get the inventory served. Thus at a first glance, inventory with positive service time may appear as a classical queueing problem. Nevertheless, this is not so since in classical queueing model it is assumed that if customers are available and server is ready to serve, then service could be rendered. This is the case as long as the availability of "additional items for service" is not required to provide service. In other words, in classical queueing theory we either ignore or assume abundance of items required for service. However, in real life, to provide service, certain items, which we call *inventory/additional item*, is required. As a result, the question of management of this inventory also come into play. Inventory with positive service time deals with this aspect as well. Thus in queueing-inventory, availability of customer(s) and server(s) ready to serve are not sufficient to provide service. Even when the above two are available, service may not take place in the absence of inventory. At a service completion epoch, the customer departs with one or more item(s) demanded. As a result, the inventory level depletes by as many units of inventory as that got served.

Research on queueing systems coupled with inventory has garnered much attention of researchers for the last three decades. The first papers of this kind are due to Melikov and Molchanov (1992) and Sigman and Simchi-Levi (1992), independently of each other. Readers may also refer to Krishnamoorthy *et al.* (2011) for a brief, but incomplete, survey of work done until 2010. Further Choi and Yoon (2016) provide a review on queueing-inventory system with phase type service time distribution applied to a very specific context such as in hospitals.

This chapter is organized as follows. In section 6.2.1, review of work done in queueing-inventory system where inventory control policies such as (s, S), (r, Q), $(S-1, S)$, randomized order size, is considered. Further we restrict to single-type inventory in this section. Section 6.2.2 provides overview of work related to queueing-inventory in which the supply of items is through a production process of a single-type inventory. The case of multitype inventory is investigated in section 6.2.3. As a subsection, we include the work reported on multiclass (priority/demanding distinct items) customers. Work on retrial queues with inventory is summarized in section 6.2.4.

Of special interest is queueing-inventory model for which explicit product form solutions are available. By this, we mean the following: suppose $N(t)$ and $I(t)$ represent, respectively, the number of customers and number of items in the system at time t. If the asymptotic distribution of $(N(t), I(t))$ can be written as the product of the distribution of the individual random variables, we say that the system admits a product form solution. When lead time is positive, the system admits a product form solution under certain strong assumptions. For example, a subset of the state space of the system is identified such that while the process is in that subset, all activities, other than lead time realization, are blocked. Work reported in queueing-inventory models yielding product form solutions will be indicated in section 6.2.1.1. Queueing systems requiring additional items for service is of recent origin. This has flavor similar to inventory with positive service time. We will discuss this in section 6.2.5. Certain possible direction for future work is given in section 6.2.6.

6.2. Queueing inventory models

Queueing inventory has its origin in Melikov and Molchanov (1992) and Sigman and Simchi-Levi (1992). These are followed by a spate of papers by Berman *et al.* (1993); Berman and Kim (1999); Berman and Sapna (2002) and several other researchers. They assume that serving inventory requires positive amount of time – deterministic or random. Among these a few deal with stochastic decomposition or even product form solutions.

To move on further, definitions and terminologies given below will be useful. We provide a few technical terms as well.

Lead time: the time lag from the epoch at which an order for replenishment of inventory is placed, until the order is materialized, is called the lead time.

(s,S)-policy: this is an inventory control policy in which the replenishment quantity is that much to bring the level back to S at the replenishment epoch. Here s is the reorder level and S is the maximum number of items stored.

***(S-1,S)**-policy*: this control policy calls for placing an order each time an item is sold out. It is also called the one-for-one ordering policy. This policy is advised for slow moving and expensive items.

(s,Q)-policy: in this control policy, the number of items ordered is fixed and is equal to $Q = S - s$. Here s is the reorder level and Q is the fixed order quantity. Q is taken larger than s to avoid perpetual order placement for replenishment.

In *randomized order size*, the decision of the order size is according to a discrete probability function p on the set $\{1, 2, ..., S\}$. The size of a replenishment order is n with probability p_n, such that $\sum_{n=1}^{S} p_n = 1$.

Production inventory: replenishment of items can also be through a production process. In this case, we call the system "Production-inventory". Usually the (s, S)-policy is adopted in production inventory: a production cycle starts at the epoch at which the inventory level drops down to s through meeting a demand or by way of deterioration. The production process stays in the "on-mode" until inventory level hits the maximum level S, at which stage the production process is "switched off".

Phase-type distribution (Neuts 1981): the continuous phase-type distributions are introduced as a natural generalization of the exponential and Erlang distributions. A phase-type distribution is obtained as the distribution of the time until absorption in a Markov chain with a finite state space and an absorbing state.

Consider a Markov process $\psi = \{X(t) : t \geq 0\}$ with a finite state space $\{1, 2, \ldots, m+1\}$ and the infinitesimal generator matrix $Q = \begin{pmatrix} T & \mathbf{T}^0 \\ \mathbf{0} & 0 \end{pmatrix}$ where T is a square matrix of order m, $\mathbf{T}^0$, a column vector and $\mathbf{0}$ the zero row vector of the same dimension. The elements of the matrices T and $\mathbf{T}^0$ satisfy $T_{ii} < 0$ for $1 \leq i \leq m$, $T_{ij} \geq 0$ for $i \neq j$; $T_i^0 \geq 0$ and $T_i^0 > 0$ for at least one i, $1 \leq i \leq m$, T_i^0 is the ith element of $\mathbf{T}^0$ and $T\mathbf{e} + \mathbf{T}^0 = \mathbf{0}$.

The initial distribution of ψ is given by the row vector $(\boldsymbol{\alpha}, \alpha_{m+1})$ where $\boldsymbol{\alpha}$ a row vector of dimension m with the property that $\boldsymbol{\alpha}\mathbf{e} + \alpha_{m+1} = 1$. The states $\{1, 2, ..., m\}$ shall be transient, while the state $m+1$ is absorbing.

Let $Z := inf\{t \geq 0 : X(t) = m+1\}$ denotes the random variable of the time until absorption in state $m+1$. The distribution of Z is called phase-type distribution (or shortly PH distribution) with representation $(\boldsymbol{\alpha}, T)$. The dimension m is called the order of the distribution. The states $\{1, ..., m\}$ are also called phases, which gives rise to the name phase-type distribution. The distribution function of Z is given by

$$F(t) := P(Z \leq t) = 1 - \alpha e^{Tt}\mathbf{e}, \text{ for all } t \geq 0$$

and the density function is

$$f(t) = \alpha e^{Tt}\mathbf{T}^0, \text{ for all } t > 0.$$

Here, the function $e^{Tt} = exp(Tt) = \sum_{n=0}^{\infty} \frac{t^n}{n!} T^n$ denotes a matrix exponential function. The Laplace–Stieltjes transform of $F(t)$ is given by

$$\phi(s) = \int_0^{\infty} e^{-st} dF(t) = \alpha_{m+1} + \alpha(sI - T)^{-1}\mathbf{T}^0$$

for all $s \in C$ with $Re(s) \geq 0$. The moments of Z are given by

$$E(Z^n) = (-1)^n n! \alpha T^{-n} \mathbf{e}$$

for all $n \in N$.

Markovian arrival process: Markovian arrival processes (MAP) introduced in Neuts (1981) is a rich class of point processes that includes many well-known processes such as Poisson, PH-renewal processes and Markov-modulated Poisson process. A significant feature of the MAP is the underlying Markovian structure that fits ideally in the context of matrix-analytic solutions to stochastic models. MAP is a generalization of the Poisson process, which keeps many useful properties of the Poisson process. For example, the memoryless property of the Poisson process is partially preserved by the MAP by conditioning on the phase of underlying Markov chain. Any stochastic counting process can be approximated arbitrarily closely by a sequence of Markovian arrival processes. MAP is a convenient tool to model both renewal and non-renewal arrivals. Chakravarthy (2001) provides an extensive survey of the batch Markovian arrival process ($BMAP$) in which arrivals are in batches where as it is in singles in MAP.

A continuous-time Markov chain $\{(N(t), I(t)), t \geq 0\}$ with state space $\{(i, j) : i \geq 0, 1 \leq j \leq m\}$ and infinitesimal generator

$$Q = \begin{pmatrix} D_0 & D_1 & & \\ & D_0 & D_1 & \\ & & \ddots & \ddots \end{pmatrix}$$

is called an MAP with matrix representation (D_0, D_1). Here, D_0 and D_1 are square matrices of order m where m is a positive integer. The diagonal elements of D_0 are negative and its off-diagonal elements are non-negative, D_1 has all its elements non-negative and $D_0 + D_1$ is an infinitesimal generator. Let $D_0 = \left(d_{ij}^{(0)}\right)$ and $D_1 = \left(d_{ij}^{(1)}\right)$, then $d_{ij}^{(0)}$ is the rate of transitions from phase i to j without an arrival, for $i \neq j$; $d_{ij}^{(1)}$ is the rate of transitions from phase i to j with an arrival and $-d_{ii}^{(0)}$ is the total rate of events in phase i. Let $N(t)$ denote the number of arrivals in $(0, t)$ and $I(t)$ the phase of the Markov chain at time t. Let $\boldsymbol{\pi}$ be the stationary probability vector of D. Then the constant $\lambda = \boldsymbol{\pi} D_1 \mathbf{e}$, referred to as *fundemental rate*, gives the expected number of arrivals per unit of time in the stationary version of the MAP.

Blocking set: if the system is assumed not to admit new arrivals as long as it is in a specific subset of the state space of the process, that subset is referred to as a blocking set for fresh arrivals. Arrivals that take place when the system is in the blocked set are assumed to be lost to the system.

Perishability: perishability is a consequence of ageing of stored items. Items perish one at a time.

In the context of airplane/train/bus seat reservation, we use terminologies such as reservation, cancellation, overbooking and common lifetime.

Reservation: purchase of an item from inventory is called reservation (e.g. reservation of a seat in bus/train/flight for a future journey).

Cancellation: a purchased item may be returned before its expiry time; this is called cancellation (e.g. canceling a reserved seat).

Common lifetime: each item on-hand may have an expiry date which in some cases is common to all; this is called the common lifetime (e.g. once a bus/train/flight departs the vacant seats have no use, those who could not board the transport system before departure miss it – at the moment of flight takeoff, next schedule is announced; batch of medicines that are manufactured together have a common expiry date). Note that on realization of common lifetime, all items remaining in the store become unusable.

Overbooking: no item is available in the system: still a finite number of reservations are permitted, this is called overbooking (e.g. customers in the waiting list).

Table 6.1 provides a birds eye view of the work reported in inventory with positive service time. We classified them based on "keywords" such as perishable inventory, postponed work, retrial queueing-inventory, production/supply chain, product form solution, multicommodity system, nature of lead time, etc. A more detailed description of a few among these is given in various sections, depending on the nature of the work reported in those papers.

6.2.1. *Single-commodity queueing-inventory systems*

As stated earlier, queueing-inventory has its origin in the works of Melikov and Molchanov (1992) and Sigman and Simchi-Levi (1992). Of these, Sigman and Simchi-Levi (1992) considered an $M/G/1$ queueing-inventory system with exponentially distributed lead time for replenishment of the inventory. They analyzed the system under light traffic. Melikov and Molchanov (1992) consider all underlying distributions to be exponential. Their analysis was through state space collapse and approximation. These are followed by a spate of papers (Berman *et al.* 1993; Berman and Kim 1999; Berman and Sapna 2002; and several other researchers). As

Keywords	References
Perishable Inventory	(Amirthakodi *et al.* 2015; Berman and Sapna 2002; Duan *et al.* 2012), (Geetha and Elango 2012; Jeganathan 2014, 2015c,d; Jeganathan *et al.* 2017; Koroliuk *et al.* 2017, 2018), (Kouki and Jouini 2015; Lawrence *et al.* 2013), (Melikov *et al.* 2017; Melikov and Shahmaliyev 2017; Melikov *et al.* 2016, 2018; Melikov and Shahmaliyev 2018; Melikov *et al.* 2019, 2017; Melikov and Shahmaliyev 2019; Shophia *et al.* 2013), (Yadavalli and Udayabaskara 2018; Yadavalli *et al.* 2017)
Postponed work	(Deepak *et al.* 2004; Devi *et al.*; Krishnamoorthy *et al.* 2016a), (Padmavathi *et al.* 2016; Sivakumar *et al.* 2012),
Orbit/retrial	(Amirthakodi *et al.* 2015; Artalejo *et al.* 2006; Cui and Wang 2013), (Jeganathan *et al.* 2013; Jeganathan 2015a; Jeganathan *et al.* 2016, 2018; Jose and Rejitha 2017; Jose and Salini 2017), (Krishnamoorthy and Shajin 2017; Krishnamoorthy *et al.* 2012; Krishnamoorthy and Shajin 2006; Krishnamoorthy *et al.* 2016b; Manikandan and Sajeev 2017), (Melikov *et al.* 2018, 2019; Melikov and Shahmaliyev 2019; Nair and Jacob 2014; Nair and Jose 2017a,b), (Padmavathi *et al.* 2015; Rejitha and Jose 2017a,b; Shajin *et al.* 2019; Shajin and Krishnamoorthy 2018), (Sivakumar 2011; Sivakumar and Arivarignan 2015; Suganya and Amirthakodi 2018), (Ushakumari 2006; Vijaya Laxmi and Soujanya 2015), (Wang 2015; Wang *et al.* 2017; Yadavalli *et al.* 2017, 2018, 2017)
Bulk arrival//service	(Chakravarthy *et al.* 2017; Ramshankar and Ramanarayanan 2014; Sundar *et al.* 2015), (Yue *et al.* 2018)
Vacation	(Divya *et al.* 2018; Divya *et al.*; Jeganathan 2014; Jeganathan *et al.* 2018), (Koroliuk *et al.* 2017, 2018; Krishnamoorthy and Viswanath 2011), (Melikov *et al.* 2017; Padmavathi *et al.* 2015, 2016; Sivakumar 2011), (Yadavalli *et al.* 2017)
Production/ supply chain	(Axsater 2013; Babak and Alireza 2011; Baek and Moon 2014, 2016; Benjaafar *et al.* 2010; Betts 2014), (Caceres *et al.* 2018; Haji *et al.* 2011; Hum *et al.* 2018; Jose and Salini 2017), (Karzana *et al.* 2017; Krishnamoorthy *et al.* 2016c; Krishnamoorthy and Viswanath 2011, 2013; Melikov and Shahmaliyev 2018), (Nair and Jose 2017a,b; Nair and Shajin 2018; Otten *et al.* 2016), (Pal *et al.* 2012; Ramshankar and Ramanarayanan 2014; Rashid *et al* 2015), (Sarkar 2012; Saidane *et al.* 2013; Shajin *et al.*; Vidyadhar and Keqi 2012), (Wang *et al.* 2013; Yue and Qin 2019; Yu and Dong 2014; Zare *et al.* 2017)
Product form Solution	(Baek and Moon 2014, 2016; Baek *et al.* 2018; Krenzler 2016; Krenzler and Daduna 2015), (Krishnamoorthy *et al.* 2016, 2015b, 2014, 2016c; Krishnamoorthy and Shajin 2017, 2006; Krishnamoorthy and Viswanath 2013; Otten *et al.* 2016), (Saffari *et al.* 2013, 2011; Schwarz *et al.* 2006, 2007), (Shajin *et al.* 2018; Shajin and Krishnamoorthy 2019; Shajin *et al.* 2016; Shajin 2019, Shajin *et al.*, Shajin and Krishnamoorthy 2018; Shajin *et al.*)

Table 6.1. *An overview of studies on inventory models with positive service time*

Keywords	References
Multicommodity	(Anbazhagan and Kumaresan 2012; Benny *et al.* 2018; Shajin 2019)
Feedback	(Amirthakodi *et al.* 2015; Amirthakodi and Sivakumar 2015), (Melikov *et al.* 2018, 2019; Suganya and Amirthakodi 2018)
Additional items Required for service	(Baek *et al.* 2017; Divya *et al.* 2018; Divya *et al.*; Hanukov *et al.* 2017), (Kazimirsky 2006; Shajin *et al.* 2019)
Two-type service	(Jeganathan 2015b)
General arrival	(Amirthakodi *et al.* 2015; key-25; Baek *et al.* 2017; Boute 2014; Divya *et al.* 2018; Divya *et al.*), (Krishnamoorthy and Shajin 2017; Krishnamoorthy and Viswanath 2011; Nair and Jose 2017a; Park *et al.* 2017), (Rejitha and Jose 2017a,b; Shajin *et al.* 2019; Shajin and Krishnamoorthy 2019; Shajin 2019; Shajin *et al.*), (Sivakumar *et al.* 2012; Sivakumar and Arivarignan 2015; Wang 2012)
General service	(Baek *et al.* 2017; Boxma 2016; Divya *et al.* 2018; Divya *et al.*; Krishnamoorthy and Shajin 2017; Krishnamoorthy and Viswanath 2011), (Nair and Jose 2017a; Rejitha and Jose 2017a,b; Shajin *et al.* 2019; Shajin *et al.*; Wang 2012)
General lead time	(Amirthakodi *et al.* 2015; Baek *et al.* 2017; Saffari *et al.* 2013; Schwarz and Daduna 2006) (Shajin *et al.* 2019; Shajin *et al.*; Sivakumar *et al.* 2012; Sivakumar and Arivarignan 2015; Ushakumari 2006)
Multiclass Customer	(Baek *et al.* 2017; Bhuvaneshwari 2018; Chen *et al.* 2012; Isotupa and Samanta 2013; Isotupa 2015), (Jeganathan *et al.* 2013; Jeganathan 2015a; Jeganathan *et al.* 2016; Liu *et al.* 2014; Melikov *et al.* 2018), (Shajin *et al.* 2019; Sivakumar and Arivarignan 2015; Wang *et al.* 2017, 2013), (Zhao and Lian 2011)
Multiserver	(Anoop and Jacob 2013; Chakravarthy *et al.*; Hum *et al.* 2018), (Krishnamoorthy *et al.* 2015a; Nair and Jacob 2014; Nair *et al.* 2015; Rajkumar *et al.* 2014), (Shajin *et al.*; Suganya and Amirthakodi 2018; Wang 2015; Wang *et al.* 2017), (Yadavalli *et al.* 2012; Yue *et al.* 2016)
Balk/impatience	(Benjaafar *et al.* 2010; Jeganathan *et al.* 2017; Krishnamoorthy and Shajin 2017), (Nair and Shajin 2018; Wang and Zhang 2017; Yadavalli *et al.* 2017)
Non-homogeneous/ compound Poisson	(Albrecher *et al.* 2017) (Babai and Jemai 2010)
Survey papers	(Choi and Yoon 2016; Krishnamoorthy *et al.* 2011)

Table 6.1 (continued). *An overview of studies on inventory models with positive service time*

stated earlier in queueing-inventory, it requires a certain positive amount of time – deterministic or random – to serve the required item to the customer. Of the above cited papers, Berman *et al.* (1993) considered a deterministic inventory model and obtained an approximation of its behavior via a dynamic programming approach. Berman and Kim (1999) provide an analysis on stochastic inventory systems, with all underlying distributions exponential. They derived the long run system state distribution and from that obtained several characteristics of the system. Berman and Sapna (2002) considered the case of arbitrarily distributed service time for a queueing-inventory system in equilibrium and analyzed its asymptotic behavior through renewal theoretic tools.

Next we explore research papers dealing with asymptotic product form solutions. This is a special case of models discussed previously. At the same time, they are of independent interest.

6.2.1.1. *Queueing-inventory system with a product form solution*

There are a few inventory models involving positive service time that have asymptotic product form solutions. By this, we mean that under certain conditions the vector process $(N(t), I(t))$, with $N(t) = \#$ of customers in the system and $I(t) = \#$ of items in the inventory at time t, has an asymptotic distribution equal to the product of the distribution of the marginals. All those models that so far provided this asymptotic independence involve a Poisson arrival process and exponentially distributed service time. However, the lead time need not have a distribution that lacks memory. The basic idea is to restrict activities in a subset of the state space of the process. For example, we can assume that no customer joins when inventory level is zero. Such subsets of the state space are referred to as blocking sets. In the example we considered for blocking set, only one activity, namely replenishment against the order placed, is permitted. These assumptions lead to an asymptotic product form solution. The importance of the result is that, despite strong correlation between the number of arrivals during the lead time and the length of the lead time, asymptotically the components of the vector process are independent. The optimal blocking set was investigated in Shajin *et al.*. We discuss below the work reported in literature that gives explicit product form solutions.

Schwarz *et al.* (2006, 2007) analyzed $M/M/1$ queueing systems with inventory. In Schwarz *et al.* (2006), stationary distribution of queue length and inventory process was derived in explicit product form for various control policies; lead time was assumed to be exponentially distributed. Customers arriving when inventory is out of stock were assumed to be lost forever. Thus, the blocking set was taken as the subset $\{(n, 0), n = 0, 1, ...\}$ of the state space of the system. In Schwarz *et al.* (2007), the authors considered a queueing network with inventory. Unlike in Schwarz *et al.* (2006), the authors do not assume that customers arriving at a node to find no inventory are lost to the system. Instead such customers were rerouted to nodes with positive inventory. Despite this, a product form solution could be arrived at for the system state. Schwarz and Daduna (2006) investigated $M/M/1$ queueing-inventory system with backordering. This means that customers who arrive when the inventory level was zero are not lost; instead they join the system. With lead time distribution exponential, they computed performance measures and derived optimality conditions under different replenishment policies. For performance measures for which no explicit solution was available, they presented an approximation scheme for all possible parameter combinations. Although the authors could not completely determine analytically the steady-state probabilities for the system state, they were able to derive functional relations between interesting probabilities and show surprising insensitivity properties of several performance measures. For the approximations, they developed an algorithm adapted to the system structure which could be easily adapted to other systems as well.

Saffari *et al.* (2013) considered an $M/M/1$ queueing system with inventory under a (r,Q) policy with generally distributed lead time and lost sales. They could also get a product form solution for the system sate probability. In another paper, (Saffari *et al.* 2011) considered a mixed exponential distribution for the lead time and obtained a product form solution.

Krishnamoorthy *et al.* (2015b) extended (Schwarz *et al.* 2007) to the case where the customer need not buy the item at a service completion epoch. Such a case arises when the item is quite expensive. The stationary distribution was shown to have a product form solution.

An (s, Q) inventory system with positive lead time and local purchase under N-policy was discussed in Krishnamoorthy *et al.* (2014). N-policy was employed in such a way that, as and when the on-hand inventory reaches $s - N$ during a lead time, a local purchase of $Q + N$ units is made, cancelling the replenishment order that is already placed. They obtained a product form solution for the state probabilities.

Krenzler (2016); Krenzler and Daduna (2015) analyzed a single-server queueing-inventory system with positive service time in a random environment. The service system and the environment interact in both directions. Whenever the environment enters a specific subset of its state space, the service process was assumed to be completely blocked and new arrivals lost. They obtained a necessary and sufficient condition for the existence of a product form steady-state distribution of the joint queueing-environment process.

Baek and Moon (2014) discussed a single-server production inventory system with lost sales in which demands are assumed to occur according to a Poisson process, and service time was assumed to be exponentially distributed. The stocks were replenished by an external order under (r, Q)-policy or an internal production. The internal production process was assumed to be a Poisson process. They derived the stationary joint distribution of the queue length and the on-hand inventory in the product form.

Otten *et al.* (2016) considered a two-echelon production-inventory system with a central supplier connected to production systems (servers) at several locations, each with a local inventory. Demand of customers arrive at each production system according to a Poisson process and was lost if the local inventory was depleted. To satisfy customers demand, a server at the production system takes exactly one unit of raw material from the associated local inventory. The central supplier manufactures raw material to replenish the local inventories, which are controlled by a continuous review base stock policy. They derived stationary distributions of joint queue length and inventory processes in explicit product form. They showed that the global search for the vector of optimal base stock levels can be reduced to a set of independent optimization problems. The explicit form of the stationary distribution enables to get additional structural insights into monotonicity properties and stability conditions.

Shajin *et al.* (2018) proposed an $M/M/1$ queue with an attached inventory system. Customers arrive at the system according to a Poisson process and are served by a single server. The stock is replenished by (s, Q)-policy, which has an exponentially distributed lead time. The service time was exponentially distributed with parameter $\alpha\mu_2$ whenever the inventory level was below $s+1$, otherwise μ_2 $(0 < \alpha \leq 1)$. This was to reduce customer loss on account of the inventory level being zero – assume that customers do not join when the inventory level is zero, thereby leading to a product form solution. An optimization problem was discussed. Also they investigated the optimal value of α by constructing a suitable cost function. In this chapter, the main tool for analysis was the matrix-geometric method.

Shajin *et al.* considered a correlated queueing-inventory system with Markovian arrival of customers, phase type distributed service time and Markovian lead time. Items in each cycle have a common lifetime. Before the realization of this, a purchased item in a cycle can be cancelled in that cycle itself, provided the inventory level has not dropped to zero. Common lifetime and intercancellation time follow independent exponential distributions. They exhaustively analyzed this system. The special case of Poisson arrival of customers and exponentially distributed service time was shown to yield a product form solution, thus extending earlier work to the case of correlated lead time.

A single-server queueing-inventory model with infinite capacity was analyzed in Shajin *et al.*. Customer demands arrived according to a Poisson process, service time and lead time were independent exponentially distributed random variables. When the inventory level enters a specific subset of its state space, new arrivals do not join the system. Optimization problems associated with the model were investigated. A special case was also discussed where, whenever the inventory level enters a complete blocking set, new arrivals do not join and the service process is also completely blocked. In this case, they obtain the system state distribution as the product of the component distributions. The optimal size of the blocking set was also investigated.

Shajin and Krishnamoorthy (2019) considered a queueing-inventory problem with Markovian arrival process and phase-type distributed service time. The lifetime of items was exponentially distributed; all items perish together that means they have a common lifetime. Reservation (purchase) and cancellation of purchased items were permitted as long as the common lifetime was not realized. In addition, overbooking items up to a certain maximum was also permitted. When the system is fully overbooked, waiting customers tend to leave the system. On realization of common lifetime, instantly S items were procured resulting in the commencement of the next cycle. In the particular case of a Poisson arrival process and exponentially distributed service time, they showed that the system admits an asymptotic product form solution. The authors adopted an algorithmic procedure to analyze the general case and for the special case probabilistic arguments.

Baek *et al.* (2018) considered an $M/M/1$ queueing-inventory model where the inventory is measured in continuous units. Customers arrive according to a Poisson process and service time was exponentially distributed. Along with the queue, there was an internal finite storage for the inventory and each service required an exponentially distributed random amount H of inventory from the storage. Therefore, a customer leaves the system with H amount of item at his/her service completion time. The inventory was replenished by an outside supplier with a random lead time under an (s, Q) inventory control policy. They assumed that the customers who arrive during stock-out periods were lost. For this queuing-inventory system, the authors obtained a product form solution.

In addition, there are a few papers on queueing-production inventory where explicit product form solutions were obtained. These are given in section 6.2.2.

6.2.1.2. *Queueing-inventory system with postponed work*

Deepak *et al.* (2004) introduced queues with postponed demands: A service system has a finite buffer and an infinite capacity pool for customers. When the buffer was full, customers join the pool. At the end of service of a customer, the customer from the head of the pool was transferred to the buffer, provided the buffer has less than a preassigned number of customers. In inventory with positive service time, the above notion was borrowed by a few researchers. Below is a brief of such papers.

Sivakumar *et al.* (2012) considered an inventory with postponed demands in discrete-time in which arrival follows a discrete Markovian arrival process; two independent phase-type distributions were introduced for lead time and selection time, respectively. Long run system behavior was investigated using the matrix-geometric method. Several performance measures associated with the system were computed.

Devi *et al.* considered the problem of controlling the selection rates of the pooled customer of a single commodity inventory system with postponed demands. The demands formed a Poisson process. The ordering policy was of (s, Q) type. The ordered items were received after a random time, which was distributed as exponential. Demands that occur during the stock-out period either entered a pool of finite size or left the system according to a Bernoulli distribution. Whenever the on-hand inventory level was positive, customers were selected one-by-one and the selection rate can be chosen from a given set. The problem was to determine a decision rule that specifies the rate of these selections as a function of the on-hand inventory level and the number of customers waiting in the pool at each instant of time to minimize the long-run total expected cost rate. The problem was modeled as a semi-Markov decision problem. The optimal policy was computed using a linear programming algorithm and the results were illustrated numerically.

6.2.1.3. *Multiserver queueing-inventory system*

Multiserver queueing-inventory, in contrast to its single-server counterpart, has not been studied extensively. A few results in this direction could be found in Krishnamoorthy *et al.* (2015a). It is of interest to note that the asymptotic independence established in single-server queueing-inventory could be extended only up to two-server cases (Yadavalli *et al.* 2012; Krishnamoorthy *et al.* 2015a; Nair *et al.* 2015).

Nair *et al.* (2015) was a remarkable deviation in the model proposed. (s, S) control policy was assumed for the system with each inventory acting as a server to the customer and going out with the customer on the latter's service completion epoch. They considered a multiserver Markovian queuing model where each server provides service only to one customer. Arrival of customers was according to a Poisson process, and whenever a customer leaves the system after getting service that server is also removed from the system. Here, the servers are considered as an inventory that will be replenished according to the standard (s, S) policy. The system was modeled as a quasi-birth–death process and an algorithmic analysis performed. The condition for ergodicity of the system, the steady-state solutions and average inventory cycle time were obtained using matrix-analytic methods. They also studied an optimization problem that minimizes the total cost induced by the waiting cost of arrivals, holding cost of the inventory of servers and ordering cost.

Yue *et al.* (2016) considered a multiserver queueing-inventory system with non-homogeneous Poisson arrivals, where the service times and lead times were exponentially distributed. The arrival process of the customers depends on the inventory level. When the on-hand inventory level decreased to zero, one item was replenished immediately with zero lead time. Essentially they adopted the notion of local purchase as discussed in Krishnamoorthy *et al.* (2014). They formulated the system as a quasi-birth–death process and obtain a stability condition of the system.

Then, the authors computed the stationary distribution of the system and some main performance measures, which were used to derive the total expected system cost. Finally, the computation of optimal (s, S) policy was illustrated by numerical examples.

Shajin *et al.* considered a multiserver queueing production inventory system with infinite capacity. Customers arrive according to a non-homogeneous Poisson process and service time follows an exponential distribution. The on-hand inventory decreases by one unit with probability δ at the moment of service completion and with probability $(1 - \delta)$ the customer leaves the system without taking the item at the service completion epoch. An emergency replenishment takes place when the inventory level decreases to zero. The production inventory system adopts an (s, S)-policy where the processing of inventory requires a positive random amount of time. The authors employed the matrix-geometric method to analyze the system. Production time for unit items was exponentially distributed. Important performance measures such as distribution of the production on time and production off time in a cycle, and mean emergency replenishment cycle time were computed.

6.2.1.4. *Queueing systems with perishable inventory*

Perishability in inventory, or for that matter in any item subject to decay, can be classified into the following categories. One classification is decay of individual items (getting decayed one at a time), like eggs in a group or fruits in a basket. In the second type, all items in a packet, like medicines with the same expiry date, become unusable immediately after the expiry date. Perishability of stocked items has great significance in inventory management since stocking large quantities may lead to heavy loss due to decay of these items. On the other hand, lack of stock or limited stock may lead to loss to the system due to unmet demands.

Models of perishable queueing systems with repeated customers were proposed in Melikov *et al.* (2016). Lifetime of stored items were assumed to be finite. Customers were served as long as inventory was available. In the absence of inventory, customers could leave the system forever or retry at a later stage. Restocking time was a positive random quantity and restocking policy was of (s, S) type. The authors developed exact and approximate methods for calculating basic characteristics of the proposed models.

The model of perishable queueing-inventory system with server vacations was studied in Koroliuk *et al.* (2018). At a service completion epoch, if no customer was found to be waiting, the server proceeded for a vacation. If on completion of the vacation, the number of customers waiting in the system was found to be below a threshold value, the server repeated the vacation until the number waiting reached the threshold. Exact and approximate methods are proposed to calculate the characteristics of the system.

A perishable queuing-inventory system with positive service time and delayed feedback was studied in Melikov *et al.* (2018). The authors assumed that the customers either leave the system with/without purchasing an item or join the orbit for decision-making. The control policy adopted was of the (s, S) type. Approximate formulas are developed to calculate the joint distributions and performance measures of the system. High accuracy of the approximate formulas is confirmed by the numerical experiments.

An important assumption in classical queueing-inventory models is that after a service completion inventory level decreases by as many units as were taken away by the customer. But in reality this condition does not always hold, because some customers may refuse to purchase the item after being served. The model with this type of customer was first studied in Krishnamoorthy *et al.* (2015a,b). Later similar models were analyzed in Melikov and Shahmaliyev (2017) and Melikov *et al.* (2017) as well.

An important class of perishable queueing-inventory models includes those in which the server takes a break depending on the current level of inventory and/or the number of orders in the system. Such models are analyzed in Koroliuk *et al.* (2017, 2018) and Melikov *et al.* (2017).

Markov models of perishable inventory systems where some customers were withdrawn after completion of the service without buying the stock were studied in Melikov and Shahmaliyev (2018). They assumed that the service time and filling supply orders were positive random variables. The inventory replenishment size was a variable and depends on the current inventory level. Exact and approximate methods for calculating the joint distribution of the inventory level and the number of orders in the system were developed. Formulas for calculating the main characteristics of the system were proposed and a related optimization problem was solved. The high level of accuracy of the formula was confirmed by numerical experiments. Perishable queueing-inventory system with positive service time and delayed feedback were studied in Melikov *et al.* (2019).

6.2.2. *Production inventory systems*

Benjaafar *et al.* (2010) considered the control of a production-inventory system with impatient customers. The authors showed that the optimal policy could be described using two thresholds: a production base-stock level that determines when production takes place and an admission threshold that determines when orders should be accepted. They described an algorithm for computing the performance of the system for any choice of base-stock level and admission threshold. In a numerical study, they compared the performance of the optimal policy against several other policies.

Krishnamoorthy and Viswanath (2011) is the first reported work on a production inventory system with positive service time; in that paper, the time for producing each item is assumed to follow a Markovian production scheme. The customer arrival process follows a Markovian arrival process and the service time to each customer has a phase-type distribution. The analysis is purely algorithmic in that study. This is followed by (Krishnamoorthy and Viswanath 2013) analyzing an (s, S) production inventory system where again the processing of inventory requires some positive amount of time. All underlying distributions were assumed to be exponential. They obtained a product form solution for the system state (three-dimensional Markov chain) using the matrix-geometric approach along with probabilistic arguments.

Vidyadhar and Keqi (2012) considered a production-inventory system where the production and demand rates were modulated by a finite state continuous-time Markov chain. Whenever the inventory position dropped to a reorder point r, an order of size q was assumed to be placed. They considered the case of stochastic lead times, where the lead times were i.i.d. exponential (μ) random variables and orders may or may not be allowed to cross and derive the distribution of the inventory level and analyze the long run holding, backlogging and ordering cost rate per unit time.

Batch demands were very common in supply chain management. Readers may refer to previous works (Saidane *et al.* 2013; Axsater 2013) and references there in for details.

Baek and Moon (2014) discussed an $M/M/1$ production inventory system with lost sales. The stocks are replenished by an external order under (r, Q)-policy or an internal production. The internal production process is assumed to be a Poisson process. They obtained a product form solution for the system state using the matrix geometric-approach.

Yu and Dong (2014) considered a two-stage production lot sizing problem, which used an inventory system with random demand arrival and proposed a numerical approach to solve the problem. For recent research on production-inventory system, readers may refer to previous works (Pal *et al.* 2012; Sarkar 2012; Wang *et al.* 2013; Krishnamoorthy *et al.* 2016c) and (Otten *et al.* 2016).

Zare *et al.* (2017) considered a two-echelon production-inventory system with a single manufacturer operating from a warehouse and a distribution center using the continuous review (R, Q) inventory policy where transportation times between the two echelons were generally distributed. They derived the optimal reorder point at the distribution center and proposed a heuristic to approximate the optimal base-stock level in the warehouse as well as the batch order size at the distribution center.

Hum *et al.* (2018) considered supply chains with multiple stages of serial or network structure. The supply chains are endogenous in the sense that they involve queues because each order's lead time is dependent on the orders already in the system. The authors define supply chain responsiveness as the probability of fulfilling customer orders within a promised lead time and study the problems of measuring and optimizing supply chain responsiveness using queueing network models. They first considered a single-server multistage serial supply chain and derived a closed-form expression for the fulfillment time distribution. For the multiserver multistage problem, the closed-form evaluation of the fulfillment time distribution becomes intractable due to the dependency of the lead times in different stages. They circumvent this difficulty by proposing a novel FCFS discipline, which enables a closed-form analysis. For the multiserver multistage Jackson-type supply chain network, to enable analysis, convert the system into an equivalent single-server single-stage system with state-dependent rates. For each case, they present detailed numerical examples for both measurement and the optimization of supply chain responsiveness.

Caceres *et al.* (2018) considered a multiperiod production inventory problem with two suppliers, where demand sizes and supplier lead time were stochastic and correlated. A discrete-time, single-item inventory system was considered, where inventory levels were reviewed periodically and managed using a base-stock policy. At the end of each period, a replenishment order was placed, which entered a queue at the buffer stage and was consequently forwarded to the first available supplier. They presented a mathematical model of this inventory system and determined optimal safety stock levels for it, in closed-form, using matrix-analytic techniques and the properties of phase-type distributions. To account for the effect of order crossovers, which occur whenever replenishment orders do not arrive in the sequence in which they were placed, the inventory shortfall distribution was analyzed.

Yue and Qin (2019) considered a production inventory system with production vacations. Customers arrive according to a Poisson process. The single production facility produces items according to an (s, S) policy and it took a vacation of random duration once the inventory level became S. It was assumed that all arriving customers were lost during the stock-out period. They first derived the stationary joint distribution of the queue length and the on-hand inventory level in product form. Then, they computed explicitly some performance measures and developed a cost function. Finally, some numerical results were presented.

6.2.3. *Multicommodity queueing-inventory system*

There are a few work reported in queueing-inventory, which discuss models involving two or more distinct commodities.

A two-commodity inventory system with a single server was considered in Benny *et al.* (2018). Customers ask for commodity$-1/2$ or both with specified probabilities at the time when taken for service. If the demanded item(s) is not available, the customer leaves the system. Thus, the resulting infinitesimal generator of the Markov chain turned out to be of the $GI/M/1$ type. The authors employed the matrix-analytic method to exhaustively analyze this system. The control policy adopted for the ith commodity is (s_i, S_i) for $i = 1, 2$. Lead time for commodity i is exponentially distributed with parameter μ_i. Arrival process is Poisson and service time is exponentially distributed with parameter depending on the type of commodity asked for. This could be easily extended to n-commodity system for $n \geq 3$.

6.2.3.1. *Queueing-inventory models with multiclass customers*

A model involving two distinct classes (priority) of customers and the necessity to use items for service provisioning was considered in an interesting paper by (Zhao and Lian 2011) as a queueing-inventory model. In this work, it was assumed that both type of customers queued up in two distinct waiting lines. The high-priority customers were considered to have non-preemptive priority over the low-priority ones. Arrival of customers follow a marked Poisson process and service time have distinct exponential distributions, depending on the priority of the customer in service. Wang (2015) extend (Zhao and Lian 2011) to a marked Markovian flow ($MMAP$) of customers. Only high-priority customers could join the queue, whereas the low-priority customers were to join an orbit of infinite capacity and retry to access the server.

Isotupa and Samanta (2013) analyzed a lost sales (s, Q) inventory system with two types of customers and stochastic lead times. Demands from each type of customer arrive according to two independent Poisson processes. A comparative study of the average cost rate for the cases where there is differentiation between service to customers based on the type of customer and the case where there is no differentiation between customers was carried out by using the concepts of rationing. They provided numerical illustrations where differentiation between customers yield lower cost and lower shortage rates for both types of customers.

Wang *et al.* (2013) considered a production service system consisting of one manufacturing facility and one service center. The finite capacity manufacturing facility produces two items to stock in a warehouse. Two classes of customers arrive at the service center. With each service completion, one unit of dedicated item in warehouse depletes exactly by one unit. To control such a system, the manager faces two scheduling problem for manufacturing facility and service center simultaneously. The first is that he/she must decide when to start or stop production; moreover, if production is warranted, which product should be produced. A similar scheduling problem for a service center was also investigated. If inventory is available and customers are waiting in queue for both types of items, the manager must decide which order should be served first. The system is assumed to be completely flexible, namely the setup time of both manufacturing facility and service center can be ignored. Customer demands of both types arrive according to independent Poisson processes. The production time of product is exponentially distributed. Inventory costs were calculated per unit time for each unit of item kept in stock. Customers of each class require exactly one item in the inventory for service and the service time is assumed to be exponentially distributed. The authors proved that the optimal service scheduling policy is an extension of the well-known optimal service scheduling policy for classical service system.

Isotupa (2015) analyzed a continuous review, lost sales $(S-1, S)$ inventory system with two demand classes high priority and low priority. The author compared two different policies – one where the two classes of customers are treated exactly alike and the other where a threshold rationing policy is used. She proved that under certain conditions there is a suboptimal rationing policy, which yields a lower cost for the supplier and higher service levels for both the high-priority and low-priority customers than the optimal policy where the two customers are treated alike.

6.2.4. *Retrial queues with inventory*

Retrial queues was introduced by Kosten (1947); until that time it was assumed that customers who could not join the system due to capacity restriction were lost forever. Instead, Kosten (1947) considered such customers as not actually lost; instead they keep on repeating their demand until the purpose was met. The papers by Anisimov (1999) and Anisimov and Artalejo (2001) could be extended to queueing-inventory set up by suitably introducing inventory into the problems discussed.

The concept of retrial in inventory was introduced by Artalejo *et al.* (2006). The authors assumed Poisson arrival of demands from outside (primary customers), negligible service time and exponentially distributed lead time under (s, S) policy. At an arrival epoch, if the inventory was found to be empty, such customers join an orbit of infinite capacity and repeat their attempts to get the inventory. The retrial rate, when n customers are in the orbit is n times the individual retrial rate (this is called the linear retrial policy). Unsatisfied demands were assumed to return to orbit. The authors derived condition for system stability. Under this condition, they computed the system state probability distribution algorithmically. In addition to deriving several performance characteristics of the system, they also illustrated these through numerical experiments. For such a system, (Ushakumari 2006) provided an exact solution for the long run system state probability distribution. Using this, she exhaustively analyzed the system.

Sivakumar (2011) introduced the concept of server vacation in retrial inventory system. He assumed distinct, but mutually independent exponential distributions for interdemand times, lead times, interretrial times and server vacation duration. He adopted a multiple vacation policy. Krishnamoorthy *et al.* (2012) discussed a queueing-inventory system with server interruption and retrial of unsatisfied customers from the orbit with infinite size.

"Search of customers in orbit" was introduced in retrial queues by (Artalejo *et al.* 2002). This concept was exploited in the paper by (Cui and Wang 2013) for a queueing-inventory system involving registration of customers proceeding to orbit.

Nair and Jacob (2014) analyzed an (s, S) inventory system with positive service time and retrial of demands by considering the inventory as servers of a multiserver queuing system. Arrival of demands was assumed to follow a homogeneous Poisson process and exponential service time distribution. On each service completion, the number of demands in the system, as well as the number of inventories (servers), got reduced by one. When all servers are busy, new arrivals join an orbit from which they try to access the service facility at an exponential rate. Using the matrix-geometric method, the authors analyze the long run behavior of the system under equilibrium.

Vijaya Laxmi and Soujanya (2015) analyzed a single-server queueing system in which demands arrive according to a Poisson process. They considered the idea of an orbit of infinite size for sending the demands that occur in the server breakdown period. The matrix-geometric solution was obtained for the steady-state probability distribution of this model. The system performance measures and cost analysis were also carried out.

Padmavathi *et al.* (2015) considered a continuous review stochastic (s, S) inventory system with Poisson demand and exponentially distributed delivery time. The demands that occur during the stock-out period or during the server vacation period enter into an orbit of infinite size. These orbiting demands retry to get satisfied by sending out signals so that the time durations were exponentially distributed. They considered two models that differ in the way that the server goes on vacation. The joint probability distribution of the inventory level, the number of demands in the orbit and the server status was obtained in the steady-state.

Krishnamoorthy and Shajin (2006) succeeded in extending product form solutions to retrial queueing-inventory systems under stringent conditions as described below. In this paper, as in Neuts and Rao (1990), the authors assumed that primary customers went directly to the orbit and from there only they retry. Further primary arrivals were assumed to not join the system when the server was found busy and also when inventory level remained zero. Under these conditions, the authors derived a product form solution for the system state.

Krishnamoorthy and Shajin (2017) considered a single-server retrial queueing-inventory system in which customers join directly to the orbit according to a Markovian arrival process. Service time of customers are independent and identical distributed phase-type random variables. Interretrial times are exponentially distributed with parameter $n\eta$ when n customers are in the orbit and η is the individual retrial rate. Unsuccessful retrial customers tend to leave the system (impatience) with positive probability. In addition, the server searches orbital customers for the next service with state-dependent probability, immediately on current service completion. The authors employed the matrix-geometric procedure to analyze the system. Several numerical experiments were provided for illustrating the performance of the system.

A single-channel queueing-inventory system with feedback and finite queue of primary customers (p-customers) with Poisson arrival was investigated in Amirthakodi and Sivakumar (2015). If the queue is full at the arrival of a p-customer, then it leaves the system without being served. The p-customer either joins the orbit for future call or leaves the system after the service completion. The orbit has finite capacity. Each customer in orbit retries for service, independently of others, after an exponentially distributed time. The system serves the repeated customers ($r-$customers) if there are no $p-$customers and/or the inventory level is zero. Service times for both types of customers are exponentially distributed, but with different parameters. Non-preemptive service policy is assumed, so that if an $r-$customer is being served at the moment of arrival of a $p-$customer, the ongoing service is not interrupted. Repetitive orbit re-joining is considered as well, that is after the service completion, an $r-$customer re-joins the orbit or leaves the system. Laplace–Stieltjes transform of waiting time was derived for both types of customers. Linear algebraic methods were employed when investigating the system. Feedback queueing-inventory models are also discussed in other works (Amirthakodi et al. 2015; Krishnamoorthy and Shajin 2017; Krishnamoorthy 2015a; Suganya and Amirthakodi 2018).

Yadavalli *et al.* (2017) considered a continuous review (s, S) inventory system at a service facility with finite homogeneous sources of demands and retrial. The lifetime of each item was assumed to be exponential. The service may get interrupted according to a Poisson process and it restarts after an exponentially distributed time. If the server is idle at the time of arrival of a customer and the inventory level is positive, then the service begins immediately. After the completion of regular service, a customer may either abandon the system forever or demand for a second service from the same server, which was multioptional. If any arriving customer finds that the server was busy or inventory level was zero, he/she either enters into the orbit with probability p or balks with probability $1 - p$. The stationary distribution of the number of customers in the system, server status and the inventory level was obtained by the matrix method. The Laplace–Stieltjes transform of the waiting time of a tagged customer was derived.

Manikandan and Sajeev (2017) considered $M/M/1/1$ queueing inventory system with retrial of unsatisfied customers. Arrivals taking place when the server is busy proceed to an orbit of infinite capacity. From the orbit, the head of the queue alone retires to access the server. Failed attempts to access an idle server with positive inventory results in the retrial customer returning to orbit. The interretrial times were independent identically distributed exponential random variables, irrespective of the number of customers in the orbit. Primary customers, who encounter an idle server without stock at its arrival epoch, leave the system for ever. They computed the condition for stability and then employed an algorithmic approach for the computation of the system steady-state probability. They derived the expected waiting time of a customer in the orbit, distribution of the time until the first customer goes to orbit and also the probability of no customer going to orbit in a given interval of time.

6.2.5. *Queues requiring additional items for service*

As stated in the introduction, in certain queueing models considered recently, the authors make the explicit assumption of the requirement of additional items for service. In telecommunication, free channels are required for message transmissions. In hospitals, to conduct certain surgical operations, special devises are necessary. In the absence of such items, certain services cannot be rendered. The work cited below gives concrete situations that need additional items for service. Either these items can be externally procured or could be prepared by the server himself. If done by the server, he/she may do so during the time when no customer is in the system or may do so while providing service. The first reported work of this kind seems to be by (Kazimirsky 2006). He studied a $BMAP/G/1$queue with infinite buffer and service time distribution depending on the number of processed items: When customers are absent from the system, the server begins to produce items that are put into the storehouse until the storehouse capacity is reached or a group of customers arrives. When a group of customers arrives, the item processing is stopped and the service of the customers begins. Customer service time depends on the amount of items in the storehouse at the beginning of the service. After the service of the customer is completed, they depart from the system and the service began with a non-empty storehouse, the number of items inside the storehouse decreases by one unit at the service completion epoch. In this queueing model, systems with the possibility of preliminary service are considered. An efficient algorithm for calculating the stationary queue length distribution was proposed and the Laplace–Stieltjes transform of the sojourn-time was derived. He also proved Little's law and an associated optimization problem was analyzed.

Hanukov *et al.* (2017), without noticing Kazimirsky (2006), considered a much simpler model. However, they examined the stochastic decomposition of the system state, which could not be arrived at in Kazimirsky (2006) due to the high complexity of the problem he considered. Hanukov *et al.* (2017) studied a single-server queue in which the service consists of two independent stages. The first stage was generic and can be performed even in the absence of customers, whereas the second requires the customer to be present. When the system was empty of customers, the server produces an inventory of the first-stage (preliminary) services, which is used to reduce customers overall sojourn times. They formulated and analyzed the resulting queueing-inventory system to derive its steady-state probability.

Divya *et al.* (2018) considered an extension of Hanukov *et al.* (2017) wherein they introduced a control policy (N- policy) for beginning a new busy period. During this idle time, the server keeps producing items for service. The number of such items accumulated is not allowed to exceed a threshold L. They assumed that the customers arrive at a single-server queueing system according to a Markovian arrival process. When the system is empty, the server goes on vacation and produces inventory for future use during this period. The inventory processing time follows phase-type distributions. These are required for the service of customers – one for each customer. The server returns from vacation when there are N customers in the system. The service time follows two distinct phase-type distributions depending on whether the processed items are available, or not, at the service commencement epoch. They analyzed the distributions of time until the number of customers hit N, the inventory level reached L, idle time and also the distribution of the number of inventory processed before the arrival of the first customer.

Divya *et al.* further extended the above work by assuming customer impatience. They investigated numerically the resulting strategic queue in equilibrium. In Divya *et al.* (2018), the authors employ algorithmic analysis for studying the system.

Baek *et al.* (2017) considered a single-server queueing system with a marked Markovian arrival process of heterogeneous customers. Type-1 customers have limited preemptive priority over type-2 customers. The system has an infinite buffer for type-2 customers and no buffer for type-1 customers. There was also a finite buffer (stock) for consumable additional items (semi-products, half-stocks, etc.), which arrive according to the Markovian arrival process. Service of a customer requires a fixed number of consumable additional items depending on the type of the customer. The service time has a phase-type distribution depending on the type of the customer. Customers in the buffer are impatient and may leave the system without service after an exponentially distributed amount of waiting time. Aiming to minimize the loss probability of type-1 customers and maximize throughput of the system, a threshold strategy of admission to service of type-2 customers was presented. Service of a type-2 customer can start only if the server was idle and the number of consumable additional items in the stock exceeds the fixed threshold. Stationary distributions of the system states and the waiting times are computed.

Shajin *et al.* (2019) extended the above work to the case of retrial by low-priority customers.

6.2.5.1. *Miscellaneous items*

Rashid *et al* (2015) presented a mathematical model for an inventory control system in which customers demands and suppliers service time were assumed to be stochastic parameters. They analyzed the model using queuing theory for a single item.

Marand *et al.* (2017) studied joint inventory control and pricing decisions for a service-inventory system. The service-inventory system was modeled as an $M/M/1$ queue in which the customer arrival rate was price dependent. The inventory of an individual item was continuously reviewed under an (r, Q) policy. The replenishment lead times of the inventory were exponentially distributed. Furthermore, customers arriving during stock-out periods were assumed to be lost.

Chakravarthy *et al.* (2017) introduced batch service into queueing-inventory models. The number of customers waiting was considered level. The service batch size could be any integer $a, a + 1, ...b$. The server stays idle when quorum for service was not available at a service completion epoch. Further for service to start, availability of items in inventory was also required. The authors exploited the quasi-Toeplitz structure of the infinitesimal generator of the underlying Markov chain to exhaustively analyze the system.

Yue *et al.* (2018) studied an $M/M/1$ queueing-inventory system with batch demands. Customers arrive in the system according to a compound Poisson process, where the size of the batch demands for each arrival was a random variable that follows a geometric distribution. The inventory was replenished according to the standard (s, S) policy. The replenishment time follows an exponential distribution. Two models were considered. In the first model, if the on-hand inventory was less than the size of the batch demands of an arrived customer, the customer takes away all the items in the inventory and a part of the customer's batch demands was lost. In the second model, if the on-hand inventory was less than the size of the batch demands of an arrived customer, the customer leaves without taking any item from the inventory. For these two models, the authors derived the condition for the existence of stationary distribution.

Wang *et al.* (2017) analyzed priority multiserver retrial queueing-inventory with the Markovian arrival process generated by a single or dual sources and exponential service times where both queueing and orbit spaces were assumed finite. The generalized stochastic Petri net models for the non-identical service rates scenario was employed. The authors established the equivalence between the two methods when service rates were identical. The generalized stochastic Petri net models were then used to investigate the impact of parameter change, including service heterogeneity on some performance measures of interest.

Concepts of reservation, cancellation and common lifetime were first introduced in queueing-inventory system by (Krishnamoorthy *et al.* 2016b) in the retrial set up; their model turns out to be of a $GI/M/1$ type since on realization of common lifetime, all waiting customers leave the system (Krishnamoorthy *et al.* 2016).

For the case of classical queueing-inventory with reservation, cancellation and common life, (Krishnamoorthy *et al.* 2016) and (Shajin *et al.* 2016) show that a product form solution exists when a suitable blocking set was introduced.

Shajin (2019) extended findings in Krishnamoorthy *et al.* (2016a,b) to overbooking for several units of time (could be days) ahead. The time taken to realize one unit time is called the common lifetime for the least among the several units of time cases considered, the second least among these will then have just two unit of time for life completion. Thus $K \to K-1 \to ... \to 2 \to 1 \to 0$, on K going to $K-1$ (on realization of one unit of time) K units of time away from the current time gets introduced. The unit of time is assumed to be phase-type distributed. Customer arrival forms a Markovian arrival process. A single-server serves inventory (one or customers) according to the first-in first-out rule for customers joining an infinite capacity waiting line. They ask for a reservation for 1 unit away, 2 units away, ..., K units away. Depending on availability the reservation or even overbooking is given. If the item was available for the requested day, then it was served to the customer. Overbooking up to a maximum fixed level is permitted for each unit of time. If for the required day such an alternative is available, the customer is served the same. If this too was not available, he/she was asked to give alternatives. If all his/her alternatives cannot be met, he/she is provided with the time (day) that it was available. If that too was not available, then he/she will have to wait; in the last case, all customers will have to wait. Service time follows independent and identically distributed exponential distribution. In addition to reservation, cancellation of a reserved item is permitted. Interoccurrence time for this was an independent exponentially distributed random variable with the appropriate parameter. At the common lifetime of the "one time unit ahead", if there are overbookings, these are transferred to the "now" K units time away inventory addition. The special case of Poisson arrival, coupled with blocking arrivals when all of $1, 2, ..., K$ unit time inventory are overbooked, is shown to yield a product form solution. For this case, an appropriate cost function is constructed and its properties are investigated numerically.

6.2.6. *Queueing-inventory: some work in progress and suggestions for future studies*

1) Consider a service provider with a single server. Customers arrive at the service station according to some probabilistic law. Service time for customers requiring only "essential items" has a specific distribution; those requiring optional items in addition to essential items have distinct service times depending on the type and number of optional items. These distributions are assumed to be independent. Further service times of customers requiring the same type of service are mutually independent.

Specifically assume that items labeled $I_1, ..., I_n$ are essential for a service, whereas items $O_1, ..., O_m$ are optional. The non-availability of the latter does not prevent the system from providing service. The items $I_1, ..., I_n$ and $O_1, ..., O_m$ are not abundantly available. As a result, the service system has to device appropriate mechanism to ensure smooth service. Further demand for these items need not materialize instantly. The lead time for different items can be distinct, especially if items are not supplied by the same vendor. Thus the problem becomes quite complex. The objectives are as follows:

a) maximization of system reliability – this amounts to computation of probability of "all essential items" available when there is at least one customer in the system;

b) the optimal values of order quantities of $I_1, ..., I_n$ and $O_1, ..., O_m$.

This requires the setting up of a cost function involving holding cost of customers as well as items in inventory, in addition to other routine costs. The main objective is to minimize the expected long run average cost per unit time.

2) Consider a single-server queueing-inventory system with the facility for advanced reservation for the next K time units (days) ahead. Customers arrive according to a probabilistic law demanding one unit of item each from the inventory of any day. It requires a random amount of time to serve each customer. The maximum of items any day can have is S. The inventoried items in the first unit of time have a common lifetime, which means that all perish together on realization of this time. The common lifetime is also a random variable. On realization of common lifetime of the first time unit, the second time unit gets shifted to the first, third to second and so on. Intercancellation time follows a random variable. The customers at the head of the queue demand an item for the ℓth day $(1 \leq \ell \leq K)$ with probability p_ℓ with $\sum_{\ell=1}^{K} p_\ell = 1$. If the demanded item is not available, the customer is offered item for the next immediate available day; he/she may reject this offer with probability p and leave the system or accept it with probability $1 - p$.

3) Analyze an (s, S)-production inventory system where the processing of inventory requires a positive random amount of time. As a result, a queue of demands is formed. Demand process is assumed to be according to a probabilistic law; duration of each service and time required to add an item to the inventory when the production is on are independent, non-identically distributed random variables. Raw materials R_1, R_2 are used for production. Both raw materials are required for production but only one unit of each item is required at a time. When the inventory level depletes to s, the production process is switched on provided both raw materials are available. The centers of raw materials adopts (r_i, R_i)-policy $(i = 1, 2)$. It takes a random amount of time for the raw materials to reach the center. No customer joins the queue when

the inventory level is zero. We investigate conditions under which this system admits stochastic decomposition.

4) A single commodity queueing-inventory with "offer" in random environment: The "offer" can be in terms of discount or other incentives to customers who purchase the item. The purpose is to boost sales. The offers in distinct environments are different. When in some environments there may not be any offer. Customers do not join the system when the inventory level is zero. Assume that environments do not change during this period of time. The duration of time the system stays in a particular environment follows exponential distribution. There are n distinct environments. (s, S) policy is followed, Lead time is exponentially distributed. The long run behavior is analyzed under the assumption of system stability. The system is seen to admit an asymptotic product form solution. An optimization problem is constructed to compute the values of s and S that minimize the average system cost. How these are affected by the environment is also investigated.

5) Queueing-inventory system with reservation of inventory items, possibility of its cancellation and overbooking for all future times up to maximum of K time frames ahead was studied in Shajin (2019). For future investigation, several possible variants of the model studied in this paper are as follows: (i) Customers asking for reservation in a specified time frame, finding no inventory readily available, may opt to leave the system with a specified probability or with complementary probability ask for other alternatives, failing which they leave the system forever. (ii) During lean season (very low traffic), sometimes airlines/trains/bus cancel schedules of certain days and accommodate passengers in departures of time frames of 1 or 2 days ahead or after. Those who do not opt for these are fully refunded. (iii) The case of multiserver is also of interest.

Literature on discrete-time inventory with positive service time is quite scarce. This seems to be an area for future work. In most investigations reviewed above, the first-order Markov chain modeling is adopted. Higher order dependence arises in many contexts. Investigating such process may yield useful results.

6.3. Acknowledgment

The authors thank the reviewers and Prof. V. Anisimov for the critical comments and suggestions.

6.4. References

Albrecher, H., Boxma, O., Essifi, R., Kuijstermans, R. (2017). A queueing model with randomized depletion of inventory. *Probab. Engrg. Inform. Sci.*, 31(1), 43–59.

Amirthakodi, M., Radhamani, V., Sivakumar, B. (2015). A perishable inventory system with service facility and feedback customers. *Ann. Oper. Res.*, 233, 25–55.

Amirthakodi, M., Sivakumar, B. (2015). An inventory system with service facility and finite orbit for feedback customers. *OPSEARCH*, 52, 225–255.

Anbazhagan, N., Kumaresan, V. (2012). Coordinated and joint ordering policies for two commodity inventory system. *Iran J. Oper. Res.*, 3(2), 94–105.

Anisimov, V.V. (1999). Averaging methods for transient regimes in overloading retrial queueing systems. *Math. Comput. Model*, 30, 65–78.

Anisimov, V.V., Artalejo, J.R. (2001). Analysis of Markov multiserver retrial queues with negative arrivals. *Queueing Syst.*, 39, 157–182.

Anoop, N.N., Jacob, M.J. (2013). An $M/M/(S,S)$ queueing-inventory system with exponential lead time and lost sales. *Math. Sci. Int. Res.*, 2(2), 276–278.

Artalejo, J.R., Joshua, V.C., Krishnamoorthy A. (2002). An $M/G/1$ retrial queue with orbital search by the server. In *Advances in Stochastic Modelling*. Artalejo, J.R., Krishnamoorthy, A. (eds). Notable Publications, Inc., NJ, 41–54.

Artalejo, J.R., Krishnamoorthy, A., Lopez-Herrero, M. J. (2006). Numerical analysis of (s,S) inventory systems with repeated attempts. *Ann. Oper. Res.*, 141, 67–83.

Axsater, S. (2013). Initiation of an inventory control system when the demand starts at a given time. *Int. J. Prod. Econ*, 143(2), 553–556.

Babai, M.Z., Jemai, Z., Dallery, Y. (2010). Analysis of order-up-to-level inventory systems with compound Poisson demand. *Eur. J. Oper. Res.*, 210(3), 552–558.

Babak H., Alireza H. (2011). A queueing-inventory system with repair center for defective items and one-for-one ordering policy. *J. Ind. Syst. Eng.*, 5(4), 230–239.

Baek, J.W., Bae, Y.H., Lee, H.W., Ahn, S. (2018). Continuous-type (s,Q)-inventory model with an attached $M/M/1$ queue and lost sales. *Perform. Evaluation*, 125, 68–79.

Baek, J. W., Dudina, O., Kim, C. (2017). A queueing system with heterogeneous impatient customers and consumable additional items. *Int. J. Appl. Math. Comput. Sci.*, 27(2), 367–384.

Baek, J.W., Moon, S.K. (2014). The $M/M/1$ queue with a production-inventory system and lost sales. *Appl. Math. Comput.*, 233, 534–544.

Baek, J.W., Moon S.K. (2016). A production-inventory system with a Markovian service queue and lost sales. *J. Korean Statist. Soc.*, 45, 14–24.

Benjaafar, S., Gayon, J.P., Tepe, S. (2010). Optimal control of a production inventory system with customer impatience. *Oper. Res. Lett.*, 38(4), 267–272.

Benny B., Chakravarthy S. R., Krishnamoorthy A. (2018). Queueing-inventory system with two commodities. *J. Indian Soc. Probab. Stat.*, 19, 437–454.

Berman, O., Kaplan, E.H., Shimshak, D.G. (1993). Deterministic approximations for inventory management at service facilities. *IIE Trans.*, 25(5), 98–104.

Berman, O., Kim, E. (1999). Stochastic models for inventory managements at service facilities. *Commun. Statist. Stochastic Models,* 15(4), 695–718.

Berman, O., Sapna, K.P. (2002). Optimal service rates of a service facility with perishable inventory items. *Naval Res. Logist.*, 49, 464–482.

Betts, J.M. (2014). Calculating target inventory levels for constrained production: A fast simulation-based approximation. *Comput. Oper. Res.*, 49, 18–27.

Bhuvaneshwari, M. (2018). Finite source inventory system with negative customers. *Int. J. Math. Trends Technol.*, 56(6), 419–429.

Boxma, O., Essifi, R., Janssen, A.J.E.M. (2016). A queueing/inventory and an insurance risk model. *Adv. Appl. Probab.*, 48(4), 1139–1160.

Boute, R.N., Disney, S.M., Lambrecht, M.R., Houdt, B.V. (2014). Coordinating lead times and safety stocks under autocorrelated demand. *Eur. J. Oper. Res.*, 232(1), 52–63.

Caceres, H., Yu, D., Nikolaev, A. (2018). Evaluating shortfall distributions in periodic inventory systems with stochastic endogenous demands and lead-times. *Ann. Oper. Res.*, 271, 405–427.

Chakravarthy, S.R. (2001). The batch Markovian arrival process: A review and future work. In *Advances in Probability Theory and Stochastic Process: Notable Publications,* Krishnamoorthy, A., Raju, N., Ramaswami, V. (eds), New Jersey, 21–49.

Chakravarthy, S.R., Maity, A., Gupta, U.C. (2017). An (s, S) inventory in a queueing system with batch service facility. *Ann. Oper. Res.*, 258(2), 263–283.

Chakravarthy, S.R., Shajin, D., Krishnamoorthy, A. (2019). Infinite server queueing-inventory models, (Accepted in *J. Indian Soc. Probab. Stat. 2019*).

Chen, H., Zhou, Z., Liu, M. (2012). A queuing-inventory system with two classes demand and subject to selected service. *J. Inf. Comput. Sci.*, 9(11), 3081–3089.

Choi K.H., Yoon B.K. (2016). A survey on the queueing inventory systems with phase-type service distributions. *QTNA '16*, December 13–15, Wellington, New Zealand.

Cui, J., Wang, J. (2013). A queueing-inventory system with registration and orbital searching processes. In *Proceedings of 2nd International Conference on Logistics, Informatics and Service Science*, Zhang, Z., Zhang, R., Zhang, J. (eds). *LISS 2012*, Springer, Berlin, Heidelberg. http://doi.org/10.1007/978-3-642-32054-5 62.

Deepak, T.G., Joshua, V.C., Krishnamoorthy A. (2004). Queues with postponed work. *TOP – Spanish J. Statist. Oper. Res.*, 12, 375–398.

Devi, P.C., Sivakumar, B., Krishnamoorthy, A. Optimal control policy of an inventory system with postponed demand. *RAIRO-Oper. Res.*, 50(1), 145–155.

Divya, V., Krishnamoorthy, A., Vishnevsky, V.M. (2018). On a queueing system with processing of service items under vacation and N-policy. *Data Communication & Computer Networks*, CCIS 919, 43–57.

Divya, V., Vishnevsky, V.M., Kozyrev, D., Krishnamoorthy, A. On a queueing system with processing of service items under vacation and N-policy with impatient customers (communicated).

Duan, Y., Li, G., Tien, J.M., Huo, J. (2012). Inventory models for perishable items with inventory level dependent demand rate. *Appl. Math. Model.*, 36, 5015–5028.

Geetha, R.M., Elango, C. (2012). Markov process for service facility systems with perishable inventory and analysis of a single server queue with reneging. *Stoch. Model: Int. J. Comput. Appl.*, 44(15), 18–23.

Haji, R., Haji, A., Saffari, M. (2011). Queueing inventory system in a two-level supply chain with one-for-one ordering policy. *J. Ind. Syst. Eng.*, 5(1), 52–62.

Hanukov, G., Avinadav, T., Chernonog, T., Spiegel, U., Yechiali, U. (2017). A queueing system with decomposed service and inventoried preliminary services. *Appl. Math. Model.*, 47, 276–293.

Hum, S.H., Parlar, M., Zhou, Y. (2018). Measurement and optimization of responsiveness in supply chain networks with queueing structures. *Eur. J. Oper. Res.*, 264(1), 106–118.

Isotupa, K.P.S. (2015). Cost analysis of an $(S-1, S)$ inventory system with two demand classes and rationing. *Ann. Oper. Res.*, 233(1), 411–421.

Isotupa, K.P.S., Samanta, S.K. (2013). A continuous review (s, Q) inventory system with priority customers and arbitrarily distributed lead times. *Math. Comput. Model.*, 57, 1259–1269.

Jeganathan, K. (2014). Perishable inventory system at service facilities with multiple server vacations and impatient customers. *J. Stat. Appl. Pro. Lett.*, 3(3), 63–73.

Jeganathan, K. (2015a). Linear retrial inventory system with second optional service under mixed priority service. *TWMS J. App. Eng. Math.*, 5(2), 249–268.

Jeganathan, K. (2015b). A stochastic inventory system with two types of services and a finite populations. *Int. J. Math. Appl.*, 3(1), 73–81.

Jeganathan, K. (2015c). A single server perishable inventory system with N additional options for service. *J. Math. Model.*, 2(2), 187–216.

Jeganathan, K. (2015d). Finite source Markovian inventory system with bonus service for certain customers. *Int. J. Adv. Appl. Math. Mech.*, 2(3), 134–143.

Jeganathan, K. (2016). Analysis of two queues in parallel with mixed priority service and a finite population. *Int. J. Oper. Res. /Nepal*, 5(1), 15–43.

Jeganathan, K., Anbazhagan, N., Kathiresan, J. (2013). A retrial inventory system with non-preemptive priority service. *Int. J. Inf. Manag. Sci.*, 24(1), 57–77.

Jeganathan, K., Kathiresan, J., Anbazhagan, N. (2016). A retrial inventory system with priority customers and second optional service. *OPSEARCH*, 53(4), 808–834.

Jeganathan, K., Reiyas, M. A., Padmasekaran, S., Lakshmanan, K. (2017). An $M/E_K/1/N$ queueing-inventory system with two service rates based on queue lengths. *Int. J. Appl. Comput. Math.*, 3(1), 357–386.

Jeganathan, K., Reiyas, M.A., Padmasekaran, S., Lakshmanan, K. (2018). Two heterogeneous servers queueing-inventory system with multiple vacations and server interruptions. *Adv. Model. Optim.*, 20(1), 113–133.

Jose, K.P., Rejitha, K.R. (2017). Two inventory models with non-identical modes of service. *Int. J. Pure Appl. Math.*, 116(4), 929–948.

Jose, K.P., Salini S.N. (2017). Analysis of two production inventory systems with buffer, retrials and different production rates. *J. Ind. Eng. Int.*, 13(3), 369–380.

Karzana, G., Razamin, R., Nerda, Z.Z. (2017). Developing a $M/G/C - FCFS$ queueing model with continuous review (R, Q) inventory system policy in a cement industry. *J. Intell. Fuzzy Syst.*, 32(6), 4059–4068.

Kazimirsky, A.V. (2006). Analysis of $BMAP/G/1$ queue with reservation of service. *Stoch. Anal. Appl.*, 24(4),703–718.

Koroliuk V.S., Melikov A.Z., Ponomarenko L.A., Rustamov A.M. (2017). Asymptotic analysis of the system with server vacations and perishable inventory. *Cybern. Syst. Anal.*, 53, 543–553.

Koroliuk, V.S., Melikov, A.Z., Ponomarenko, L.A., Rustamov, A.M. (2018). Models of perishable queueing-inventory systems with server vacations. *Cybern. Syst. Anal.*, 54(1), 31–44.

Kosten, L. (1947). On the influence of repeated calls in the theory of probabilities of blocking. *De Ingenieur*, 59(1), 1–25.

Kouki, C., Jouini, O. (2015). On the effect of lifetime variability on the performance of inventory systems. *Int. J. Prod. Econ.*, 167, 23–34.

Krenzler, R. (2016). Queueing systems in a random environment. Doctoral Thesis, Hamburg University.

Krenzler, R., Daduna, H. (2015). Loss systems in a random environment steady-state analysis. *Queueing Syst.*, 80(1), 127–153.

Krishnamoorthy, A., Benny, B., Shajin, D. (2016). A revisit to queueing-inventory system with reservation, cancellation and common lifetime. *OPSEARCH*, 54(2), 336–350.

Krishnamoorthy, A., Lakshmy, B., Manikandan, R. (2011). A survey on inventory models with positive service time. *OPSEARCH*, 48(2), 153–169.

Krishnamoorthy, A., Manikandan, R., Shajin, D. (2015a). Analysis of multi-server queueing-inventory system. *Adv. Oper. Res.*, 2015, 1–16.

Krishnamoorthy, A., Manikandan, R., Lakshmy, B. (2015b). A revisit to queueing-inventory system with positive service time. *Ann. Oper. Res.*, 233, 221–236.

Krishnamoorthy, A., Varghese, R., Lakshmy, B. (2014). An (s, Q) inventory system with positive lead time and service time under N-policy. *Calcutta Stat. Assoc. Bull.*, 66(3–4), 241–260.

Krishnamoorthy, A., Nair, S.S., Narayanan, V.C. (2012). An inventory model with server interruptions and retrials. *Oper. Res. Int. J.*, 12, 151–171.

Krishnamoorthy, A., Shajin, D. (2006). Stochastic decomposition in retrial queueing inventory system. *QTNA '16,* December 13–15, Wellington, New Zealand.

Krishnamoorthy, A., Shajin, D. (2017). $MAP/PH/1$ retrial queueing-inventory system with orbital search and reneging of customers. *International Conference on Analytical and Computational Methods in Probability Theory and its Applications (ACMPT),* RUDN University, Russia.

Krishnamoorthy, A., Shajin, D., Lakshmy, B. (2016a). $GI/M/1$ type queueing-inventory systems with postponed work, reservation, cancellation and common lifetime. *Indian J. Pure Appl. Math.*, 47(2), 357–388.

Krishnamoorthy, A., Shajin, D., Lakshmy, B. (2016b). On a queueing-inventory with reservation, cancellation, common lifetime and retrial. *Ann. Oper. Res.*, 247(1), 365–389.

Krishnamoorthy, A., Shajin, D., Lakshmy, B. (2016c). Product form solution for some queueing-inventory supply chain problem. *OPSEARCH*, 53(1), 85–102.

Krishnamoorthy, A., Viswanath, C.N. (2011). Production inventory with service time and vacation to the server. *IMA J. Manag. Math.*, 22(1), 33–45.

Krishnamoorthy, A., Viswanath, C.N. (2013). Stochastic decomposition in production inventory with service time. *Eur. J. Oper. Res.*, 228, 358–366.

Lawrence, A.S., Sivakumar, B., Arivarignan, G. (2013). A perishable inventory system with service facility and finite source. *Appl. Math. Model.*, 37(7), 4771–4786.

Liu, M., Feng, M., Wong, C.Y. (2014). Flexible service policies for a Markov inventory system with two demand classes. *Int. J. Prod. Econ.*, 151, 180–185.

Manikandan, R., Sajeev, S.N. (2017). $M/M/1/1$ queueing-inventory system with retrial of unsatisfied customers. *Commun. Appl. Anal.*, 21(2), 217–236.

Marand, A.J., Li, H., Thorstenson, A. (2017). Joint inventory control and pricing in a service-inventory system. *Int. J. Prod. Econ.*, 209, 78–91.

Melikov, A., Krishnamoorthy, A., Shahmaliyev, M. (2018). Perishable queuing inventory systems with delayed feedback. In *Information Technologies and Mathematical Modelling. Queueing Theory and Applications* 2018 (ITMM 2018), Communications in Computer and Information Science, Dudin, A., Nazarov, A., Moiseev, A. (eds). Springer, Cham, 912.

Melikov, A., Krishnamoorthy, A., Shahmaliyev, M. (2019). Numerical analysis and long run total cost optimization of perishable queuing inventory systems with delayed feedback. *Queueing Models Serv. Manag.*, 2(1), 83–111.

Melikov, A.Z., Molchanov, A.A. (1992). Stock optimization in transportation/storage systems. *Cybern. Syst. Anal.*, 28(3), 484–487.

Melikov, A.Z., Ponomarenko, L.A., Aliyev, I.A. (2018). Markov models of systems with demands of two types and different restocking policies. *Cybern. Syst. Anal.*, 54(6), 900–917.

Melikov, A.Z., Ponomarenko, L.A., Shahmaliyev, M.O. (2016). Models of perishable queueing-inventory system with repeated customers. *Journal of Automation and Information Sciences*, 48(6), 22–38.

Melikov, A.Z., Ponomarenko, L.A., Shahmaliyev, M.O. (2017). Analysis of perishable queuing-inventory systems with different types of requests. *J. Automat. Inform. Sci.*, 49, 42–60.

Melikov, A.Z., Rustamov, A.M., Ponomarenko, L.A. (2017). Approximate analysis of a queueing-inventory system with early and delayed server vacations. *Automation and Remote Control*, 78(11), 1991–2003.

Melikov, A., Shahmaliyev, M. (2017). A perishable queueing-inventory system with positive service time and $(S-1, S)$ replenishment policy. *Commun. Comput. Inform. Sci.*, 800, 83–96.

Melikov, A., Shahmaliyev, M. O. (2018). Markov models of inventory management systems with a positive service time. *J. Comput. Sys. Sci. Int.*, 57(5), 766–783.

Melikov, A.Z., Shahmaliyev, M.O. (2019). Queueing system $M/M/1/\infty$ with perishable inventory and repeated claims. *Autom. Remote Control*, 80(1), 38–50.

Nair, A.N., Jacob, M.J. (2014). (s, S) Inventory with positive service time and retrial of demands: An approach through multi-server queues. *Hindawi Publishing Corporation ISRN Oper. Res.*, 2014, 1–6.

Nair, A.N., Jacob, M.J., Krishnamoorthy, A. (2015). The multi-server $M/M/(s, S)$ queueing inventory system. *Ann. Oper. Res.*, 233, 321–333.

Nair, S.S., Jose, K.P. (2017a). A $MAP/PH/1$ production inventory model with varying service rates. *Int. J. Pure Appl. Math.*, 117(12), 373–381.

Nair, S.S., Jose, K.P. (2017b). Analysis of a production inventory system with two modes of service using matrix analytic method. *Far East J. Math. Sci.*, 101(12), 2703–2720.

Nair, A.S., Shajin, D. (2018). Customer impatience: A supply chain model. *Int. J. Pure Appl. Math.*, 119(12), 2779–2786.

Neuts, M.F. (1981). *Matrix-Geometric Solutions in Stochastic Models: An Algorithmic Approach*. The Johns Hopkins University Press, Baltimore, MD [1994 version is Dover Edition].

Neuts, M.F., Rao, B.M. (1990). Numerical investigation of a multiserver retrial model. *Queueing Syst.*, 7, 169–190.

Otten, S., Krenzler, R., Daduna, H. (2016). Models for integrated production-inventory systems: Steady state and cost analysis. *Int. J. Prod. Res.*, 54(20), 6174–6191.

Padmavathi, A., Sivakumar, B., Arivarignan, G. (2015). A retrial inventory system with single and modified multiple vacation for server. *Ann. Oper. Res.*, 233, 335–364.

Padmavathi, I., Shophia, L. A., Sivakumar, B. (2016). A finite-source inventory system with postponed demands and modified M vacation policy. *OPSEARCH*, 53(1), 41–62.

Pal, B., Sana, S.S., Chaudhuri, K. (2012). Three-layer supply chain – a production-inventory model for reworkable items. *Appl. Math. Comput.*, 219(2), 530–543.

Park, J., Baek, J.W., Kim, Y.B. (2017). Simulation study on the lost sales inventory systems with attached service queue. *Int. J. Industrial Eng. Theory*, 24(2).

Rajkumar, M., Sivakumar, B., Arivarignan, G. (2014). An infinite waiting hall at a multi-server inventory system. *Int. J. Inventory Res.*, 2(3), 189–221.

Ramshankar, R., Ramanarayanan, R. (2014). Stochastic analysis of bulk production with bulk and catastrophic sales system using matrix geometric approach. *Int. J. Comput. Appl.*, 107(7), 16–21.

Rashid, R., Hoseini, S.F., Gholamian, M.R., Feizabandi, M. (2015). Application of queuing theory in production-inventory optimization. *J. Ind. Eng. Int.*, 11, 485-494.

Rejitha, K.R., Jose, K.P. (2017a). A stochastic inventory system with two modes of service and retrial of customers. *OPSEARCH*, 55, 134–149.

Rejitha, K.R., Jose, K.P. (2017b). A queuing-inventory system with MAP, retrials and different replenishment rates. *Int. J. Pure Appl. Math.*, 117(11), 289–297.

Saffari, M., Asmussen, S., Haji, R. (2013). The $M/M/1$ queue with inventory, lost sale and general lead times. *Queueing Syst.*, 75, 65–77.

Saffari, M., Haji, R., Hassanzadeh, F. (2011). A queueing system with inventory and mixed exponentially distributed lead times. *Int. J. Adv. Manuf. Technol.*, 53, 1231–1237.

Saidane, S., Babai, M., Aguir, M. (2013). On the performance of the base-stock inventory system under a compound Erlang demand distribution. *Comput. Ind.l Eng.*, 66(3), 548–554.

Sarkar, B. (2012). An inventory model with reliability in an imperfect production process. *Appl. Math. Comput.*, 218(9), 4881–4891.

Schwarz, M., Daduna, H. (2006). Queueing systems with inventory management with random lead times and with backordering. *Math. Meth. Oper. Res.*, 64, 383–414.

Schwarz, M., Sauer, C., Daduna, H., Kulik, R., Szekli, R. (2006). $M/M/1$ queueing systems with inventory. *Queueing Syst.*, 54, 55–78.

Schwarz, M., Wichelhaus, C., Daduna, H. (2007). Product form models for queueing networks with an inventory. *Stoch. Models*, 23(4), 627–663.

Shajin, D., Benny, B., Deepak, T.G., Krishnamoorthy, A. (2016). A relook at queueing-inventory system with reservation, cancellation and common lifetime. *Commun. Appl. Anal.*, 20, 545–574.

Shajin, D., Benny, B., Razumchik, R.V., Krishnamoorthy, A. (2018). Discrete product inventory control system with positive service time and two operation modes. *J. Autom. Remote Control*, 79(9), 1593–1608.

Shajin, D., Dudin, A.N., Dudina, O., Krishnamoorthy, A. (2019). A two-priority single server retrial queue with additional items. *J. Ind. Manag. Optim.*, doi: 10.3934/jimo2019085.

Shajin, D., Krishnamoorthy, A. (2018). Stochastic decomposition in retrial queueing-inventory system. *RAIRO-Oper. Res.* doi:10.1051/ro/2018118.

Shajin D., Krishnamoorthy A. (2019). On a queueing-inventory system with impatient customers, advanced reservation, cancellation, overbooking and common lifetime. *Oper. Res. Int. J.*, doi: 10.1007/s12351-019-00475-3.

Shajin, D., Krishnamoorthy, A., Dudin, A.N., Joshua, V.C., Jacob, V. (2019). On a queueing-inventory system with advanced reservation and cancellation for the next K time units ahead: the case of overbooking. *Queueing Syst.*, 1–35.

Shajin, D., Krishnamoorthy, A., Manikandan, R. On a queueing-inventory system with common lifetime and Markovian leadtime process (Communicated).

Shajin, D., Krishnamoorthy, A., Melikov, A.Z. On partial and complete blocking set of states in queueing-inventory model. *Appl. Comput. Math.*, 18(2), 189–201.

Shajin, D., Krishnamoorthy, A., Melikov, A.Z. Multi-server Queueing Production Inventory System with Emergency Replenishment (Communicated).

Shophia, L.A., Sivakumar, B., Arivarignan, G. (2013). A perishable inventory system with service facility and finite source. *Appl. Math. Model.*, 37, 4771–4786.

Sigman, K., Simchi-Levi, D. (1992). Light traffic heuristic for an $M/G/1$ queue with limited inventory. *Ann. Oper. Res.*, 40, 371–380.

Sivakumar, B. (2011). An inventory system with retrial demands and multiple server vacation. *Qual. Technol. Quant. Manag.*, 8(2), 125–146.

Sivakumar, B., Jayaraman, R., Arivarignan, G. (2012). A discrete time inventory system with postponed demands. *J. Comput. Appl. Math.*, 236, 3073–3083.

Sivakumar, B., Arivarignan, G. (2015). An inventory system with two types of customers and retrial demands. *Int. J. Syst. Sci.: Oper. Logist.*, 2(2), 90–112.

Srivathsan, S., Viswanathan, S. (2017). A queueing-based optimization model for planning inventory of repaired components in a service center. *Comput. Ind. Eng.*, 106, 373–385.

Suganya, C., Amirthakodi, M. (2018). An inventory system with two heterogeneous servers and orbital search for feedback customers. *Int. J. Math. Model. Numer. Optim.*, 8(4).

Sundar, V., Rama, G., Ramshankar, R., Sandhya, R. (2015). Ramanarayanan R., Bulk Demand (S, s) Inventory System with PH Distributions. *IOSR J. Eng.*, 5(1), 24–33.

Ushakumari, P.V. (2006). On (s, S) inventory system with random lead time and repeated demands. *J. Appl. Math. Stoch. Anal.*, 1–22.

Vidyadhar, K., Keqi, Y. (2012). Production-inventory systems in stochastic environment and stochastic lead times. *Queueing Syst.*, 70, 207–231.

Vijaya Laxmi, P., Soujanya, M.L. (2015). Retrial inventory system with negative customers and service interruptions. *OPSEARCH*, 52(2), 212–224.

Wang, F.F. (2012). A generic modeling framework for queueing-inventory systems with a single removable server. *J. Chinese Inst. Ind. Eng.*, 29(1), 1–15.

Wang, F.F. (2015). Approximation and optimization of a multi-server impatient retrial inventory-queueing system with two demand classes, *Qual. Technol. Quant. Manag.*, 12(3), 267–290.

Wang, F.F., Bhagat, A., Chang, T.M. (2017). Analysis of priority multi-server retrial queueing-inventory systems with MAP arrivals and exponential services. *OPSEARCH*, 54, 44–66.

Wang, J., Zhang, X. (2017). Optimal pricing in a service-inventory system with delay-sensitive customers and lost sales. *Int. J. Prod. Res.*, 55(22), 6883–6902.

Wang, K., Jiang, Z., Xie, W. (2013). Optimal scheduling policies for a production service system with two items. 7th *IFAC Conference on Manufacturing Modelling, Management and Control, International Federation of Automatic Control,* Saint Petersburg, Russia.

Yadavalli, V.S.S., Jeganathan, K., Krishna Sundar, D., Saravanan, S. (2017). (s, Q) Inventory system with delayed working vacations. *J. Control Syst. Eng.*, 5(1), 18–36.

Yadavalli, V.S.S., Jeganathan, K., Venkatesan, T., Padmasekaran, S., Kingsly, S.J. (2017). A retrial queueing-inventory system with J-additional options for service and finite source. *ORiON*, 33(2), 105–135.

Yadavalli, V.S.S., Kathiresan, J., Anbazhagan, N. (2018). A continuous review inventory system with retrial customers and two-stage service, *Appl. Math. Inf. Sci.*, 12(2), 441–449.

Yadavalli, V.S.S., Sivakumar, B., Arivarignan, G., Adetunji, O. (2012). A finite source multi-server inventory system with service facility. *Comput. Ind. Eng.*, 63, 739–753.

Yadavalli, V.S.S., Udayabaskara, S. (2018). A single product perishable inventory system with compulsory waiting period for reordering and with stochastic lead-time. *Eur. J. Ind. Eng.*, 12(5), 682–707.

Yu, A.J., Dong, Y. (2014). A numerical solution for a two-stage production and inventory system with random demand arrivals. *Comput. Oper. Res.*, 44, 13–21.

Yue, D., Qin, Y.J. (2019). A production inventory system with service time and production vacations. *Syst. Sci. Syst. Eng.*, https://doi.org/10.1007/s11518-018-5402-8

Yue, D., Zhao, G., Qin, Y. (2018). An $M/M/1$ queueing-inventory system with geometric batch demands and lost sales. *J. Syst. Sci. Complex.*, 31(4), 1024–1041.

Yue, D., Zhao, G., Yue, W. (2016). Analysis of a multi-server queueing-inventory system with non-homogeneous poisson arrivals. *QTNA '16 December 13-15*, Wellington, New Zealand.

Zare, A.G., Mehrizi, H.A., Berman, O. (2017). Exact analysis of the (R, Q) inventory policy in a two-echelon production-inventory system. *Oper. Res. Lett.*, 45(4), 308–314.

Zhao, N., Lian, Z. (2011). A queueing-inventory system with two classes of customers. *Int. J. Prod. Econ.*, 129(1), 225–231.

7

A Stability Analysis Method of Regenerative Queueing Systems

Evsey MOROZOV[1,2] and Bart STEYAERT[3]

[1] *Institute of Applied Mathematical Research, Russian Academy of Sciences, Petrozavodsk, Russia*

[2] *Petrozavodsk State University, Russia*

[3] *Department of TELIN, Ghent University, Belgium*

This chapter is devoted to a stability analysis of regenerative queueing systems based on a renewal analysis technique. This approach allows us to obtain simple proofs for stability conditions for the stochastic processes that are considered. In order to illustrate the method, we specifically focus on the stability conditions of the well-known classical continuous-time single-server $GI/G/1$ queueing system, and the $GI/G/m$ multiserver system. For the latter system, we consider the zero-delayed case, as well as the delayed case, meaning that the system initially is non-empty. Some applications of the regenerative method with respect to the stability analysis of other, more complicated, queueing models are briefly discussed as well.

7.1. Introduction

The stability analysis of stochastic models for telecommunication systems is one of the topics intensively studied nowadays. This analysis is, as a rule, a difficult problem requiring a refined mathematical technique, especially when we endeavor beyond the framework of Markovian models. Stability analysis allows us to delimit

Queueing Theory 2,
coordinated by Vladimir ANISIMOV, Nikolaos LIMNIOS.

the region of the system parameters where the stationary regime of the processes describing the dynamics of the associated queuing system exists. The presented method is devoted to the stability analysis of queueing systems possessing the *regeneration property*. This means that a trajectory of the random processes describing the behavior of the system is divided into independent and identically distributed (i.i.d.) *regeneration cycles* at the consecutive *regeneration epochs,* which constitute an *embedded renewal process* (Asmussen 2003; Glynn 1994; Kalashnikov 1994; Serfozo 2009; Sigman and Wolff 1993). The dynamics of a wide class of queueing models describing modern communication and computing systems can be expressed in terms of such regenerative processes (Asmussen 2003; Borovkov 2000; Serfozo 2009).

An important feature of the presented approach is a separation of the predefined conditions on the *negative drift condition* and the *regeneration condition.* The former condition requires that the mean increment of the workload or queue size process outside a bounded set B must be negative. This in turn implies that the process itself *does not go to infinity* and thus visits the bounded set B infinitely often. As an alternative, the latter (regeneration) condition is used to show that, starting in the set B, the process starts a new regeneration cycle within a finite time interval with a positive probability. Then we use a key asymptotic result from renewal theory to establish the system's stability. More precisely, we use a characterization of the limiting behavior of the remaining regeneration time, or, in other words, the remaining renewal time of the regeneration process. This characterization claims that if the length of the remaining renewal time does not go to infinity in probability, then the mean length of the regeneration period (the mean distance between the adjacent regeneration instances) is finite. Such a regenerative process is called *positive recurrent,* in analogy with a Markov process (Asmussen 2003; Feller 1971; Serfozo 2009). As it follows from the asymptotic theory of regenerative processes, it turns out that positive recurrence (under a mild technical assumption) implies the existence of the stationary distribution of the process under consideration (Asmussen 2003; Serfozo 2009). This observation allows us to further reduce the stability analysis by verifying that, under predefined negative drift and regeneration conditions, the queueing process *does not go to infinity in probability.* In many cases, this verification turns out to be much easier than a direct proof of the finiteness of the mean regeneration period.

It is also worth mentioning that, as we will show, an important role in our analysis is played by the tightness of the remaining service time process (Morozov 1997b). The regenerative approach leads to a unique and a transparent method to establish sufficient stability conditions for a wide class of communication and computing systems, including new models describing retrial systems and systems with optical buffers. Moreover, as we show below, this approach provides a simple proof of the necessary stability conditions as well using the strong laws of large numbers (SLLN) and the corresponding system (balance) equations.

Although the stability conditions of the classical queueing systems that we present below are well known, their analysis allows us to demonstrate our method in great detail, and outlines an "algorithm" on how to act in the stability analysis of other models. We would like to emphasize that an important novel element of the analysis is that we also investigate several models under an *arbitrary initial state*. In this case, the length of the first regeneration cycle (if finite) has a different distribution than a generic cycle (starting at a regeneration point). Also in this case, positive recurrence assumes finiteness with probability (w.p.) 1 of the first regeneration period. The main underlying idea here is the following. Under a negative drift condition, we prove that the basic regenerative process hits a bounded set B *infinitely often* w.p.1, while, under the regeneration condition, we show that the mean number of such visits is *finite within each regeneration cycle*. This immediately implies that the number of regeneration cycles is not less than two, and hence the first period is finite w.p.1. In this regard, we would also like to mention the work in Sigman *et al.* (1994).

In the subsequent sections, we demonstrate the main steps of the regenerative stability analysis by considering the classical $GI/G/1$ and $GI/G/m$ queueing systems. The stability conditions of these systems are indeed well known (see, for instance, Asmussen 2003; Kiefer and Wolfowitz 1995), and the proofs that are developed below illustrate in detail the main generic steps of the regenerative stability analysis method. Moreover, these steps reappear under various forms when we consider other, more complicated queueing models. Although a single-server queueing system is a particular case of the corresponding multiserver system, we believe that it is instructive to increase the complexity on a step-by-step basis in the development of the regenerative stability analysis method, and therefore to consider the cases $m = 1$ and $m > 1$ separately. For the same reason, we make a distinction between the zero-delayed multiserver model, and the one with an arbitrary initial state. Also, some results with respect to instability of these systems are included, and we conclude this chapter with a literature survey of related systems, whose stability has been successfully investigated by the regenerative approach.

Although this text mainly is oriented toward the mathematical aspects of the regenerative stability analysis method, the authors believe that the material is also interesting for the engineering community working in the telecommunication area.

7.2. Preliminaries

We will not present a comprehensive introduction to the theory of regenerative processes, which can be found in various previous works (Asmussen 2003; Serfozo 2009; Smith 1955, among others). Instead, we confine this section to a few basic definitions and results which are used in the analysis developed in the subsequent sections below.

Consider either a continuous-time stochastic process $\{X(t),\ t \geq 0\}$, or a discrete-time process $\{X_n,\ n \geq 0\}$, both defined on a metric space (for instance, the m-dimensional Euclidean space R^m). The basic discrete-time processes that occur in our analysis turn out to be embedded into a continuous-time process at random instants $\{\tau_n\}$, which, by assumption, constitute a renewal process that generates the arrivals into a queueing system, and therefore $X_n = X(\tau_n^-)$. In other words, X_n is the state of the process $X(t)$ observed by the nth arrival. With respect to the stochastic processes that we consider below, it is standard to assume that the paths of continuous-time processes are *right-continuous with left-hand limits* (Asmussen 2003; Serfozo 2009; Sigman and Wolff 1993).

Consider the continuous-time case, and assume that there exist instances $0 = T_0 < T_1 < T_2 < \cdots$, such that $T_n \to \infty$ w.p.1, and assume that:

i) $\{X(T_n + s),\ s \geq 0\}$ is independent of T_n and the pre-history, $\{X(s),\ s < T_n\},\ n \geq 0$;

ii) the distribution of $\{X(T_n + s),\ s \geq 0\}$ is independent of $n \geq 1$.

The process $\{X(t)\}$ is then called a *regenerative process*. The epochs $\{T_n,\ n \geq 0\}$ are called *regeneration points*, and the nth *regeneration cycle* of the process is defined as the following random element (of the state space of the paths)

$$G_n := (\{X(t)\} : T_n \leq t < T_{n+1};\ T_{n+1} - T_n),\ n \geq 0. \qquad [7.1]$$

For $n \geq 1$, by construction, these cycles (and regeneration periods $T_{n+1} - T_n$) are independent and identically distributed (i.i.d.). The process is called *delayed* if the first cycle G_0 has a different distribution than the other cycles, and is called *zero-delayed* otherwise. In the zero-delayed case, $T_1 =_{st} T_2 - T_1$ (where $=_{st}$ represents a *stochastic equality*), and we denote by T a generic regeneration period. The definitions for discrete-time process are similar, and we denote by θ_n the regeneration points for this case, and θ a generic regeneration period.

Define the *remaining regeneration time* at instant $t \geq 0$ as

$$T(t) := \min_{k \geq 1} (\, T_k - t : T_k - t > 0\,), \qquad [7.2]$$

and, at instant $n \geq 0$,

$$\theta(n) := \min_{k \geq 1} (\, \theta_k - n : \theta_k - n > 0\,), \qquad [7.3]$$

respectively. Note that in the zero-delayed case, $\theta(0) =_{st} \theta,\ T(0) =_{st} T$.

One of the key elements of the approach that we present below is the following *dichotomy* describing the asymptotic behavior of $\theta(n)$ and $T(t)$ (Feller 1971). Assume $T(0) < \infty$ w.p.1, then, as $t \to \infty$,

$$T(t) \Rightarrow \infty \text{ if and only if } \mathsf{E}T = \infty. \qquad [7.4]$$

Similarly, if $\theta_1 < \infty$ w.p.1, then, as $n \to \infty$,

$$\theta(n) \Rightarrow \infty \text{ if and only if } \mathsf{E}\theta = \infty. \qquad [7.5]$$

In particular, if $\mathsf{E}T = \infty$, then for each $x \geq 0$,

$$\lim_{t\to\infty} \mathsf{P}(T(t) > x) = 1, \qquad [7.6]$$

and, similarly, $\mathsf{E}\theta = \infty$ implies that, for each $k \geq 0$,

$$\mathsf{P}(\theta(n) > k) \to 1, \; n \to \infty, \qquad [7.7]$$

We would like to point out that that an elegant proof of [7.4] (provided period T is *non-lattice*) can be found as a by-product of the analysis developed in Dynkin (1996) by means of reduction of the problem to *Blackwell's renewal theorem* (Asmussen 2003).

Next, consider a regenerative queueing process $\{X(t), t \geq 0\}$ with arrival instants $\{\tau_n\}$, and the corresponding embedded process $\{X_n =: X(\tau_n^-), n \geq 0\}$. An important feature of a regenerative process is that the regeneration property is preserved under an arbitrary measurable mapping f. That is, the process $\{f(X(t))\}$, respectively, the process $\{f(X_n)\}$, is regenerative with the same embedded renewal process of regenerations $\{T_n\}$, and $\{\theta_n\}$, respectively. Moreover, the underlying process $\{X(t)\}$ may be multidimensional (for details, see Asmussen 2003; Serfozo 2009). If $\mathsf{E}T < \infty$ ($\mathsf{E}\theta < \infty$) and $T_1 < \infty$ ($\theta_1 < \infty$) w.p.1, then the process $\{X(t)\}$ ($\{X_n\}$) is called *positive recurrent.*

The following statement is a variant of the *basic limit theorem* for the positive recurrent regenerative process (Asmussen 2003; Serfozo 2009). Denote by $M = \sup\{|f(X(t)| : 0 \leq t < T\}$ and suppose T, M and TM have finite means. If θ is *aperiodic* and T *non-lattice*, then the following limits exist

$$\lim_{n\to\infty} \mathsf{E}(f(X_n)) = \frac{\mathsf{E}\sum_{k=0}^{\theta-1} f(X_k)}{\mathsf{E}\,\theta}; \qquad [7.8]$$

$$\lim_{t\to\infty} \mathsf{E}(f(X(t)) = \frac{\mathsf{E}\Big(\int_0^T f(X(t))dt\Big)}{\mathsf{E}\,T}. \qquad [7.9]$$

In the above expressions, the E-operator denotes expectation with respect to distribution of the process at a regeneration point, corresponding to the zero-delayed case. (We represent this distribution by φ and the corresponding expectation by E_φ in section 6.6.1.) For our analysis, the following particular case of the basic limit regenerative theorem, where function f equals indicator function 1, is crucial (Asmussen 2003; Serfozo 2009):

$$\lim_{n\to\infty} \mathsf{P}(X_n \in \cdot) = \frac{\mathsf{E}\Big(\sum_{k=0}^{\theta-1} \mathsf{1}(X_k \in \cdot)\Big)}{\mathsf{E}\,\theta} =: \pi(\cdot), \quad [7.10]$$

$$\lim_{t\to\infty} \mathsf{P}(X(t) \in \cdot) = \frac{\mathsf{E}\Big(\int_0^T \mathsf{1}(X(t) \in \cdot)dt\Big)}{\mathsf{E}\,T} =: \hat{\pi}(\cdot). \quad [7.11]$$

Consequently, the finiteness of the mean regeneration period $\mathsf{E}T$ and $\mathsf{E}\theta$ is the key requirement for positive recurrence, i.e. for the existence of the stationary distribution π, and/or $\hat{\pi}$, of the regenerative process.

7.3. The single-server system

We consider a $GI/G/1$ system, which is fed by a renewal input with arrival instants $\{\tau_n,\, n \geq 0\}$, and i.i.d. interarrival times, and we let τ represent a generic interarrival time with finite mean $\mathsf{E}\tau$; hence, the customer arrival rate equals $1/\mathsf{E}\tau$. Also, let $\{S_n,\, n \geq 0\}$ be the set of i.i.d. service times, with finite mean $\mathsf{E}S$, that are assumed to be independent of the arrival process.

The primary quantities of interest for this system, respectively, are $\nu(t)$, the number of customers in the system, and $W(t)$, the workload (i.e. the remaining time required to serve the $\nu(t)$ customers), at time instant t, $t \geq 0$.

In addition, we also consider the sequences of random variables $\{\nu_n, n \geq 0\}$ and $\{W_n, n \geq 0\}$, with $\nu_n := \nu(\tau_n^-)$ and $W_n := W(\tau_n^-)$, respectively. Note that the discrete-time processes ν_n and W_n regenerate at the arrival instants with indices

$$\theta_{n+1} := \min_{k\geq 1} \{\, k > \theta_n \,:\, \nu_k = W_k = 0\,\}, \quad n \geq 0, \quad \theta_0 =: 0\,. \quad [7.12]$$

On the other hand, the regeneration instants T_n of the continuous-time processes $\nu(t)$ and $W(t)$ similarly satisfy

$$T_{n+1} := \min_{k\geq 1} \{\, \tau_k > T_n \,:\, \nu_k = W_k = 0\,\}, \quad n \geq 0,\ T_0 =: 0. \quad [7.13]$$

For further use, we also define the remaining interarrival time process as

$$\tau(t) = \min_{k\geq 1} \{\, \tau_k - t : \tau_k - t \geq 0\,\},\ t \geq 0. \quad [7.14]$$

Note that the following relation exists between the quantities θ_n and T_n:

$$T_n = \tau_{\theta_n}, \quad n \geq 0. \qquad [7.15]$$

Consider a generic regeneration period $T =_{st} T_{n+1} - T_n$, and similarly, let θ be a generic time index difference between regeneration points, $\theta =_{st} \theta_{n+1} - \theta_n$, $n \geq 1$. Note that these generic variables are connected by the following useful stochastic equality

$$T =_{st} \tau_\theta.$$

Clearly, by Wald's identity (Billingsley 1995), the following relation holds

$$\mathsf{E}T = \mathsf{E}\tau\, \mathsf{E}\theta, \qquad [7.16]$$

provided that these expectations exist. Also, if $\tau_0 = 0$ and $\nu_0 = 0$, then the corresponding queueing process is called *zero-delayed*, in which case the first regeneration periods T_1 and θ_1 are stochastically equivalent to the generic periods T and θ, respectively, that is $\theta =_{st} \theta_1$, $T =_{st} T_1$. In view of [7.16], $\mathsf{E}\,\theta < \infty$ holds if and only if $\mathsf{E}\,T < \infty$. If the queueing process is zero-delayed, we say that the process has a *zero initial state*. We recall that the renewal process $\{\theta_n\}$ is positive recurrent if

$$\theta_1 < \infty \quad \text{w. p. 1 and} \quad \mathsf{E}\,\theta < \infty, \qquad [7.17]$$

and the process $\{T_n\}$ is positive recurrent if

$$T_1 < \infty \quad \text{w. p. 1 and} \quad \mathsf{E}T < \infty. \qquad [7.18]$$

Under a general (non-zero) initial state, it is possible that $\nu(0) > 0$, in which case the queueing process is called *delayed*, and to have positive recurrence, besides $\mathsf{E}T < \infty$, we need also $T_1 < \infty$ w.p.1. Note that, from [7.15], $T_1 < \infty$ holds if and only if $\theta_1 < \infty$. Consequently, for any system under consideration, it suffices to prove positive recurrence for either the continuous-time or the discrete-time processes. We call a *queueing system positive recurrent* if a regenerative process describing the dynamics of the system (for instance, workload or queue size) is so.

In view of the previous definitions, a generic regeneration period θ equals the number of arrivals in the queueing system $GI/G/1$ (or $GI/G/m$) during the corresponding continuous-time (generic) regeneration period T.

Therefore, in order to prove positive recurrence in the zero-delayed discrete-time case, due to [7.7], it suffices to show that there exist a deterministic sequence $n_k \to \infty$, and constants $C < \infty$, $\delta > 0$, such that

$$\inf_{k\geq 1} \mathsf{P}(\theta(n_k) \leq C) \geq \delta, \qquad [7.19]$$

implying $\mathsf{E}\theta < \infty$, while in the delayed case, we additionally must establish that the inequality $\theta_1 < \infty$ w.p.1 holds as well. (Alternatively, we could deal with $T(t)$ and T_1.)

Below we demonstrate that this approach turns out to be very effective to establish sufficient stability conditions, and even stability criteria, of many queueing models. Denote by $\mathbf{X}(t) = \{W(t),\, \nu(t),\, \tau(t)\}$, $t \geq 0$, and $\mathbf{X}_n = \{W_n,\, \nu_n\}$, $n \geq 0$ a *basic* continuous-time process and the embedded process, respectively, where we take into account that $\tau(\tau_n) = 0$. We note that these processes have regenerations defined by [7.13] and [7.12], respectively. Actually, depending on the purpose, we may consider any suitable process with regenerations [7.13] or [7.12].

In what follows, the initial value of any process is assumed to be taken at time instant 0^-, which corresponds to the state of the process observed by customer 0.

THEOREM 7.1.– If the traffic intensity satisfies

$$\rho =: \mathsf{E}S/\mathsf{E}\tau < 1\,, \qquad [7.20]$$

then the processes $\{\mathbf{X}(t)\}$ and $\{\mathbf{X}_n\}$ are positive recurrent regardless of the initial states, and the process $\{\mathbf{X}_n\}$ has a stationary distribution as $n \to \infty$. If furthermore the interarrival time τ is non-lattice, then the process $\{\mathbf{X}(t)\}$ has a stationary distribution as $t \to \infty$.

PROOF. We first introduce the idle time $I(t)$ of the server in the interval $[0,\, t]$, defined as

$$I(t) = \int_0^t 1(\nu(s) = 0)\, ds,$$

and $A(t)$, the number of arrivals in the interval $[0,\, t)$. Also define the total workload that has arrived in the interval $[0, t)$ as

$$V(t) = \sum_{n=0}^{A(t)} S_n$$

Obviously, we may write

$$W(0) + V(t) = t - I(t) + W(t) \geq t - I(t),$$

and hence,

$$I(t) \geq t - V(t) - W(0) = t - V(t) + o(t),\ t \to \infty.$$

Invoking the SLLN property, we obtain

$$\lim_{t\to\infty} \frac{V(t)}{A(t)} \to \mathsf{E}\,S, \quad \lim_{t\to\infty} \frac{A(t)}{t} \to 1/\mathsf{E}\tau, \qquad [7.21]$$

and may therefore write

$$\liminf_{t\to\infty} \frac{I(t)}{t} \geq 1 - \lim_{t\to\infty} \frac{V(t)}{A(t)} \frac{A(t)}{t} = 1 - \rho > 0. \qquad [7.22]$$

In particular, the summary idle time of the server in the interval $[0,\ t]$ satisfies

$$I(t) \to \infty \quad \text{w. p. 1 as } t \to \infty. \qquad [7.23]$$

Because $I(t)/t \leq 1, t > 0$, then it follows from [7.22] by Fatou's lemma (Billingsley 1995) that

$$\liminf_{t\to\infty} \frac{\mathsf{E}\,I(t)}{t} \geq \mathsf{E} \liminf_{t\to\infty} \frac{I(t)}{t} = 1 - \rho. \qquad [7.24]$$

Then, applying Fubini's theorem (Billingsley 1995) to interchange the expectation and integral operators, we obtain

$$\liminf_{t\to\infty} \frac{\mathsf{E}\,I(t)}{t} = \liminf_{t\to\infty} \frac{1}{t} \int_0^t \mathsf{P}(\nu(s) = 0)ds \geq 1 - \rho > 0. \qquad [7.25]$$

It is easy to see that [7.25] contradicts the convergence of $\mathsf{P}(\nu(t) = 0) \to 0$, and hence there exist a deterministic (non-random) sequence $z_i \to \infty$ and a constant $\delta > 0$ such that

$$\inf_i \mathsf{P}(\nu(z_i) = 0) \geq \delta. \qquad [7.26]$$

It has been shown in Morozov (1997b) that the remaining interarrival time process $\tau(t)$, defined in [7.14], is tight, and therefore, for $\delta > 0$ satisfying [7.26], a constant $C < \infty$ exists such that

$$\inf_{t\geq 0} \mathsf{P}(\tau(t) \leq C) \geq 1 - \frac{\delta}{2}. \tag{7.27}$$

Since the instants $\{z_i\}$ are non-random, then [7.27] holds when the parameter t is replaced by $\{z_i\}$. Note that conditioning on the event $\{\nu(z_i) = 0\}$, the first arrival after instant t generates a regeneration. As a result, we find that

$$\begin{aligned}\inf_i \mathsf{P}(T(z_i) \leq C) &\geq \inf_i \mathsf{P}\Big(\nu(z_i) = 0,\, \tau(z_i) \leq C\Big)\\ &\geq \inf_i \mathsf{P}(\nu(z_i) = 0) - \mathsf{P}(\tau(z_i) > C) \geq \frac{\delta}{2}.\end{aligned} \tag{7.28}$$

Evidently, C and δ are both independent of the sequence z_i. Consequently, from [7.28], then $T(t) \not\Rightarrow \infty$ holds, implying that $\mathsf{E}T < \infty$ and $\mathsf{E}\theta < \infty$ in view of [7.16] and [7.4]. Moreover, in the zero-delayed case, $T_1 =_{st} T < \infty$ w. p. 1, while in the delayed case, from [7.23], the initial busy period satisfies $u_0 = \inf(t > 0 : W(t) = 0) < \infty$ w.p. 1. Because a regeneration occurs within the interval $[0,\, u_0 + \tau(\, u_0\,)]$, then

$$T_1 \leq u_0 + \tau(\, u_0\,) < \infty,\ \theta_1 \leq A(T_1) < \infty \quad \text{w. p.1}, \tag{7.29}$$

holds as well. In addition, using assumption [7.20], then $\mathsf{E}(\tau - S) > 0$, implying $\mathsf{P}(\tau - S > 0) > 0$, and thus the process $\{\theta_n\}$ is aperiodic, leading to

$$\mathsf{P}(\theta = 1) = \mathsf{P}(\tau > S) > 0, \tag{7.30}$$

and hence, the process $\{\mathbf{X}_n\}$ has a stationary distribution as $n \to \infty$. If the interarrival time τ is non-lattice, then the regeneration period T is non-lattice as well (see [7.15]), and $\{\mathbf{X}(t)\}$ has a stationary distribution as $t \to \infty$.

7.4. The zero-delayed multiserver system

Consider the FCFS m-server $GI/G/m$ queueing system with equivalent servers, the same input process as above, and i.i.d. service times $\{S_n\}$ with mean $\mathsf{E}S$. Denote by $W_i(t)$ the remaining workload in server i at instant t^-, and let $W(t) := \sum_{i=1}^m W_i(t)$. Although in general $W(0) \geq 0$, in theorem 7.2, we only prove the finiteness of the mean generic regeneration period $\mathsf{E}T$, which implies the positive recurrence of the zero-delayed system, because in this case $T_1 =_{st} T$. The finiteness (w.p.1) of the first regeneration period T_1 for the delayed case is more challenging and will be treated in the next section.

Let $\nu(t)$ be the number of customers in the system at instant t^-. Consider also the process $\mathbf{Y}(t) = \{\nu(t), W(t), \tau(t)\}$, $t \geq 0$. Note that the instants $\{T_n\}$, defined in [7.13], are regenerations of the process $\{\mathbf{Y}(t)\}$, and also of the process $\{\nu(t), \mathbf{W}(t), \tau(t)\}$, where $\mathbf{W}(t) = (W_1(t), \dots, W_m(t))$. In the following statement, we prefer to work with the process $\{\mathbf{Y}(t)\}$.

THEOREM 7.2.– Assume an arbitrary initial state $\mathbf{Y}(0) = \mathbf{y}_0$, and that conditions

$$\rho =: \mathsf{E}S/\mathsf{E}\tau < m, \qquad [7.31]$$

and

$$\mathsf{P}(\tau > S) > 0, \qquad [7.32]$$

hold. Then $\mathsf{E}T < \infty$.

PROOF.– For any server i, denote by $I_i(t)$ its idle time in the interval $[0, t]$, and let $I(t)$ be the total amount of time in $[0, t]$ during which not all servers are busy (i.e. at least one server is idle). Then, for each $i = 1, \dots, m$,

$$I(t) = \int_0^t \mathsf{1}(\nu(s) < m)\, ds \geq \int_0^t \mathsf{1}(W_i(s) = 0)\, ds = I_i(t), \quad t \geq 0. \qquad [7.33]$$

Note that $I(t) \geq \max_i I_i(t)$, and hence, $mI(t) \geq \sum_{i=1}^m I_i(t)$. Then we can write down the following relation

$$W(0) + V(t) = tm - \sum_{i=1}^{m} I_i(t) + W(t) \geq m\,t - m\,I(t), \qquad [7.34]$$

implying, as above,

$$\liminf_{t\to\infty} \frac{I(t)}{t} \geq 1 - \frac{\rho}{m} =: \delta_0 > 0, \qquad [7.35]$$

and

$$\liminf_{t\to\infty} \frac{\mathsf{E}I(t)}{t} = \liminf_{t\to\infty} \frac{1}{t} \int_0^t \mathsf{P}(\nu(s) = 0) ds \geq \delta_0. \qquad [7.36]$$

Then the inequality $\limsup_{t\to\infty} \mathsf{P}(\nu(t) < m) > 0$ holds, and consequently

$$\inf_i \mathsf{P}(\nu(z_i) < m) \geq \delta, \qquad [7.37]$$

for a non-random sequence $z_i \to \infty$ and some $\delta > 0$.

Next, denote by $S_i(t)$ the remaining service time in server i at instant t^-, then the summary process $\{\sum_{i=1}^m S_i(t),\ t \geq 0\}$ is tight (Morozov 1997b). Note that $W(t) = \sum_{i=1}^m S_i(t)$ is conditioned on the event $\{\nu(t) < m\}$, and that the remaining interarrival time process $\{\tau(t)\}$ is also tight, because it is the remaining time of the standard positive recurrent renewal process. Also define the set $\mathbf{B}_D = [0,\ m) \times [0,\ D] \times [0,\ D]$, and the event

$$\mathcal{E}_i(D) = \Big\{\mathbf{Y}(z_i) \in \mathbf{B}_D\Big\},$$

for a constant D and a fixed (arbitrary) time instant z_i satisfying [7.37]. Due to the tightness of $\{\tau(t)\}$ and [7.37], we can choose the constant D sufficiently large such that

$$\inf_i \mathsf{P}(\mathcal{E}_i(D)) \geq \frac{\delta}{2}, \qquad [7.38]$$

where, for the events $\mathcal{A}$, $\mathcal{B}$, we apply the inequality $\mathsf{P}(\mathcal{AB}) \geq \mathsf{P}(\mathcal{A}) - \mathsf{P}(\bar{\mathcal{B}})$, and $\bar{\mathcal{B}}$ denotes the complement of the event $\mathcal{B}$. (Indeed, $\mathsf{P}(\cdot) = \mathsf{P}_{\mathbf{y}_0}(\cdot)$ in [7.38], but if it is not required, we omit in the notation the dependence of the probability measure on the initial state of the process.) Also, due to [7.32] and $\mathsf{E}\,\tau < \infty$, there exist constants $\delta_0,\ \delta_1,\ C \in (0,\ \infty)$ such that

$$\mathsf{P}(C \geq \tau > S + \delta_0) \geq \delta_1. \qquad [7.39]$$

Then, denote by $L = \lceil D/\delta_0 \rceil$ and, for a fixed arbitrary z_i, denote by $n_i = \min(k : \tau_k \geq z_i)$, the index of the first arrival after instant z_i. Also denote by $\tau^{(n)} = \tau_{n+1} - \tau_n$ the n-th interarrival time, $n \geq 0$, and introduce the event

$$\mathcal{A}_i(L) = \bigcap_{k=0}^{L-1} \Big\{C \geq \tau^{(n_i+k)} > S_{n_i+k} + \delta_0\Big\}, \qquad [7.40]$$

which consisted of independent events. The event [7.40] implies that (starting from the time instant τ_{n_i}), every interarrival time is upper bounded by the constant C and exceeds the service time of the new customer by no less than δ_0. As a result, during each interarrival time, the workload decreases by no less than δ_0 as long as the system is not completely empty. Hence, we can expect that an appropriate series of such events leads to an empty state of the system and regeneration occurs. Indeed, conditioned on the event $\mathcal{E}_i(D) \cap \mathcal{A}_i(L)$, a customer arrives in the time interval $[z_i, z_i + D + LC]$, which observes an empty system, and hence, a regeneration occurs in this interval. Since $\mathsf{P}(\mathcal{A}_i(L)) \geq \delta_1^L$, then the following lower bound, uniform in z_i and i, holds:

$$\mathsf{P}\Big(T(z_i) \leq D + LC\Big) \geq \delta \, \frac{\delta_1^L}{2} > 0, \qquad [7.41]$$

implying that $\mathsf{E}T < \infty$, which concludes the proof. □

REMARK 7.1.– We may summarize the analysis leading to [7.41] as follows. The *regeneration* condition [7.32], via condition [7.39], allows us to establish that if the process $\mathbf{Y}(z_i) = \mathbf{y} \in \mathbf{B}_D$, then the length of the remaining regeneration time $T(z_i)$ is upper bounded by a constant $C_0 := D + LC$ with a probability, which is lower bounded by δ_1^L. Although both C_0 and δ_1^L in general depend on the set $\mathbf{B}$, the following uniform inequality holds (also for all z_i):

$$\inf_{\mathbf{y}\in\mathbf{B}_D} \mathsf{P}_{\mathbf{y}}(T(z_i) \leq C_0) = \inf_{\mathbf{y}\in\mathbf{B}_D} \mathsf{P}(T(z_i) \leq C|\mathbf{Y}(z_i) = \mathbf{y}) \geq \delta_1^L. \qquad [7.42]$$

However, for the analysis of the multiserver system with an arbitrary initial state, given below, we need to obtain a similar uniform bound for the discrete remaining regeneration time $\theta(n)$. Denote by $\mathbf{Y}(\tau_n) = \mathbf{Y}_n$, and note that, since $z_i \leq \tau_{n_i}$, then $\mathbf{Y}(z_i) \geq \mathbf{Y}_{n_i}$ in the component-wise sense. It then follows from above that, conditioned on the event $\mathcal{E}_i(D)$, the event $\mathcal{A}_i(L)$ implies a regeneration after *at most* L arrivals, i.e.

$$\mathsf{P}(T(z_i) \leq D + LC\,|\mathcal{E}_i(D) \cap \mathcal{A}_i(L)) = \mathsf{P}(\theta(n_i) \leq L\,|\mathcal{E}_i(D) \cap \mathcal{A}_i(L)). \qquad [7.43]$$

Specifically, if $\mathbf{Y}(z_i) = \mathbf{y} \in \mathbf{B}_D$, then, for all n_i, the remaining regeneration time $\theta(n_i)$ is uniformly upper bounded by (see [7.42])

$$\inf_{\mathbf{y}\in\mathbf{B}_D} \mathsf{P}_{\mathbf{y}}(\theta(n_i) \leq L) = \inf_{\mathbf{y}\in\mathbf{B}_D} \mathsf{P}(\theta(n_i) \leq L|\mathbf{Y}_{n_i} = \mathbf{y}) \geq \delta_1^L. \qquad [7.44]$$

It is evident that this conclusion holds for any bounded set $\mathbf{B}$, possibly with different constants δ_1, L.

REMARK 7.2.– It is easy to check that, for an m-server system, condition [7.31] in general only implies the condition $\mathsf{P}(\tau > S/m) > 0$, while condition [7.32] may not hold. In the latter case, classical regenerations do not exist. Nonetheless, it is still possible to develop the stability analysis under condition [7.31] only, relying on the so-called one-dependent regeneration property (Borovkov 1984, 1987; Sigman 1990). However, this topic goes beyond the scope of the analysis, which is being developed in this book.

7.5. The delayed multiserver system: finiteness of the first regeneration period

It follows from [7.35] that $I(t)$, the time during which the system is *partially empty* (i.e. at least one server is idle) in the interval $[0, t]$, satisfies $I(t) \to \infty$ w. p. 1, as $t \to \infty$. However, in general, this does not imply the finiteness of the first regeneration period, since $I(t)$ may not contain regeneration points.

To establish the finiteness (w.p.1) of the first regeneration period, we present an approach that has been presented in Morozov (1997a) and also effectively used in the stability analysis of many queueing systems. (We mention some of these systems in section 7.6.1.) We split the proof into two steps. First, we show that the number of arrivals that see a basic process in a bounded set increases unlimitedly as $t \to \infty$. Second, we show that, within the first regeneration cycle, the basic process visits any bounded set at most a finite number of times w.p.1. It then follows that the number of regeneration cycles is not less than two, and thus, $\theta_1 < \infty$, and $T_1 < \infty$ w.p.1.

First, denote by $\mathbf{S}(t) = (S_1(t), \dots, S_m(t))$, the vector of the remaining service times, and consider the process $\{\mathbf{Z}(t), t \geq 0\}$, where $\mathbf{Z}(t) = (\nu(t), \mathbf{S}(t))$. Also, denote by $\mathbf{Z}_n = (\nu_n, \mathbf{S}(\tau_n))$, $n \geq 0$. The processes $\{\mathbf{Z}(t)\}$ and $\{\mathbf{Z}_n\}$, which regenerate at the instances [7.13] and [7.12], respectively, are required for the following analysis.

THEOREM 7.3.– If assumptions [7.31] and [7.32] hold, then positive recurrence of the processes $\{\mathbf{Z}(t)\}$ and $\{\mathbf{Z}_n\}$ will occur for an arbitrary initial state $\mathbf{Y}(0) = \mathbf{y}_0$. In particular,

$$\theta_1 < \infty, \quad T_1 < \infty \quad \text{w.p.1.} \qquad [7.45]$$

PROOF.– In the light of theorem 7.2, for positive recurrence of the considered process, it is sufficient to merely prove [7.45].

Step 1. We assume that there exist independent, i.i.d. basic service times sequences $\{S_n^{(i)}, n \geq 0\}$, with a common generic element S, which are used to generate the consecutive service times in server i. Moreover, for each fixed i, we consider a renewal process

$$\hat{Z}_0^{(i)} = S_i(0),\ \hat{Z}_n^{(i)} = S_i(0) + S_1^{(i)} + \cdots + S_n^{(i)},\ n \geq 1,$$

where $S_i(0)$ is the initial remaining service time in server i corresponding to the state $\mathbf{Z}(0)$, and the service times $S_n^{(i)}$ are those used in the *real service process* in server i. (We use the "hat" notation to distinguish this renewal process from the original service process.) Then the remaining renewal time $\hat{S}_i(t)$ at instant t in the process $\{\hat{Z}_n^{(i)}, n \geq 0\}$ is defined as

$$\hat{S}_i(t) = \min_n(\hat{Z}_n^{(i)} - t : \hat{Z}_n^{(i)} - t \geq 0). \qquad [7.46]$$

Although the actual remaining service times $S_i(t)$ and $\ddot{S}_i(t)$ are in general are not comparable, taking into account the possible *empty periods* of the servers, then for an arbitrary value $x \geq 0$ we obtain the following inequalities:

$$\int_0^t \mathsf{1}(S_i(u) \leq x)du \geq \int_0^t \mathsf{1}(\hat{S}_i(u) \leq x)du,\ \ t \geq 0,\ i = 1, \ldots, m. \qquad [7.47]$$

For each x, the process $I(\hat{S}_i(u) \geq x),\ u \geq 0$, is regenerative with generic regeneration period S, and $\int_0^t I(\hat{S}_i(u) \geq x)du,\ t \geq 0$, is the corresponding cumulative process. It then follows from regeneration theory that, w.p.1,

$$\lim_{t\to\infty} \frac{1}{t} \int_0^t \mathsf{1}(\hat{S}_i(u) \leq x)du = \frac{1}{\mathsf{E}S} \int_0^x \mathsf{P}(S > u)du$$
$$=: F_e(x),\ x \geq 0,\ i = 1, \ldots, m, \qquad [7.48]$$

where, because of $\mathsf{E}S < \infty$, the distribution F_e of the stationary remaining renewal time satisfies $F_e(x) \to 1$ as $x \to \infty$. In particular, for given value $\delta_0 = 1 - \rho/m > 0$, we can choose a constant $M = M(\delta_0) \in (0, \infty)$ such that

$$F_e(M) \geq 1 - \frac{\delta_0}{4(m-1)}. \qquad [7.49]$$

Now, define the set

$$\mathbf{B}_M = [0, M] \times \cdots \times [0, M] \in \mathsf{R}_+^m.$$

Then it follows from [7.47] and [7.49] that

$$\liminf_{t\to\infty} \frac{1}{t} \int_0^t \mathsf{1}(\mathbf{S}(u) \in \mathbf{B}_M)\, du$$

$$\geq \liminf_{t\to\infty} \frac{1}{t} \int_0^t \mathsf{1}(S_1(u) \leq M) du - \limsup_{t\to\infty} \frac{1}{t} \int_0^t \sum_{i=2}^{m} \mathsf{1}(S_i(u) > M) du$$

$$\geq F_e(M) - (m-1)(1 - F_e(M)) \geq 1 - \delta_0/2. \tag{7.50}$$

Therefore, we obtain

$$\limsup_{t\to\infty} \frac{1}{t} \int_0^t \mathsf{1}(\mathbf{S}(u) \notin \mathbf{B}_M)\, du \leq \frac{\delta_0}{2}. \tag{7.51}$$

Consider the set $\mathbf{B}_0 = [0,\, m-1] \times \mathbf{B_M}$ and denote by

$$\mathcal{M}_0(t) = \int_0^t \mathsf{1}(\mathbf{Z}(u) \in \mathbf{B}_0)\, du,$$

the time the process $\mathbf{Z}$ is in the set $\mathbf{B_0}$ during the interval $[0,\, t]$. Note that any regenerative process, for instance, $\{\nu(t)\}$, $\{W(t)\}$ or $\{\mathbf{Z}(t)\}$ can be chosen as the basic process in the stability analysis. Evidently, relation [7.35] remains true for an arbitrary initial value of the corresponding basic process. Note that the term $W(0^-)$ in relation [7.34] can be expressed via the components of vector $\mathbf{Z}(0)$ as follows:

$$W(0) = \sum_{i=1}^{m} S_i(0) + \sum_{k=1}^{(\nu(0)-m)^+} S_k^*, \tag{7.52}$$

where the random variables S_k^* represent the i.i.d. service times of the customers present in the system at instant 0^- (and are distributed as S). Relying on the inequality

$$\mathsf{1}(\mathbf{Z}(u) \in \mathbf{B}_0) \geq \mathsf{1}(\nu(u) < m) - \mathsf{1}(\mathbf{S}(u) \notin \mathbf{B}_M),$$

we obtain from [7.35] and [7.51] that

$$\liminf_{t\to\infty} \frac{\mathcal{M}_0(t)}{t} \geq \liminf_{t\to\infty} \frac{1}{t} \int_0^t \mathsf{1}(\nu(u) < m) du$$

$$- \limsup_{t\to\infty} \frac{1}{t} \int_0^t \mathsf{1}(\mathbf{S}(u) \notin \mathbf{B}_M)\, du \geq \frac{\delta_0}{2}. \tag{7.53}$$

Let

$$G_0(t) = \sum_{k=0}^{A(t)} 1(\mathbf{Z}_k \in \mathbf{B}_0),$$

represent the number of arrivals within interval $[0, t)$, which meet the process $\{\mathbf{Z}_k\}$ in the set $\mathbf{B}_0$. It follows from [7.53] that $\mathcal{M}_0(t) \to \infty$, and moreover, using the definition of the n-th interarrival time $\tau^{(n)} = \tau_{n+1} - \tau_n$, we obtain

$$\mathcal{M}_0(t) \leq G_0(t) \max_{0 \leq n \leq A(t)} \tau^{(n)}. \qquad [7.54]$$

Because $\mathsf{E}\tau < \infty$, then by (Smith 1955),

$$\max_{0 \leq n \leq A(t)} \tau^{(n)} = o(t), \quad t \to \infty, \qquad [7.55]$$

and it thus follows from [7.53] to [7.55] that

$$G_0(t) \to \infty, \;\; t \to \infty, \;\; \text{w. p. 1.} \qquad [7.56]$$

For the following analysis, we return to the process $\{\mathbf{Y}_k\}$. Because the remaining interarrival time $\tau(\tau_k) = 0$ for each arrival instant τ_k, then, considering the set

$$\mathbf{B}^* = [0, m) \times [0, mM] \times \{0\},$$

where constant M was defined in [7.49], we obtain the following inclusion

$$\{\mathbf{Z}_k \in \mathbf{B}_0\} \subseteq \{\mathbf{Y}_k \in \mathbf{B}^*\}.$$

Hence, w.p.1, as $t \to \infty$,

$$\sum_{k=0}^{A(t)} 1(\mathbf{Y}_k \in \mathbf{B}^*) \geq G_0(t) \to \infty. \qquad [7.57]$$

In other words, the summary number of visits by the process $\{\mathbf{Y}_k\}$ to the set $\mathbf{B}^*$ grows unlimitedly w.p.1 as time increases, for an arbitrary initial state $\mathbf{Y}_0 = \mathbf{y}_0$.

Step 2. Now we prove that the mean number of visits to the set $\mathbf{B}^*$ by the process $\{\mathbf{Y}_k\}$ within the same regeneration cycle is indeed finite, which evidently contradicts

[7.57]. Consequently, we obtain the required result: the first regeneration period is finite w.p.1.

Note that the instant z_i in [7.43] is arbitrary, and in turn, the bound [7.44] holds for each instant n, i.e.

$$\inf_n \inf_{\mathbf{y}\in\mathbf{B}^*} \mathsf{P}_{\mathbf{y}}(\theta(n) \le L) \ge \varepsilon, \qquad [7.58]$$

for $L = \lceil mM/\delta_0 \rceil$, $\varepsilon = \delta_1^L$. Then, applying condition [7.58], we arrive at the bound

$$\mathsf{E}_{\mathbf{y}_0} \sum_{k=0}^{\theta_1 - 1} \mathsf{1}(\mathbf{Y}_k \in \mathbf{B}^*) \le \frac{L}{\varepsilon}, \qquad [7.59]$$

which holds for each initial state $\mathbf{y}_0$. As a result, the number of such visits within the first regeneration cycle satisfies $\sum_{k=0}^{\theta_1-1} \mathsf{1}(\mathbf{Y}_k \in \mathbf{B}^*) < \infty$ w.p.1, for each initial state $\mathbf{y}_0$. Combined with [7.57], this implies that the total number of regeneration cycles is *no less than two*, in turn implying that $\theta_1 < \infty$, and hence $T_1 < \infty$, w.p.1. □

REMARK 7.3.– It follows from the above analysis that the number of regeneration cycles is indeed infinite w.p. 1, or equivalently, $\theta_n < \infty$ for each $n \ge 1$. (In this regard, see also Sigman *et al.* (1994).)

7.6. Instability

In this section, we briefly discuss the *instability condition* of the basic $GI/G/m$ queueing system. In particular, in the following statement we demonstrate that the regenerative stability analysis allows us to study the so-called *boundary case* $\rho = m$, in which case the popular fluid stability analysis does not work (Dai 1995).

THEOREM 7.4.–

i) Assume $\rho - m =: \varepsilon > 0$. Then irrespective of the initial state $\nu(0)$, $W(0)$, w.p.1, the following inequalities hold:

$$\liminf_{t\to\infty} \frac{\nu(t)}{t} \ge \frac{\varepsilon}{\mathsf{E}S}, \quad \liminf_{t\to\infty} \frac{W(t)}{t} \ge \varepsilon. \qquad [7.60]$$

ii) If $\rho = m$ and $W(0) = 0$ (corresponding to a zero initial state), then $\mathsf{E}T = \infty$.

Thus, under condition $\rho > m$, both the queue size and the workload grow linearly in time w.p.1, regardless of the initial state of the system, and we call this case *strong* instability. On the other hand, under condition $\rho = m$, both processes unlimitedly grow *in probability*, which will be referred to as *weak instability*.

PROOF.–

i) Assume $\rho > m$ and denote by $\hat{D}_i(t)$ the number of customers, which can be served by server i in the interval $[0, t)$, provided it is permanently working, and let $\hat{D}_i(0) = 1$. In the construction of the process $\{\hat{D}_i(t)\}$, we use the same service times as in real service process in server i, $i = 1, \ldots, m$. Consider the process $\hat{D}(t) = \sum_i \hat{D}_i(t)$. As before, we define $A(t)$ and $D(t)$, respectively, as the total number of arrivals and departures from the system in the interval $[0, t)$. Evidently, we can write down the following relations:

$$\nu(t) = \nu(0) + A(t) - D(t) \geq A(t) - \hat{D}(t),\ t \geq 0. \tag{7.61}$$

Since $\hat{D}(t)/t \to m/\mathsf{E}S$, then we immediately obtain the first inequality in [7.60], after dividing by t and taking the corresponding limit of both hand sides in [7.61]. Next, denote by $\mathcal{G}(t)$ the set of customers waiting in the queue at instant t^-, so the capacity of this set,

$$|\mathcal{G}(t)| = (\nu(t) - m)^+ = Q(t)\,,$$

is the queue size at instant t^-. Note that, from expression [7.61], we can derive that

$$\liminf_{t\to\infty} \frac{Q(t)}{t} \geq \frac{\varepsilon}{\mathsf{E}S}.$$

Then the second inequality in [7.60] results from the inequality

$$W(t) \geq \sum_{k\in\mathcal{G}(t)} S_k, \tag{7.62}$$

which implies, by invoking the SLLN property

$$\begin{aligned}\liminf_{t\to\infty} \frac{W(t)}{t} &\geq \liminf_{t\to\infty} \frac{\sum_{k\in\mathcal{G}(t)} S_k}{Q(t)} \frac{Q(t)}{t} \\ &= \mathsf{E}S \liminf_{t\to\infty} \frac{Q(t)}{t}.\end{aligned} \tag{7.63}$$

Of course, this also implies that $\mathsf{E}T = \infty$.

ii) Assume that $\rho = m$. Because, by assumption, $\nu(0) = 0$, this corresponds to the zero-delayed case, $T_1 =_{st} T$. Assume, by contradiction, that $\mathsf{E}T < \infty$ holds. Then $W(t) = o(t)$, $t \to \infty$ would hold as well. Denote by $B_i(t)$ the length of the busy time of server i in the interval $[0, t]$. Then it follows from the system equation $V(t) = W(t) + \sum_i B_i(t)$ and the SLLN property that

$$\rho = \lim_{t\to\infty} \frac{\sum_i B_i(t)}{t} < m\,. \qquad [7.64]$$

This contradiction implies that indeed $\mathsf{E}T = \infty$.

Thus, we can conclude that, under assumption [7.32], condition $\rho < m$ is the *stability criterion* of the (zero-delayed) $GI/G/m$ queueing system.

REMARK 7.4.– Indeed, under assumption [7.32], condition $\rho < m$ is stability criterion for an *arbitrary initial state* (Kiefer and Wolfowitz 1995; Morozov 1997b).

The following statement contains further results related to the boundary case.

THEOREM 7.5.– If $\rho = m$, then, as $t \to \infty$,

$$\nu(t) \Rightarrow \infty, \ W(t) \Rightarrow \infty, \qquad [7.65]$$

$$\mathsf{E}\nu(t) = o(t), \ \mathsf{E}W(t) = o(t). \qquad [7.66]$$

Moreover, if $\nu(t) \Rightarrow \infty$, then

$$W(t) \Rightarrow \infty \text{ and } \rho \geq m. \qquad [7.67]$$

Similarly, if $W(t) \Rightarrow \infty$, then

$$\nu(t) \Rightarrow \infty \text{ and } \rho \geq m. \qquad [7.68]$$

PROOF.– First, consider the case $\rho = m$. Suppose now that $\nu(t) \not\Rightarrow \infty$. Then [7.37] holds and $\mathsf{E}T < \infty$ follows. Then, invoking

$$V(t) = W(t) + mt - I(t), \qquad [7.69]$$

we obtain in the limit the contradiction $\rho < m$, see also [7.35] and [7.64]. This contradiction leads to the conclusion that $\nu(t) \Rightarrow \infty$. Moreover, for each $x \geq 0$ and integer N, we find

$$\begin{aligned} \mathsf{P}(W(t) > x) &\geq \mathsf{P}(\sum_{k=1}^{N} S_k > x,\ \nu(t) - m > N) \\ &\geq \mathsf{P}(\sum_{k=1}^{N} S_k > x) - \mathsf{P}(\nu(t) \leq N + m). \end{aligned} \qquad [7.70]$$

For any value x and arbitrary $\varepsilon > 0$, we can choose N such that $\mathsf{P}(\sum_{k=1}^{N} S_k > x) \geq 1 - \varepsilon/2$, and then choose t_0 such that $\mathsf{P}(\nu(t) \leq N + m) \leq \varepsilon/2$ for $t \geq t_0$. This implies that $\mathsf{P}(W(t) > x) \geq 1 - \varepsilon$, $t \geq t_0$, and because x and ε are arbitrary, [7.65] follows. Then the mean fraction of summary idle time is given by

$$\frac{1}{t}\mathsf{E}I(t) = \frac{1}{t}\int_0^t \mathsf{P}(W(u) = 0)du \to 0, \quad t \to \infty,$$

and, because of

$$\lim_{t\to\infty} \frac{1}{t}V(t) = \mathsf{E}S/\mathsf{E}\tau = \rho, \qquad [7.71]$$

see [7.21], it follows from $\mathsf{E}I(t) = \mathsf{E}V(t) - mt - \mathsf{E}W(t) \geq 0$, that

$$\limsup_{t\to\infty} \frac{1}{t}\mathsf{E}W(t) \leq \lim_{t\to\infty} \frac{1}{t}\mathsf{E}V(t) - m = 0. \qquad [7.72]$$

Because

$$W(t) \geq_{st} \sum_{k=1}^{(\nu(t)-m)^+} S_k,$$

and, since $\nu(t)$ and the service times of the customers that are waiting in the queue are independent random variables, we have by Wald's identity,

$$\mathsf{E}W(t) \geq \mathsf{E}S\,\mathsf{E}(\nu(t) - m)^+,$$

implying $\mathsf{E}\nu(t) = o(t)$.

We have shown above that $\nu(t) \Rightarrow \infty$ implies $W(t) \Rightarrow \infty$. Assume that $W(t) \Rightarrow \infty$ but $\nu(t) \not\Rightarrow \infty$. Then, similar as in Theorem 6.2, we arrive at $\mathsf{E}T < \infty$, further leading to $\rho < m$. This contradiction shows that $\nu(t) \Rightarrow \infty$, indeed implying $\mathsf{E}I(t) = o(t)$. Finally, the inequality $\rho \geq m$ then follows from [7.72] and [7.69], and it is written as

$$\mathsf{E}W(t) = \mathsf{E}V(t) - mt + o(t) \geq 0. \qquad \square$$

7.6.1. *Some comments on the method*

We conclude by giving some additional comments to the method described above. Consider a (one-dimensional) discrete-time regenerative process $X = \{X_n\}$ with regeneration instances $\{\theta_k\}$, which describes a queueing system at the arrival instants of a renewal input. Based on the result [7.59], we may conclude that, for each bounded set B, the expected number of visits to B by the regenerative process X, satisfying condition like [7.44], is finite. Denote by φ the distribution of the process X_{θ_k}, i.e. observed at regeneration points; we call φ the *regeneration measure*. In most cases, the measure φ is degenerate and concentrated at the point 0. However, this not always true. In this regard, it is worth mentioning that Harris Markov chains represent a wide class of regenerative processes with non-degenerate regenerative measure φ (see (Asmussen 2003; Serfozo 2009)). However, assume that this is indeed the case, and consider a state x belonging to the domain $\mathcal{D}_\varphi$ of the measure φ, and let $X_{\theta_k} = x$. Then, by analogy to expression [7.59], we can define a finite measure

$$\phi_x(B) := \mathsf{E}_x\Big(\sum_{k=0}^{\theta-1} \mathsf{1}(X_k \in B)\Big) = \sum_{k=0}^{\infty} \mathsf{P}_x(X_k \in B,\, \theta > k) \leq C, \qquad [7.73]$$

for some constant $C < \infty$. It is useful to note that

$$\lim_{n\to\infty} \mathsf{P}_x(X_n \in B,\, \theta > n) = 0, \;\; n \to \infty. \qquad [7.74]$$

Furthermore, because the regenerative process $\{X_n\}$ starts anew at θ_k, the *regeneration measure* completely describes the state of the process at each regeneration instant θ_k. In this case, for each bounded set B, we can define the *stationary finite measure*

$$\phi_\varphi(B) := \int_{x\in\mathcal{D}_\varphi} \phi_x(B)\varphi(dx), \qquad [7.75]$$

which equals the mean number of visits to the set B during a generic regeneration period.

The comments above show that if a regenerative process obeys a regeneration condition such as [7.32], then *each bounded set B has finite measure [7.73]* for each initial state x. This further implies that the mean number of visits to any bounded set B is finite *irrespective of whether or not the process is positive recurrent.* Obviously, if the process is positive recurrent, this measure is finite, since

$$\phi_\varphi(B) \le \phi_\varphi(\mathsf{R}_+) = \mathsf{E}_\varphi\,\theta < \infty.$$

Otherwise, if the process is *not positive recurrent*, then our result shows that the process *goes to infinity*. Indeed, consider a (zero-delayed, for simplicity) *non-positive recurrent* process $\{X_n\}$, in which case

$$\mathsf{E}_\varphi\theta = \infty. \qquad [7.76]$$

It further may be either *null recurrent*, in which case, in addition, $\theta < \infty$ w.p.1, or *terminating*, for which $\mathsf{P}(\theta = \infty) > 0$. Assume that, for such a process, for some $x \in \mathcal{D}_\varphi$ and a bounded set B, $\mathsf{P}_x(X_n \in B) \not\to 0$, implying $\inf_i \mathsf{P}_x(X_{n_i} \in B) \ge \varepsilon$ for a deterministic sequence $n_i \to \infty$ and some $\varepsilon > 0$. Then, as above, we can establish that $\mathsf{E}_\varphi\,\theta < \infty$, and this contradiction shows that, for a non-positive recurrent process, and for any bounded set B and any $x \in \mathcal{D}_\varphi$, the following convergence holds:

$$\mathsf{P}_x(X_n \in B) \to 0,\ n \to \infty. \qquad [7.77]$$

Indeed, it is easy to show, also by contradiction, that under [7.76], convergence [7.77] holds for an *arbitrary initial state x*. In this regard, we note that if $\{X_n\}$ is a null-recurrent Harris Markov chain with a general state space, then [7.77] holds as well (Asmussen 2003; Serfozo 2009).

We would also like to highlight the following remarkable feature of the presented approach. As a careful reader may have observed, in order to establish positive recurrence, we indeed verify a very *weak* requirement: namely that the process does *not go to infinity*. At first glance, this requirement seems absolutely not sufficient to obtain such a nice property as positive recurrence, because we could not exclude *a priori* a phenomenon such as oscillations of the process over time with no visits to the regeneration state. However, such undesirable effect is removed by the regeneration condition, and as a result, the asymptotic behavior of the regenerative process is indeed completely defined by the characterization we widely apply in the analysis: either the process *goes to infinity* or it is *positive recurrent.*

We also mention that in many cases the regenerative analysis of the positive recurrent systems allows us to obtain, in a short and elegant way, the explicit expressions for some important stationary performance measures. We mention here the well-known Pollaczeck–Khinchine and Little's formulae, the loss probability in a finite buffer system, etc., but this topic goes beyond the scope of this chapter. (A lot of material on this topic can be found in Asmussen (2003) and Serfozo (2009).)

In conclusion, we would like to point out an analogy between our method and the Lyapunov function method (Bremaud 1998; Down 2005), which however is only applied to Markov processes. As a rule, proving stability, we deal either with the (one-dimensional) basic process directly, or with the sum of its components (in the multidimensional case), which plays the role of a Lyapunov function in our analysis. Indeed, we can show that this function (under some predefined conditions) has a negative mean increment. This explains to some extent why the sufficient stability conditions that are deduced in this way turn out to be the stability criteria, or are very close to the stability criteria. At the same time, considering the process itself as a Lyapunov function, we avoid the typical difficulties that arise in the construction of such a function for different Markov processes related to queueing theory. In the regenerative method that we apply, the finiteness of the mean return time to a *regeneration state* is explicitly included in the definition of positive recurrence. For a given (non-Markov) queueing process, a regeneration state is often unique, and for this reason the regenerative approach requires an additional step (realized by means of the regeneration condition) to show that the regeneration state is visited with a positive probability from an arbitrary point of the bounded set.

7.7. Related research

Now we briefly discuss the stability analysis of other queueing systems that have been successfully investigated by the regenerative approach described above.

First, we mention an m-server system with non-identical servers considered in Morozov (1993). The input is a renewal process with rate λ, and server i has service rate μ_i. This system, unlike classical queueing models, does not possess the monotonicity property, and it makes the corresponding stability analysis particularly difficult. To overcome the difficulty caused by the non-monotonicity, the regenerative stability analysis is based in this case on a comparison theorem, which finally allows us to prove that condition

$$\lambda < \sum_{i=1}^{m} \mu_i, \qquad [7.78]$$

is the stability criterion for this system. The regenerative *instability* analysis of this system can be found in Morozov (2002). One more generalization of the classic multiserver system is a multiclass system with *server-dependent* and *class-dependent* service. Such a model is studied in Morozov (2014a), where the stability criterion is obtained as well. A related stability result can be found in Foss and Chernova (1998). Sufficient stability conditions of a very general multiserver system with a dependence between the input process, the service time process and the workload have been presented in Morozov (2014b).

Another important class of models whose stability has been investigated by the regenerative method are related to optical buffer systems, which are considered in the papers (Rogiest *et al.* 2010; Morozov *et al.* 2012). A specific problem arises with the buffering of optical bursts of data (or customers, in queueing terms). Since light cannot be frozen, optical buffering can be implemented by delaying the light with a set of fiber delay lines (FDLs). An interesting feature of these models is that, in order to obtain sufficient stability conditions, we must take into account the remaining time that the subsequent transmitted customers must spend in their assigned FDL while the server is idle after the previous departure. As a result, some of the server capacity remains unused. Nevertheless, in the framework of the regenerative analysis that we propose, it is possible to establish tight sufficient stability conditions, which include the *distance* between the longest FDLs. The key point of the analysis is to consider the system approaching a saturated regime (when, roughly speaking, the FDLs with very long lengths are occupied).

The same idea (i.e. to consider a saturated regime) is also used in the regenerative stability analysis of a wide class of *retrial models* (Avrachenkov and Morozov 2014; Avrachenkov *et al.* 2018; Morozov 2007; Morozov and Phung-Duc 2017). In such models, an arriving class-i customer joins a virtual orbit (with infinite capacity) when it encounters a busy server upon arrival, and then tries to access server again after some delay that depends on the specifics of the model under consideration. Among many existing variations of the retrial policy, the following main types are typically considered: the *classical retrial policy*, when the retrial rate depends (as a rule, linearly) on the number of orbital customers acting independently (Morozov 2007; Morozov and Phung-Duc 2017), and the *constant retrial rate discipline*, when the retrial rate does not depend on the orbit's content (Avrachenkov and Morozov 2014; Avrachenkov *et al.* 2018). It is intuitively clear that under the classic discipline, the stability condition coincides with the stability criterion of a classical buffered system since, when the system approaches saturation, the retrial system behavior approaches that of the corresponding buffered system. We call such a discipline *asymptotically work conserving*, because the lost capacity of the server, caused by the idle time occurring after each departure, vanishes as the number of orbital customers increases (Morozov 2007). On the other hand, in the constant retrial rate case, this lost capacity must be taken into account in the formulation of the stability conditions.

Some very recent studies deal with multiclass systems with so-called *coupled orbits*, where the retrial rate of a given orbit depends on the binary state (idle/busy) of other orbits (Morozov and Dimitriou 2017; Morozov and Morozova 2018). This model reflects the effect of interference in modern mobile communication systems.

Using a coupling-based regenerative approach and the PASTA property (Wolff 1982), the necessary and sufficient stability conditions of such models are found as well.

As an additional example of the effective application of the regenerative stability analysis, we would also like to mention the discrete-time queueing model with interruptions studied in Morozov *et al.* (2011). The discrete nature of the model allows us to consider more complicated inputs, being a superposition of independent (discrete) renewal processes. In this case, in order to construct a synchronization of the renewal instants of the aggregated input, the discrete structure of the renewal intervals of the component processes is used. Further analysis as always is based on a negative drift and regeneration condition, and is rather straightforward.

Finally, we mention a system with regenerative input presented in previous works (Morozov 1997a, 2004; Morozov and Delgado 2009). This model is motivated by studying acyclic queueing networks because, unlike renewal process, regenerative input preserves the regeneration property crossing the (non-saturated) nodes of the network (Morozov 1991; Sigman 1988a; Morozov 2004). A specific (non-restrictive) condition that we require to be fulfilled in this model is that the sequence of interarrival times of the regenerative input becomes renewal with a positive probability. This requirement is in fact a regeneration condition, allowing us to unload the system by means of a series of independent events, as in [7.40], to reach a regeneration state in a limited time. Then, under a standard negative drift assumption, this system turns out to be positive recurrent. In this regard, we would like to mention some recent papers (Afanasyeva and Tkachenko 2014; Afanasyeva 2019) devoted to stability analysis of queueing systems, where a somewhat different concept of regenerative input is considered.

In order to advocate even further the importance and usefulness of the regenerative approach, we would like to point out that a wide class of Markov chains (i.e. *recurrent Harris chains*) possesses a regeneration property, and in this case the regenerative stability analysis can be reformulated in terms of positive recurrence of the corresponding Markov chain, again by means of the finiteness of the mean regeneration period. Although in this chapter we consider only classic regeneration processes (with i.i.d. regeneration cycles), the interpretation of stability in the terms of positive Harris recurrence turns out to be meaningful in the case of *one-dependent* regeneration, when a dependence between two adjacent cycles is allowed while the cycle lengths stay independent (Morozov 1994; Sigman and Wolff 1993; Sigman 1988a,b, 1990). We call this property *weak regeneration* in Morozov (2004). Some more specific aspects of weakly regenerative queueing systems and networks can be found in previous works (Morozov 1991; Nummelin 1981).

When discussing the regenerative approach, we do need to mention the important problem of estimating by simulation the performance quantities of the regenerative queueing systems under consideration. The regenerative performance analysis, which is mainly based on the limiting relations [7.8] and [7.9], does not allow to obtain stationary performance measures in an explicit form, except for a few important results such as Little's formula as mentioned above, and in some of the more intricate models, the stability conditions may even contain certain parameters that must be estimated in advance, see for instance (Tezcan 2013). Regenerative process theory is among the most powerful and elegant ideas used in simulation output analysis (Asmussen and Glynn 2007). Its main advantage is based on the possibility to apply well-developed statistical procedures to correlated data that model the traffic of present broadband communications systems. The regenerative structure greatly enhances the ability of the simulator to apply output analysis and variance reduction techniques, yielding substantial advantages over non-regenerative systems by exploiting the refined structural properties of the process.

Summarizing the discussion of the regenerative approach, we would like to emphasize one more time the key role of the renewal structure of the regenerations, allowing us to develop the stability analysis method described above. It is important to mention that the proof of positive recurrence beyond renewal theory is generally a tough problem.

Stability analysis of queueing models is of course not restricted to the regenerative approach. A detailed and quite general presentation of the stability analysis of stochastic models can be found in Borovkov (1984, 1987, 2000). In particular, the stability (ergodicity) of general stochastic processes describing the behavior of telecommunication networks is considered in Borovkov (2000), and a detailed presentation of stability theory of general Markov chains, which is useful from the point of view of the problems treated in this chapter, can be found in Meyn and Tweedie (1993).

Another alternative and effective method for the stability analysis of a wide class of queueing systems, including regenerative systems, is the *fluid approximation* (Chen 1995; Dai 1995). In this method, the original stochastic model is replaced by its deterministic analogue, called the *fluid model*. The stability of the original system is then deduced from the stability of the fluid model (Dai 1995). We would like to emphasize that the regenerative approach, unlike the fluid approximation, allows us to capture the so-called boundary case, which in case of classic $GI/G/1$ queueing system is expressed by the known relation for the traffic intensity $\rho = 1$. On the other hand, the fluid approach can cover the stability analysis of queueing systems with multiple independent renewal inputs, in which case the regenerative approach faces considerable difficulties. A comprehensive overview of various stability methods of stochastic models can also be found in Foss and Konstantopoulos (2004).

Although the literature devoted to the stability of queueing models is vast, there is not yet a detailed discussion of the regenerative stability analysis method (an exclusion is (Morozov and Delgado 2009)), and we believe that the work presented here fills this gap at least in part.

7.8. Acknowledgments

The authors wish to express their thanks to the referee for a number of valuable comments which, we hope, allowed us to improve the presentation of the material.

The research of EM is in part supported by Russian Foundation for Basic Research, projects 18-07-00147, 18-07-00156 and 19-07-00303.

7.9. References

Afanasyeva, L.G., Tkachenko, A.V. (2014). Multichannel queueing systems with regenerative input flow. *Theory Probab. Appl.*, 58(2), 174–192.

Afanasyeva, L.G. (2019). Asymptotic analysis of queueing models based on synchronization method. *Methodology and Computing in Applied Probability*. https://doi.org/10.1007/s11009-019-09694-9.

Asmussen, S. (2003). *Applied Probability and Queue*, 2nd ed. Springer-Verlag, New York.

Asmussen, S., Glynn, P.W. (2007). *Stochastic Simulation: Algorithms and Analysis*. Springer, New York.

Avrachenkov, K., Morozov, E. (2014). Stability analysis of GI/GI/c/K retrial queue with constant retrial rate. *Math. Methods Oper. Res.*, 79, 273–291.

Avrachenkov, K., Morozov, E., Steyaert, B. (2016). Sufficient stability conditions for multi-class constant retrial rate systems. *QUESTA*, 81, 149–171.

Billingsley, P. (1995). *Probability and Measure*, 3rd ed. John Wiley & Sons, New York.

Borovkov, A. (1984). *Asymptotic Methods in Queueing Theory*. John Wiley & Sons, New York.

Borovkov, A. (1987). Limit theorems for queueing networks I. *Theory Probab. Appl.*, 31(3), 413–427.

Borovkov, A. (2000). *Ergodicity and Stability of Stochstic Processes*. John Wiley & Sons, New York.

Bremaud, P. (1998). *Markov Chains: Gibbs Fields, Monte Carlo Simulation, and Queues*. Springer-Verlag, New York.

Chen, H. (1995). Fluid approximation and stability of multiclass queueing networks: Work-conserving disciplines. *Ann. Appl. Probab.*, 5(3), 637–665.

Dai, J. (1995). On positive Harris recurrence of multiclass queueing networks: A unified approach via fluid limit models. *Ann. Appl. Probab.*, 5(1), 49–77.

Down, D. (2005). A survey of Markovian methods for stability of networks. *LN*, 199, 490–504.

Dynkin, E.B. (1996). Some limit theorems for sums of independent random variables with infinite mathematical expectations. *Selected Translations Math. Stat. Prob.*, 1, 171–189.

Feller, W. (1971). *An Introduction to Probability Theory, Volume II*. John Wiley & Sons, New York.

Foss, S., Chernova, N. (1998). On the stability of a partially accessible multi-station queue with state-dependent routing. *QUESTA*, 29, 55–73.

Foss, S., Konstantopoulos, T. (2004). An overview on some stochastic stability methods. *Journal of the Operations Research Society of Japan*, 47(4), 275–303.

Glynn, P. (1994). Some topics in regenerative steady-state simulation. *Acta Appl. Math.*, 34, 225–236.

Kalashnikov, V. (1994). *Topics on Regenerative Processes*. CRC Press, Boca Raton, FL.

Kiefer, J., Wolfowitz, J. (1955). On the theory of queues with many servers. *Trans. Amer. Math. Soc.*, 78, 1–18.

Meyn, S.P., Tweedie, R.L. (1993). *Markov Chains and Stochastic Stability*. Springer-Verlag, London.

Morozov, E. (1991). Conservation of a regenerative stream in an acyclic net. *Journal of Soviet Mathematics*, 57(4), 3302–3305.

Morozov, E. (1993). A comparison theorem for queueing system with non-identical channels. *Lecture Notes in Mathematics, Stability Problems for Stochastic Models*, 1546, 129–133.

Morozov, E. (1994). Wide sense regenerative processes with applications to multi-channel queues and networks. *Acta Appl. Math.*, 34, 189–212.

Morozov, E. (1997a). The stability of non-homogeneous queueing system with regenerative input. *J. Mathemat. Sci.*, 89, 407–421.

Morozov, E. (1997b). The tightness in the ergodic analysis of regenerative queueing processes. *QUESTA*, 27, 179–203.

Morozov, E. (2002). Instability of nonhomogeneous queueing networks. *J. Mathematic. Sci.*, 112, 4155–4167.

Morozov, E. (2004). Weak regeneration in modeling of queueing processes. *QUESTA*, 46, 295–315.

Morozov, E. (2007). A multiserver retrial queue: Regenerative stability analysis. *QUESTA*, 56, 157–168.

Morozov, E. (2014a). Stability criterion of a general multiserver multiclass queueing system. *Proceedings of 29th International Symposium on Computer and Information Sciences*, Krakow, Poland, 27–28.

Morozov, E. (2014b). Stability analysis of a general state-dependent multiserver queue. *J. Mathematic. Sci.*, 200(4), 462–472.

Morozov, E., Delgado, R. (2009). Stability analysis of regenerative queues. *Autom. Remote Control*, 70, 1977–1991.

Morozov, E., Dimitriou, I. (2017). Stability analysis of a multiclass retrial system with coupled orbit queues. *LNCS*, 10497, 85–98.

Morozov, E., Fiems, D., Bruneel, H. (2011). Stability analysis of multiserver discrete-time queueing systems with renewal-type server interruptions. *Perf. Eval.*, 68, 1261–1275.

Morozov, E., Morozova, T. (2018). Analysis of a generalized retrial system with coupled orbits. *Proceedings of 23rd Conference of the FRUCT Association*, 253–260.

Morozov, E., Phung-Duc, T. (2017). Stability analysis of a multiclass retrial system with classical retrial policy. *Perf. Eval.*, 112, 15–26.

Morozov, E., Rogiest, W., De Turck, K., Fiems, D. (2012). Stability of multiwavelength optical buffers with delay-oriented scheduling. *Europ. Trans. on Telecomm.*, 23(3), 217–226.

Nummelin, E. (1981). Regeneration in tandem queues. *Adv. Appl. Prob.*, 13, 221–230.

Rogiest, W., Morozov, E., Fiems, D., Laevens, K., Bruneel, H. (2010). Stability of single-wavelength optical buffers. *Europ. Trans. on Telecomm.*, 21, 202–212.

Serfozo, R.F. (2009). *Basics of Applied Stochastic processes*. Springer-Verlag, Heidelberg.

Sigman, K. (1988a). Regeneration in tandem queues with multiserver stations. *J. Appl. Prob.*, 25, 391–403.

Sigman, K. (1988b). Queues as Harris recurrent Markov chains. *QUESTA*, 3, 179–198.

Sigman, K. (1990). One-dependent regenerative processes and queues in continuous time. *Math. Oper. Res.*, 15, 175–189.

Sigman, K., Thorisson, H., Wolff, R.W. (1994). A note on the existence of regeneration times. *Appl. Prob.*, 31, 1116–1122.

Sigman, K., Wolff, R.W. (1993). A review of regenerative processes. *SIAM Review*, 35, 269–288.

Smith, W.L. (1955). Regenerative stochastic processes. *Proc. Royal Soc., Series A*, 232, 6–31.

Tezcan, T. (2013). Stability analysis of N-model systems under a static priority rule. *QUESTA*, 73, 235–259.

Wolff, R.W. (1982). Poisson arrivals see time averages. *Oper. Res.*, 30(2), 223–231.

8

Transient Analysis of Markovian Queueing Systems: a Survey with Focus on Closed-forms and Uniformization

Gerardo RUBINO[1]

[1]*Inria, Rennes, France*

Analyzing the transient behavior of a queueing system is much harder than studying its steady-state, the difference being basically that of moving from a linear system to a linear *differential* system. However, a huge amount of efforts has been made for the former problem from all kinds of points of view: trials to find closed-forms of the main state distributions, algorithms for numerical evaluations, approximations of different types, exploration of other transient metrics than the basic state distributions, etc.

In this chapter, we focus on the first two elements, the derivation of closed-forms for the main transient state distributions and the development of numerical techniques. This chapter is organized as a survey, and the main goal is to position and to underline the role of the uniformization technique, for both finding closed-forms and for developing efficient numerical evaluation procedures. In some cases, we extend the discussion to other related transient metrics that are relevant for applications.

Queueing Theory 2,
coordinated by Vladimir ANISIMOV, Nikolaos LIMNIOS.

8.1. Introduction

Queueing models are the most frequently used when analyzing the performance of a resource sharing system, and they are applied on a daily basis in many different areas. As for other dynamical systems, most of their analysis is done in steady-state and for several reasons (see the nice developments in the Introduction of Whittle (1986)). Studying the system in equilibrium can provide qualitative insight into its structure and its limits, and also about the connections between the different parameters that characterize it. Moreover, in several areas, the related systems get very close to steady-state in short time delays for human standards (often in minutes for computing facilities or telecommunication networks).

However, if we are interested in the behavior of a system long before equilibrium, or if we do not know how far we will be from equilibrium at some specific point in time, only the transient behavior of the system matters. Another situation of interest here is when the parameters' values of a model must be changed, to take into account, for instance, a change in the environment; just after the change, the behavior of the system is naturally captured by a transient analysis. This is not just a theoretical consideration; anyone that needs to evaluate the performance of a system will in general try to identify time regions where the parameters characterizing the environment do not change too much, and will look at both the transient distributions at the beginning of that time area, and then at steady-state, for the case where we are far enough from the changing zone. The problem is that transient analysis is much harder than studying the asymptotic behavior, and the amount of known results and related algorithmic tools is significantly less than for the steady-state case. Of course, when we rely on simulation, analyzing the system in the transient regime is even simpler than in steady-state, since the latter corresponds to look at what happens when time goes to infinity.

In this chapter, we will describe some elements of the transient behavior of queues when they are in their transient phase, limiting the focus on Markovian models, because they are by far the most frequently used in performance or dependability evaluation studies. We will also concentrate the effort on analytical results when available, because they are richer from the mathematical point of view and also because they show structural aspects of the behavior of the models, which are harder to see when applying numerical techniques or simulators. We will also look at algorithmic tools specialized to specific classes of models.

We will only consider here queues and not networks of queues, because even if in steady-state there are many results concerning Markovian networks, mainly around the concept of product form, there is nothing similar for the transient regime. Boucherie and Taylor (1993) showed that to have a product-form transient distribution, the queueing network must be composed of $./M/\infty$ nodes, plus supplementary conditions. For material related to this negative result, see also Boucherie (1993) (where the transient behavior of the important Engset model is explored). It is shown that only in particular cases (uninteresting in practice) such a product-form transient distribution of a network of queues exists. Readers can also refer to Lebah and Pellaumail (1993) for an analysis of Jackson networks in the transient regime in the positive case (nodes with an infinite number of servers).

There are many related topics that have been excluded from this text to avoid the immediate increase in its size. These include a large part of the material concerning the numerical analysis of Markovian queueing models, where the idea is to numerically solve the Kolmogorov differential equations, the development of approximations instead of exact analysis, including mean field approaches, the consideration of non-stationary arrival processes of state-dependent models, the study of other transient metrics than the basic distributions (for instance, different cases of hitting times, or important queueing metrics such as waiting or response times – their moments, their distributions, etc.), the time to reach equilibrium in the transient phase, etc. Another world we avoided is that of discrete-time queues, simply because of their limited use compared to the continuous-time queues. As stated before, we concentrate on analytical results related to the fundamental state distributions, and, in particular, on closed-forms. From the methodological viewpoint, we will often focus on the use of the uniformization technique that allows us to attack the problem in discrete-time.

This chapter is organized as follows. Section 8.2 presents the general framework of the chapter, and introduces the uniformization technique that moves a problem specified in continuous-time to a discrete-time setting. In section 8.3, we introduce some queueing models where closed-forms for the standard state distribution are not hard to obtain, and then, the most fundamental system, the $M/M/1$ model, together with its bounded version, the $M/M/1/H$ queue. Here, we basically describe the historical approaches that led to the first closed-form expressions of the associated transient distributions. In section 8.4, a more recent way of deriving the transient distribution of the $M/M/1$ model is presented, and some new material is added to the previously published work. This approach is based on uniformization, and it leads to a particularly simple and useful closed-form expression. Then, in section 8.5 another technique based on the concept of duality, described following (Anderson 1991), is presented. It allows us to obtain again the same kind of expression as in section 8.4 (since the initial step is again uniformization), but following a very different path. The method extends to other Markovian queueing systems. Section 8.6 briefly describes other models, metrics and results related to the main stream of the chapter.

8.2. Basics on Markovian queues

In this section, we establish the Markovian framework and the main notation. The goal is to present then the uniformization procedure, which plays an important role in analyzing transient distributions of basic queueing systems.

8.2.1. *Markov models*

Let us resume here the main elements we need in Markov models, and choose some global notation. Let us use the term "process" for a Markov model in continuous-time and "chain" if time is discrete. We will only consider discrete state spaces here.

Since we are dealing with queues, we will focus on homogeneous continuous-time Markov models on a finite or infinite denumerable state space S (often $\mathbb{N}$, or a segment $\{0, 1, \ldots, H\}$ of positive integers). If X is such a model, its infinitesimal generator will be denoted by A and the transition rate matrix by Q. The transition rate from state i to state $j \neq i$ will then be $Q_{i,j} = A_{i,j}$. The departure rate from state i is $d_i = -A_{i,i} = \sum_j Q_{i,j}$[1]. For any $i \in S$, $Q_{i,i} = 0$. In matrix terms, if D is the diagonal matrix with $D_{i,i} = d_i$, then $A = Q - D$. When $d_i = 0$, we say that state i is absorbing. If X represents the state of a queueing model, in general it has no such absorption state; process X is then irreducible. In basic queues, the analysis focuses on the number of customers in the system; in such cases, we will use $N(t)$ instead of $X(t)$.

Let us denote by Y the discrete-time Markov chain canonically embedded into X at X's jump times. If the departure rate from state i is $d_i > 0$, then the transition probability to move to state j coming from state i, $j \neq i$, is $P_{i,j} = Q_{i,j}/d_i$. If $d_i = 0$, that is, if state i is absorbing, then we set $P_{i,j} = 0$ for all $j \neq i$ and $P_{i,i} = 1$. In matrix notation and for an irreducible model, we write $P = D^{-1}Q$, where D^{-1} is the diagonal matrix whose element (i, i) is $\left(D^{-1}\right)_{i,i} = 1/d_i$. Chain Y shares its initial distribution with process X.

Matrix $\left(\mathbb{P}(X(t) = j \mid X(0) = i)\right)_{i,j \in S}$ is the transition function of process X, and we will denote it by $P(t)$. When the state space is finite, matrix e^{At} exists for any $t \in \mathbb{R}_{\geq 0}$ and $e^{At} = P(t)$. In the infinite case, previous matrix exponential does not always exists, but if it does, the same equality holds. A sufficient condition for its existence is that the set of departure rates $\{ d_i, i \in S \}$ is upper bounded (see below).

In this chapter, we will denote $p_x(t) = \mathbb{P}(X(t) = x)$ or $p_x(t) = \mathbb{P}(N(t) = x)$. The distribution of $X(t)$ (or $N(t)$) will be denoted $p(t)$ and will be seen as a row

1. "Pathological" Markov processes where this is not true can be constructed. We ignore these cases, our models will always be stable and conservative. Readers can see Anderson (1991) or Kijima (1997) for details.

vector, the usual convention in the Markovian world. The steady-state distribution for the process, assuming its existence, will be π, also a row vector, with the element indexed by state x and denoted byπ_x. The Chapman–Kolmogorov equations are $p'(t) = p(t)A$ and the equilibrium or balance equations can be written as$\pi A = 0$. When dealing with queues, and only to simplify the presentation, we will in general assume that the queue is empty at the time origin, that is, $\mathbb{P}(N(0) = 0) = 1$. In the provided references, readers can find the details for different initial conditions.

8.2.1.1. *Some notation for basic queues*

When discussing the $M/M/1$ or the $M/M/1/H$ models, the arrival rate will always be $\lambda > 0$ and the service rate will be $\mu > 0$. Then, the transition probability of moving from state i to state $i+1$ is $p = \lambda/(\lambda + \mu)$ and of moving from i to $i-1$, for $i \geq 1$, is $q = \mu/(\lambda + \mu) = 1 - p$; the load is $\varrho = \lambda/\mu = p/q$. The canonical Markov process associated with these models is $\{ N(t), t \in \mathbb{R}_{\geq 0} \}$, where $N(t)$ is the number of customers in the system at time t.

8.2.2. *Uniformization*

If the space-state S is finite, or if it is infinite denumerable with bounded departure rates, we say that X is uniform or uniformizable. In this case, the matrix exponential e^{At} exists. Taking any positive real number Λ satisfying $\sup_i d_i \leq \Lambda$, we can construct a new process as follows. First, we define a new matrix $U = I + A/\Lambda$, where I is the identity matrix. It is easy to see that U is stochastic. We can then build a new discrete-time Markov chain on the same state space S, with the same initial distribution as X and whose transition probability matrix is U; let us call Z this new chain, called the *uniformized* chain built from X with *uniformization rate* Λ. We call this procedure *uniformization*, or *randomization*, or *Jensen's method* because it was Arne Jensen who first proposed it as a procedure (Jensen 1953). Observe that X is irreducible $\iff Z$ is irreducible.

From $U = I + A/\Lambda$, write $A = -\Lambda(I - U)$, multiplying by t and taking exponentials; we get

$$\mathrm{e}^{At} = \mathrm{e}^{-\Lambda t(I-U)} = \mathrm{e}^{-\Lambda tI}\mathrm{e}^{\Lambda tU} = \mathrm{e}^{-\Lambda t}\mathrm{e}^{\Lambda tU}. \qquad [8.1]$$

The last equality comes from the fact that I commutes with any matrix M and thus, so do matrices aI and bM for any coefficients a and b. The last is an immediate property of the matrix exponential function. Writing the Taylor series of previous matrix function of t at $t = 0$, we obtain

$$\mathrm{e}^{At} = \sum_{n\geq 0} \mathrm{e}^{-\Lambda t}\frac{(\Lambda tU)^n}{n!} = \sum_{n\geq 0} \mathrm{e}^{-\Lambda t}\frac{(\Lambda t)^n}{n!}U^n. \qquad [8.2]$$

Looking at probability distributions as row vectors and by denoting α the initial distribution of X (and thus, also of Z), $p(t)$ the distribution of $X(t)$, $t \in \mathbb{R}_{\geq 0}$, and $q(n)$ the distribution of $Z(n)$, $n \in \mathbb{N}$, we get the vectorial representation

$$p(t) = \sum_{n \geq 0} \mathrm{e}^{-\Lambda t} \frac{(\Lambda t)^n}{n!} q(n). \qquad [8.3]$$

In scalar form, for any $j \in S$,

$$p_j(t) = \sum_{n \geq 0} \mathrm{e}^{-\Lambda t} \frac{(\Lambda t)^n}{n!} q_j(n), \qquad [8.4]$$

with, concerning X, $p(t) = \big(\ldots p_j(t) \ldots\big)$ where $p_j(t) = \mathbb{P}(X(t) = j)$, and similarly, $q(n) = \big(\ldots q_j(n) \ldots\big)$, where $q_j(n) = \mathbb{P}(Z(n) = j)$.

REMARK 8.1.– (The power of uniformization). Uniformization is a very fruitful and powerful way of working with Markovian models. It is fruitful because it moves a problem specified in continuous-time into a new problem that evolves in discrete-time, that is, a linear differential problem into a linear difference problem. Moreover, the transformation has a very simple probabilistic interpretation, as briefly described before, that has led to many new results in many published works. It is powerful because when we need numerical values, the obtained series involves only positive terms and the only operators involved are sums and products; this leads to stable algorithms. Next remark emphasizes a last important feature of this transformation, extremely useful for numerical evaluations.

REMARK 8.2.– (Error control in uniformization). Assume we need to know, say, $p_j(t)$, for some fixed $j \in \mathbb{N}$, with an absolute error less than ε. If we truncate the series at index $n = N$, the absolute error is

$$err_N(t) := p_j(t) - \sum_{0 \leq n \leq N} \mathrm{e}^{-\Lambda t} \frac{(\Lambda t)^n}{n!} q_j(n) = \sum_{n > N} \mathrm{e}^{-\Lambda t} \frac{(\Lambda t)^n}{n!} q_j(n) \geq 0. \qquad [8.5]$$

Now, since $q_j(n) < 1$,

$$0 \leq err_N(t) < \sum_{n > N} \mathrm{e}^{-\Lambda t} \frac{(\Lambda t)^n}{n!} < \varepsilon \iff \sum_{0 \leq n \leq N} \mathrm{e}^{-\Lambda t} \frac{(\Lambda t)^n}{n!} > 1 - \varepsilon.$$

So, given ε, we compute

$$N_{\text{min}} = \min \left\{ N \in \mathbb{N} \;\middle|\; \sum_{0 \leq n \leq N} \mathrm{e}^{-\Lambda t} \frac{(\Lambda t)^n}{n!} > 1 - \varepsilon \right\}. \qquad [8.6]$$

Previous observations then state that summing the first $N_{\min}$ terms of the series in [8.4] guarantees an approximation of $p_j(t)$ with absolute error less than ε. The key observation here is that this is not only elementary, but (i) computing $N_{\min}$ is an extremely fast procedure, and (ii) we can do it beforehand, and thus evaluate the cost of the whole numerical evaluation before starting it.

Just for completeness, another way of connecting chain Z with the original process X is the following. If $\{ N(t), t \in \mathbb{R}_{\geq 0} \}$ is the counting process of a Poisson process independent of Z and with rate Λ, then we have $Z(N(.)) \equiv_{\text{st}} X(.)$, where '$\equiv_{\text{st}}$' means "stochastically equivalent" (the same joint distributions at any number of arbitrary points in $\mathbb{R}_{\geq 0}$).

There are many references concerning this important tool of the Markovian universe. See, for instance, (de Souza e Silva and Gail 2001) and the many references therein.

8.3. First examples

In this section, we will describe the transient regime of some basic models for which obtaining transient distributions is simple and/or well known. In the infinite state case, the main difficulty in the analysis of these models is the fact that we are dealing with an infinite linear differential system with an infinite number of unknowns.

8.3.1. *The Ehrenfest model in continuous-time*

Let us start with the very famous and simple model proposed in 1907 by Tatiana and Paul Ehrenfest for the analysis of the second principle of thermodynamics. We have two boxes numbered 1 and 0 and H particles. Initially, all particles are in box 1, and the particles move randomly from one box to the other. Let us say that $N(t)$ is the number of particles in box 1 at time t (so, $N(0) = H$). The system's dynamics corresponds to the birth and death process $\{ N(t),\ t \in \mathbb{R}_{\geq 0} \}$, with non-null rates $Q_{j,j-1} = j\lambda$ for $1 \leq j \leq H$, and $Q_{j,j+1} = (H-j)\lambda$ for $0 \leq j \leq H-1$.

This process can be also represented as the sum $X_1(t) + \cdots + X_H(t)$, where the $X_i(t)$s are i.i.d., and each term is the state of the two-state Markov process given in Figure 8.1, with the initial condition $X_1(0) = 1$.

Observe that the number of particles in box 1 at t is also the number of customers at time t in a finite-source queue with state-dependent arrival rate $\lambda_i = (H-i)\lambda$, $i = 0, 1, \ldots, H-1$ (that is, a queue fed by H Exponential sources, each with the same rate λ) and state-dependent service rate $\mu_j = j\lambda$, $j = 1, 2, \ldots, H$ (that is, a queue with H servers, each with, again, the same rate λ), the $M/M/H//H$ model in Kendall's notation.

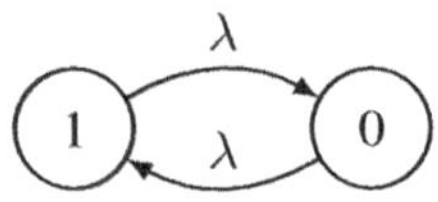

Figure 8.1. *The evolution of an individual particle in the continuous-time Ehrenfest model. When the initial state is state 1, the probability for the particle to be in state 1 at time* t *is* $p(t) = \left(1 + \mathrm{e}^{-2\lambda t}\right)/2$

From $\mathbb{P}(X_i(t) = 1) = p(t)$ for the ith particle, we immediately obtain for any $j \in \{0, 1, 2, \dots, H\}$, the expression

$$p_j(t) = \frac{1}{2^H} \binom{H}{j} \left(1 + \mathrm{e}^{-2\lambda t}\right)^j \left(1 - \mathrm{e}^{-2\lambda t}\right)^{H-j}.$$

The mean number of units in, say, box 1, is then $H\left(1 + \mathrm{e}^{-2\lambda t}\right)/2$.

8.3.2. The $M/M/\infty$ model

This is probably the simplest queueing model with an infinite state space. The family of distributions $\{\, p(t), t \in \mathbb{R}_{\geq 0} \,\}$ satisfies the system $p'(t) = p(t)A$ that is expressed as follows. For $j \geq 1$,

$$p_j'(t) = -(\lambda + j\mu)p_j(t) + \lambda p_{j-1}(t) + (j+1)\mu p_{j+1}(t)$$

and if $j = 0$,

$$p_0'(t) = -\lambda p_0(t) + \mu p_1(t).$$

The derivation of the solution to this infinite differential system is pretty simple. See, for instance, the probabilistic proof in Kulkarni (2017, page 183), or see the 2nd edition of (Gross *et al.* 2018) (in the References we mention the current 5th edition of the book, but the derivation details are in the 2nd edition). Assume, to simplify the discussion, that the system is empty at time 0. The result is that the transient distribution is Poisson: for any $j \in \mathbb{N}$,

$$p_j(t) = \mathrm{e}^{-\psi(t)} \frac{\psi(t)^j}{j!}, \qquad \text{where } \psi(t) = \varrho\left(1 - \mathrm{e}^{-\mu t}\right) \text{ and } \varrho = \frac{\lambda}{\mu}.$$

It follows that the mean number of customers in the system at time t is

$$\mathbb{E}(N(t)) = \psi(t) = \varrho\left(1 - \mathrm{e}^{-\mu t}\right).$$

8.3.3. *A queue with no server and catastrophes*

Even if generally speaking, it is hard to attack a system of infinite differential equations looking for closed-form solutions, the $M/M/\infty$ model is not the only case where this task can be successfully performed and using elementary techniques only. Consider a system where units arrive according to a Poisson process with rate λ, and where the service is exponentially distributed with rate μ, but when it ends, the whole buffer is emptied. Equivalently, think of a queue with no server and catastrophes or breakdowns, also called by some authors a birth process with catastrophes, where the latter arrive according to a new Poisson process independent of the arrival one, with rate μ. It still corresponds to a queueing system storing units, and such that after an exponentially distributed amount of time, once the system is occupied with at least one unit, all units staying there are instantaneously sent to some other place to be consumed (or destroyed). Assume, as usual, that the system is empty at time 0. Figure 8.2 illustrates this behavior.

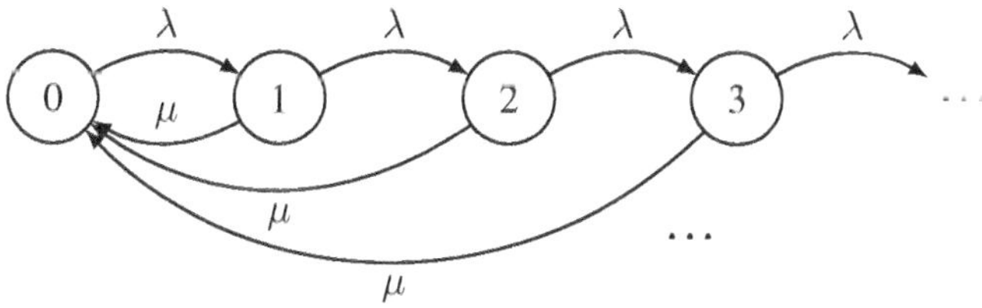

Figure 8.2. *An example where a closed-form for the transient behavior is easy to derive*

In this example, the Chapman–Kolmogorov equations are $p_0'(t) = -\lambda p_0(t) + \mu(1 - p_0(t))$ and for $j \geq 1$, $p_j'(t) = -(\lambda + \mu)p_j(t) + \lambda p_{j-1}(t)$. Denote by $\widetilde{f}$ the Laplace transform of function f of the time variable t, and by s the Laplace transform variable. Taking Laplace transforms in the Chapman–Kolmogorov equations, the corresponding linear system is easily solved. We get, for any $j \geq 0$,

$$\widetilde{p}_j(s) = \left(\frac{\lambda}{\lambda + \mu + s}\right)^j \frac{\mu + s}{s(\lambda + \mu + s)}.$$

For the inversion of these transforms, let us set first the notation

$$\varrho = \frac{\lambda}{\mu}, \qquad p = \frac{\lambda}{\lambda + \mu}, \qquad \mathcal{P}_M(t) = \sum_{m=0}^{M} e^{-(\lambda+\mu)t} \frac{(\lambda + \mu)^m t^m}{m!}.$$

We then obtain first

$$p_0(t) = \frac{1 + \varrho e^{-(\lambda+\mu)t}}{1+\varrho} = 1 - p + pe^{-(\lambda+\mu)t},$$

and for $j \geq 1$,

$$p_j(t) = \pi_j - p^j \mathcal{P}_{j-1}(t) + p^{j+1}\mathcal{P}_j(t),$$

where the steady-state distribution π is given, for any $j \in \mathbb{N}$, by

$$\pi_j = \frac{1}{1+\varrho}\Big(\frac{\varrho}{1+\varrho}\Big)^j = (1-p)p^j.$$

If we want to compute the mean number of units at time t, $m(t) = \mathbb{E}\big(N(t)\big) = \sum_{j\geq 1} jp_j(t)$, it is better to stay in the Laplace world, and compute

$$\widetilde{m}(s) = \sum_{j\geq 1} j\widetilde{p}_j(s) = \frac{\lambda}{s(\mu+s)} = \varrho\Big(\frac{1}{s} - \frac{1}{\mu+s}\Big),$$

leading immediately, by inversion of this transform, to $m(t) = \varrho\big(1 - e^{-\mu t}\big)$. Observe that this expectation is the same as for the $M/M/\infty$ model.

8.3.4. *The fundamental* $M/M/1$ *model*

This is the most fundamental queue and one of the most frequently used queueing models. Its transient regime has been the object of a large amount of publications, with many variations around similar approaches and many interesting results. Its derivation is more difficult than for the $M/M/\infty$ model. In this survey, we will first focus on some of the main historical developments. Then, we will describe our own contributions following the approach we believe is the richest, the uniformization-based approach.

Start by writing the associated Chapman–Kolmogorov equations: for $j \geq 1$,

$$p'_j(t) = -(\lambda+\mu)p_j(t) + \lambda p_{j-1}(t) + \mu p_{j+1}(t) \qquad [8.7]$$

and for the empty queue,

$$p'_0(t) = -\lambda p_0(t) + \mu p_1(t). \qquad [8.8]$$

Researchers looked for a solution to these equations for years, and the breakthrough arrived in the period 1953–1956, where different works were published, proposing the first expressions of the $p_j(t)$ probabilities, all in terms of Bessel functions.

The first analytical results on the transient regime of the $M/M/1$ models appear to be those of Clarke in 1953, in an unpublished work that only appeared as an internal technical report of the University of Michigan (Clarke 1953). In that text, the author obtains the classic solution in terms of modified Bessel functions of the first kind[2] making a change in variable and solving the new differential system (now, of the hyperbolic type) in terms of the solutions to Volterra integral equations.

In 1954 two papers appeared, Ledermann and Reuter (1954) and Bailey (1954), providing the same type of expressions of the transient distribution of the $M/M/1$ queue. In the first work, the authors analyze the spectrum of general birth and death processes (including the case of processes where state 0 is made absorbing, useful for instance to study the busy periods of queues). The second paper follows a completely different approach, by applying z-transforms first, then the Laplace transform, and obtaining a closed-form for the double transform function. The latter is then inverted, leading to the same expression as before based on modified Bessel functions. We will present it in some detail below. Then we have Morse (1955), in which the same type of representation of the transient distribution appears but not in a constructive manner, basically showing that it satisfies the differential equations. Closing this series of papers of the 1950s, we have Champernowne (1956), who follows a purely combinatorial path to derive again the same type of expression.

Let us provide now some elements of the solving procedure in Bailey (1954). This is probably the most frequently described method for deriving the transient distribution of the $M/M/1$ queue (for instance, in textbooks). Consider the generating function of the distribution $p(t)$ (or z-transform)

$$G(z,t) \stackrel{\text{def}}{=} \sum_{j\geq 0} p_j(t) z^j,$$

2. The modified Bessel function of the first kind, with index or order $k \in \mathbb{N}$, is the function of z

$$z \mapsto Ik(z) = \sum n \geq 0 \frac{1}{n!(k+n)!}\left(\frac{z}{2}\right)k + 2n.$$

defined at least on the domain $\{|z| \leq 1\}$ for any $t \in \mathbb{R}_{\geq 0}$. If we multiply the differential equation associated with state j by z^j and sum on j over $\mathbb{N}$, we obtain the equation

$$\frac{\partial G(z,t)}{\partial t} = -(\lambda+\mu)G(z,t) + \mu p_0(t) + \mu\frac{G(z,t) - p_0(t)}{z} + \lambda z G(z,t),$$

with the initial condition $G(z,0) = z^i$ if $p_i(0) = 1$. As before, let us concentrate on the case of an empty queue at time 0, that is, on the border condition $G(z,0) = 1$. Observe that since $|G(z,t)| \leq 1$ when $|z| \leq 1$, for all $t \in \mathbb{R}_{\geq 0}$, the convergence region of the Laplace transform of $\partial G(z,t)/\partial t$ is, at least, $\{\operatorname{Re}(s) > 0\}$, for all $t \geq 0$. Taking then Laplace transforms on both sides of previous equation, we obtain

$$s\widetilde{G}(z,s) - G(z,0) = -(\lambda+\mu)\widetilde{G}(z,s) + \mu\widetilde{p}_0(s)$$
$$+\mu\frac{\widetilde{G}(z,s) - \widetilde{p}_0(t)}{z} + \lambda z\widetilde{G}(z,s),$$

leading, using $G(z,0) = 1$, to

$$\widetilde{G}(z,s) = \frac{\mu\widetilde{p}_0(s)(1-z) - z}{\lambda z^2 - (\lambda+\mu+s)z + \mu}.$$

At this point, the difficulty is the unknown transform $\widetilde{p}_0$ appearing in the numerator. But look at the denominator as a polynomial in z. We first observe that $\lambda z^2 - (\lambda+\mu+s)z + \mu = 0$ has the two solutions

$$r_1(s) = \frac{\lambda+\mu+s-\sqrt{(\lambda+\mu+s)^2 - 4\lambda\mu}}{2\lambda},$$
$$r_2(s) = \frac{\lambda+\mu+s+\sqrt{(\lambda+\mu+s)^2 - 4\lambda\mu}}{2\lambda}.$$

Some elementary algebra then shows that $r_1(s) \neq r_2(s)$ because $\lambda\mu > 0$, and that $|r_1(s)| < 1 < |r_2(s)|$. Since function $\widetilde{G}$ is analytic inside the unit disc in z, the numerator of the obtained expression for this function must also be zero when $z = r_1(s)$ allowing solving for $\widetilde{p}_0$. This is the technical finding[3] that led to obtaining formal results following this direct approach (for instance, in Ledermann and Reuter (1954, end of page 366) it is said that the generating function method cannot provide

3. Some authors prefer to use Rouché's theorem to prove that the denominator in the expression of G^* has only one zero inside the unit disc.

a path toward the solution, showing that an idea for finding function $\widetilde{p}_0$ was missing, until this breakthrough arrived). The closed-form for $\widetilde{p}_0(s)$ is

$$\widetilde{p}_0(s) = \frac{r_1(s)}{\mu(1 - r_1(s))},$$

from which we get a closed-form for $\widetilde{G}$.

This function is then inverted in s and developed at 0 as a Taylor series, leading to the well-known expression

$$p_j(t) = \mathrm{e}^{-(\lambda+\mu)t}\Big[\varrho^{j/2} I_j + \varrho^{(j+1)/2} I_{j+1} + (1-\varrho)\varrho^j \sum_{k=j+2}^{\infty} \varrho^{-k/2} I_k\Big] \qquad [8.9]$$

where all appearing modified Bessel functions (I_j, I_{j+1}, I_k) are all taken at the same point $2t\sqrt{\lambda\mu}$, and where $\varrho = \lambda/\mu$. This is the expression most often found in textbooks, for instance. Numerically speaking, it is not an efficient way of obtaining the transient probabilities (Abate and Whitt 1989). We claim that this position is taken today by uniformization-based approaches (see below).

8.3.4.1. *A different approach*

In the book by Sharma (1990), the author proposed an approach (used by him and coauthors in several papers) where the idea is to study the joint process $\big(A(t), D(t)\big)_{t\in\mathbb{R}_{\geq 0}}$, with $A(t)$ = number of arrivals to the queue in $[0, t]$ and $D(t)$ = number of departures in the same period (an idea coming from (Pegden and Rosenshine 1982); see also (Towsley 1987)). Assuming again that the system is empty at the beginning, if $p_{a,d}(t) = \mathbb{P}(A(t) = a, D(t) = d)$, we have, for any $j \in \mathbb{N}$,

$$p_j(t) = \mathbb{P}(N(t) = j) = \sum_{\ell\geq 0} p_{j+\ell,\ell}(t).$$

The author starts by writing the system of differential equations satisfied by the distribution of this bi-dimensional Markov process; for instance, for $a > d \geq 1$,

$$p'_{a,d}(t) = -(\lambda+\mu)p_{a,d}(t) + \lambda p_{a-1,d}(t) + \mu p_{a,d-1}(t).$$

Taking Laplace transforms, solving for them, inverting the transforms, and making usage of some technical methods such as hypergeometric series to obtain integrals, the result is a new type of expression. For any state $j \in \mathbb{N}$,

$$p_j(t) = (1-\varrho)\varrho^j + \varrho^j \mathrm{e}^{-(\lambda+\mu)t} \sum_{\ell=0}^{\infty} \frac{(\lambda t)^\ell}{\ell!} \sum_{m=0}^{j+\ell} (\ell - m)\frac{(\mu t)^{m-1}}{m!}, \qquad [8.10]$$

where $\varrho = \lambda/\mu$. As we can see, the obtained representation is simpler than the previous one based on Bessel functions. However, observe that the factor $(\ell - m)$ inside the series at the r.h.s. can be negative, thus leading to numerical instabilities in some cases. See also Conolly and Langaris (1993) for related material.

8.3.5. $M/M/1$ *with bounded waiting room: the* $M/M/1/H$ *model*

First of all, observe that for this finite system, since the number of free unknowns is H, when H is very small, we can obtain closed-forms by just asking a tool such as `Maple` or `Mathematica` to symbolically solve the corresponding differential equations. For instance, for the very first values and just solving the equations by hand, we have the following expressions:

– for the $M/M/1/1$,

$$p_0(t) = \frac{\mu}{\lambda+\mu}\left(1 - \mathrm{e}^{-(\lambda+\mu)t}\right),$$

$$p_1(t) = \frac{\lambda}{\lambda+\mu} + \frac{\mu}{\lambda+\mu}\mathrm{e}^{-(\lambda+\mu)t}.$$

– for the $M/M/1/2$,

$$p_0(t) = \frac{1}{D}\left(\mu^2 + \frac{r_2}{2}e^{-r_1 t} + \frac{r_1}{2}e^{-r_2 t}\right),$$

$$p_1(t) = \frac{\lambda}{\mu D}\left(\mu^2 + \frac{\lambda(\sqrt{\lambda\mu} - \mu^2)}{2}\mathrm{e}^{-r_1 t} - \frac{\lambda(\sqrt{\lambda\mu} + \mu^2)}{2}\mathrm{e}^{-r_2 t}\right),$$

$$p_2(t) = \frac{\lambda}{\mu D}\left(\lambda\mu - \frac{r_2\sqrt{\lambda\mu}}{2}e^{-r_1 t} + \frac{r_1\sqrt{\lambda\mu}}{2}e^{-r_2 t}\right),$$

where $0 < r_1 = \lambda+\mu-\sqrt{\lambda\mu} < r_2 = \lambda+\mu+\sqrt{\lambda\mu}$ and $D = r_1 r_2 = \lambda^2 + \lambda\mu + \mu^2$.

In the $M/M/1/H$ model, we deal with a finite Markov process, which suggests to try spectral methods to analyze transients. This has of course been done; we will illustrate the approach following the well-known book by Takács (1962), for the historical importance of his contribution. The idea is simply to find the eigenvalues of the transition matrix of the process, that have size $H+1$. The $H+1$ eigenvalues values are

$$\{0\} \cup \left\{2\sqrt{\lambda\mu}\cos\left(\frac{h\pi}{H+1}\right) - (\lambda+\mu), \quad h = 1, \ldots, H\right\}.$$

Then, the two matrices B and B^{-1} such that $A = BDB^{-1}$ with D diagonal, are computed, and finally, e^{At} is obtained through the expression $B\mathrm{e}^{Dt}B^{-1}$. In scalar form, the given expression of the transient distribution is the following (as usual, starting with the queue empty at time 0): if $\varrho \neq 1$,

$$p_j(t) = \frac{1-\varrho}{1-\varrho^{H+1}}\varrho^j - \frac{2\varrho^{j/2}}{H+1}\sum_{1\leq h\leq H}\left\{\frac{e^{-(\lambda+\mu-2\sqrt{\lambda\mu}\cos(\frac{h\pi}{H+1}))t}}{1+\varrho-2\sqrt{\varrho}\cos(\frac{h\pi}{H+1})}\right. \quad [8.11]$$

$$\left.\times\sqrt{\varrho}\sin(\frac{h\pi}{H+1})\left[\sin(\frac{jh\pi}{H+1})-\sqrt{\varrho}\sin(\frac{(j+1)h\pi}{H+1})\right]\right\},$$

and if $\varrho = 1$,

$$p_j(t) = \frac{1}{H+1} - \frac{1}{H+1}\sum_{1\leq h\leq H}\left\{\frac{e^{-2\lambda(1-t)\cos(\frac{h\pi}{H+1})}}{1-\cos(\frac{h\pi}{H+1})}\right. \quad [8.12]$$

$$\left.\times\sin(\frac{h\pi}{H+1})\left[\sin(\frac{jh\pi}{H+1})-\sin(\frac{(j+1)h\pi}{H+1})\right]\right\}.$$

It is worth mentioning that in another well-known book (Morse 1958), the same expression is obtained following a slightly different approach based on trigonometric representations.

8.3.5.1. *The $M/M/1$ as the limit of the $M/M/1/H$ when $H \to \infty$*

Takács (1962) also obtains the transient distribution of the unbounded model, that is, the $M/M/1$ queue, basically by making $H \to \infty$ in [8.12] and [8.19]. The obtained representation is as follows: for any state $j \in \mathbb{N}$ and starting from the empty queue at time 0,

$$p_j(t) = (1-\varrho)\varrho^j 1(\varrho < 1) - R_j(t), \quad [8.13]$$

with $1(C) = 1$ iff the condition or predicate C is true, 0 otherwise, and where the transient part $R_j(t)$ has the integral representation

$$R_j(t) = \frac{2e^{-(\lambda+\mu)t}\varrho^{\frac{j}{2}}}{\pi}\int_0^\pi \frac{e^{2t\sqrt{\lambda\mu}\cos(y)}}{1-2\sqrt{\lambda\mu}\cos(y)+\varrho}$$

$$\times\sqrt{\varrho}\sin(y)\left[\sin(jy)-\sqrt{\varrho}\sin((j+1)y)\right]dy. \quad [8.14]$$

8.3.6. *Comments*

As mentioned previously, a survey-oriented paper on a topic such as this implies choices, given the impressive amount of publications. Let us add here a few more references to diminish the set of forgotten papers and works.

Concerning the $M/M/1$ model, in van de Coevering (1995) we find integral representations of the first two moments of the basic occupation process at t, with a correction of the formulas given in Takács (1962), correction provided by Takács himself to the author. In the two-page paper Parthasarathy (1987), the author derives an expression of the transient distribution in terms of the classic modified Bessel functions of the first kind, but in a very direct way, using a smart change of variable and generating functions.

Concerning the spectral approach in the analysis of the bounded $M/M/1/H$ case, several variants and improvements have been proposed by different authors. See, for instance, the books Cohen (1982), Bhattacharya and Waymire (2009), Sharma (1990) and the many references therein.

The list of papers working on different aspects of the transient behavior of the $M/M/1$ model and variants is huge, so it is hard to get close to exhaustivity. Let us just complete a list of book references with abundant material on the topic: we have already mentioned many (Morse 1958; Takács 1962; Anderson 1991; Cohen 1982), but we can also add the following works (Bailey 1964; Prabhu 1965; Saaty 1982; Karlin and Taylor 1981; Robert 2003, Chapters 5 and 6), etc.

Considering the model presented in section 8.3.3, there are many papers dealing with different variants and/or generalizations, some of them dedicated to transient behaviors (for instance, Swift 2001; Economou and Fakinos 2008). In section 8.5.5, the $M/M/1/H$ model with catastrophes is considered.

8.4. An uniformization-based path for the $M/M/1$ with matrix generating functions

In this section, we describe a path we followed in part of our past work for finding the transient distribution of the $M/M/1$ model, published in Leguesdron (1993). The starting point here is uniformization to move the problem to discrete-time. In the paper we obtain the following expression for the transient distribution of the $M/M/1$ model, when we start with an empty system: for any $j \in \mathbb{N}$,

$$p_j(t) = \varrho^j \sum_{n \geq j} e^{-(\lambda+\mu)t} \frac{(\lambda + \mu)^n t^n}{n!} C_{n,j}, \qquad [8.15]$$

where the coefficient $C_{n,j}$ represents the following *finite* sum

$$C_{n,j} = \sum_{k=0}^{\lfloor \frac{n-j}{2} \rfloor} \frac{n+1-2k}{n+1} \binom{n+1}{k} p^k q^{n-k}, \qquad [8.16]$$

with $p = \lambda/(\lambda+\mu)$, $q = \mu/(\lambda+\mu) = 1-p$ and $\varrho = \lambda/\mu = p/q$.

This is a typical uniformization-based representation, coming with the associated powerful benefits: a probabilistic interpretation (the number $C_{n,j}$ is the probability that the uniformized discrete-time Markov chain is at state j at time n, see Figure 8.3), a numerically stable procedure (there are only sums and products of positive numbers, furthermore bounded since we basically deal with probabilities) and, for practical purposes, a before-hand error bound on the computation (see section 8.2.2).

Again, compare [8.16] with the classic expression [8.9] given in terms of Bessel functions, or with [8.10], or with the pair [8.13] and [8.14]. The other expressions do not offer the nice properties mentioned in a general way in remark 8.1.

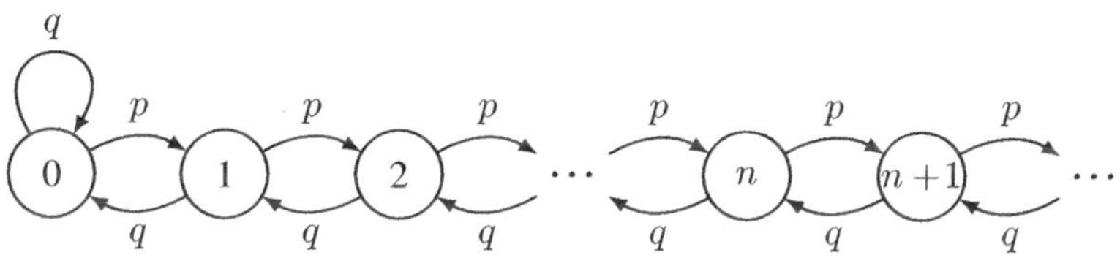

Figure 8.3. *Uniformization of the $M/M/1$ canonical process w.r.t. the uniformization rate $\lambda+\mu$; $p = \lambda/(\lambda+\mu)$ and $q = 1-p = \mu/(\lambda+\mu)$*

The approach of Leguesdron (1993) goes through the following steps. First of all, the development is observed in the algebra of infinite matrices equipped with the norm $||M||_\infty = \sup_i \sum_j |M_{i,j}|$, indexed on $\mathbb{N}$. The set of infinite matrices with a finite norm is a Banach non-commutative algebra, and for any matrix M in this space, its powers M^n, $n \in \mathbb{N}$, and its exponential e^M are well defined and belong to the same space $\mathcal{M}$. Following the uniformization approach, we reduce the evaluation of $P(t) = e^{At}$ to that of U^n, where $U = I + A/(\lambda+\mu)$. This evaluation is done through the analysis of the matrix generating function of the sequence of powers of U, exploiting the properties of $\mathcal{M}$ and a sort of product-form appearing in one of the key matrices of the analysis. The matrix generating function is obtained in closed-form, leading immediately to the elements of U^n also in closed-form, first for the case of $i = 0$.

8.4.1. *General case*

The presentation here completes that of Leguesdron (1993). First of all, for any two integers i, j in the model's state space (this is valid for the $M/M/1$ and for the $M/M/1/H$ models) and any time $t \in \mathbb{R}_{\geq 0}$, we have

$$P_{i,j}(t) = \varrho^{j-i} P_{j,i}(t). \qquad [8.17]$$

This is an immediate consequence of reversibility, valid even in the unstable $\varrho \geq 1$ case (see, Abate et al. 1991, relation (10)). In particular, it means that we can limit ourselves to exhibiting the transition function $P_{i,j}(t)$ only for $i \leq j$.

In Leguesdron (1993), U is written as the sum $U = pR + qL + qK$, where matrix R is the right-shift operator $R_{i,j} = I_{i+1,j}$, matrix L is the left-shift operator $L_{i,j} = I_{i,j+1}$ and K has only one non-zero element, its corner: $K_{0,0} = 1$ (thus, $K = I - L \cdot R$). Let us denote $\widehat{U} = pR + qL$. If we denote by $G_M(z) = \sum_{n \geq 0} M^n z^n$ the generating function of the sequence of powers of $M \in \mathcal{M}$ at point z, the paper shows (relation (16) in Leguesdron (1993)) that

$$\left(G_U(z)\right)_{i,j} = \left(G_{\widehat{U}}(z)\right)_{i,j} + \frac{1}{\varrho^{i+1}} \left(G_U(z)\right)_{0,i+j+1}.$$

This means that once the case of $P_{0,h}(t)$ is solved, that is, once $\left(G_U(z)\right)_{0,i+j+1}$ is known, all we need is to get a closed-form expression of $\left(G_{\widehat{U}}(z)\right)_{i,j}$. This is done in Leguesdron (1993) at the end of section 8.3.2, for the case of $i \leq j$. Make first the change of notation $j = i + d$, $d \geq 0$. Then, the only non-null elements of the form $\left(\widehat{U}^m\right)_{i,i+d}$ are

$$\left(\widehat{U}^{2n+d}\right)_{i,j} = \begin{cases} p^{n+d} q^n \binom{2n+d}{n} & \text{if } n \leq i, \\ p^{n+d} q^n \left[\binom{2n+d}{n} - \binom{2n+d}{n-i-1} \right] & \text{if } n \geq i+1. \end{cases} \qquad [8.18]$$

Let us go here until the end of the process, in order to exhibit an explicit expression of the transient distribution of the model for any $i \leq j$. Making the change of variable $2n + d = m$, then looking again for the non-zero elements, and going back to the coefficient $[z^n]G_{\widehat{U}}$ (the coefficient of z^n in the series expansion of $G_{\widehat{U}}$), we obtain, after some algebra, the following uniformization-based expressions, given separately for d even and d odd:

– even case: if $j - i = 2h$, $h \geq 0$,

$$P_{i,i+2h}(t) = \sum_{\ell=h}^{h+i} \mathrm{e}^{-(\lambda+\mu)t} \frac{((\lambda+\mu)t)^{2\ell}}{(2\ell)!} \binom{2\ell}{\ell-h} p^{\ell+h} q^{\ell-h} \qquad [8.19]$$

$$+ \sum_{\ell \geq h+i+1} \mathrm{e}^{-(\lambda+\mu)t} \frac{((\lambda+\mu)t)^{2\ell}}{(2\ell)!} \left[\binom{2\ell}{\ell-h} - \binom{2\ell}{\ell-h-i-1} \right] p^{\ell+h} q^{\ell-h}$$

$$+ \varrho^{i+2h} \sum_{n \geq 2i+2h+1} \mathrm{e}^{-(\lambda+\mu)t} \frac{((\lambda+\mu)t)^{n}}{n!} C_{n,2i+2h+1};$$

– odd case: if $j - i = 2h + 1$, $h \geq 0$,

$$P_{i,i+2h+1}(t) = \sum_{\ell=h}^{h+i} \mathrm{e}^{-(\lambda+\mu)t} \frac{((\lambda+\mu)t)^{2\ell+1}}{(2\ell+1)!} \binom{2\ell+1}{\ell-h} p^{\ell+h} q^{\ell-h} \qquad [8.20]$$

$$+ \sum_{\ell \geq h+i+1} \mathrm{e}^{-(\lambda+\mu)t} \frac{((\lambda+\mu)t)^{2\ell+1}}{(2\ell+1)!}$$

$$\times \left[\binom{2\ell+1}{\ell-h} - \binom{2\ell+1}{\ell-h-i-1} \right] p^{\ell+h+1} q^{\ell-h}$$

$$+ \varrho^{i+2h+1} \sum_{n \geq 2i+2h+2} \mathrm{e}^{-(\lambda+\mu)t} \frac{((\lambda+\mu)t)^{n}}{n!} C_{n,2i+2h+2}.$$

Again, this is an uniformization-based expression, that is, a Poissonian sum (this is also called the Poisson generating function of the sequence $(U^n)_{i,i+d}$ in n, for fixed i, d, at point $(\lambda + \mu)t$, and it is also basically what is called an exponential generating function – the factor $\mathrm{e}^{-(\lambda+\mu)t}$ must be removed to use this terminology). This means that all the nice properties described before (numerical stability, probabilistic interpretation, error control) hold here.

8.4.2. *Mean number of customers at time* t *in the* $M/M/1$

Let us show how to use the obtained result to evaluate the mean number of customers in the $M/M/1$ system at time t. First of all, observe that from a general point of view, if we model a queue using an uniform Markov process $\{N(t), t \in \mathbb{R}_{\geq 0}\}$ where $N(t)$ is the number of customers in the queue at time t, then, from the general uniformization representation

$$p_j(t) = \sum_{n \geq 0} \mathrm{e}^{-\Lambda t} \frac{(\Lambda t)^n}{n!} q_j(n),$$

we have

$$\begin{aligned}\mathbb{E}(N(t)) &= \sum_{j\geq 1} j \sum_{n\geq 0} \mathrm{e}^{-\Lambda t}\frac{(\Lambda t)^n}{n!} q_j(n)\\ &= \sum_{n\geq 0} \mathrm{e}^{-\Lambda t}\frac{(\Lambda t)^n}{n!} \sum_{j\geq 1} j q_j(n)\\ &= \sum_{n\geq 0} \mathrm{e}^{-\Lambda t}\frac{(\Lambda t)^n}{n!} \mathbb{E}(Y(n)).\end{aligned}$$

This shows that we can use the uniformization approach to obtain $\mathbb{E}(N(t))$ for any model of this type (thus including all the $M/M/$ basic ones, for instance).

Coming now back at the specific case of the $M/M/1$ model, and starting from an empty queue, we saw that

$$\mathbb{P}(N(t)=j) = p_j(t) = \sum_{n\geq j} \mathrm{e}^{-\Lambda t}\frac{(\Lambda t)^n}{n!} \varrho^j c_{n,j},$$

where

$$c_{n,j} = \sum_{k=0}^{\lfloor \frac{n-j}{2} \rfloor} \varphi_{n,k} \qquad \text{and} \qquad \varphi_{n,k} = \frac{n+1-2k}{n+1}\binom{n+1}{k} p^k q^{n-k},$$

with $\Lambda = \lambda+\mu$, $p=\lambda/\Lambda$, $q=\mu/\Lambda = 1-p$ and $\varrho = \lambda/\mu = p/q$.

Then,

$$\begin{aligned}\mathbb{E}(N(t)) &= \sum_{j\geq 1} j p_j(t)\\ &= \sum_{j\geq 1} j\varrho^j \sum_{n\geq j} \mathrm{e}^{-\Lambda t}\frac{(\Lambda t)^n}{n!} c_{n,j}\\ &= \sum_{n\geq 1} \mathrm{e}^{-\Lambda t}\frac{(\Lambda t)^n}{n!} d_n,\end{aligned}$$

where d_n is the finite sum

$$d_n = \sum_{1\leq j\leq n} j\varrho^j c_{n,j}.$$

We already saw that we can use this uniformization-based approach to numerically evaluate $\mathbb{E}(N(t))$ in a pretty general setting. Here, we can specialize the truncation level and the effective computations, by observing first that $\varrho^j c_{n,j} < 1$ allows us to bound the number d_n by the number $n(n+1)/2$. Then, if we want to numerically evaluate $\mathbb{E}(N(t))$ by truncating previous series at level N, we observe that the absolute error err_N satisfies

$$err_N < \sum_{n>N} \mathrm{e}^{-\Lambda t} \frac{(\Lambda t)^n}{n!} \frac{n(n+1)}{2}.$$

Using the notation $\mathcal{P}_H(x) = \sum_{0 \leq h \leq H} \mathrm{e}^{-x} x^h / h!$ introduced in section 8.3.3, we can write, after some algebra,

$$\begin{aligned} \sum_{n>N} \mathrm{e}^{-\Lambda t} \frac{(\Lambda t)^n}{n!} \frac{n(n+1)}{2} &= \frac{(\Lambda t)^2}{2} \left[1 - \mathcal{P}_{N-2}(\Lambda t)\right] + \frac{3\Lambda t}{2} \left[1 - \mathcal{P}_{N-1}(\Lambda t)\right] \\ &< \left[1 - \mathcal{P}_{N-2}(\Lambda t)\right] \frac{\Lambda t(3 + \Lambda t)}{2}, \end{aligned}$$

which shows that the standard uniformization approach can be applied as for the probability distributions, given that $\mathcal{P}_\infty(x) = 1$, for instance using N^* defined as

$$N^* = \min\left\{ N \in \mathbb{N} \;\middle|\; \mathcal{P}_{N-2}(\Lambda t) > 1 - \frac{2\varepsilon}{\Lambda t(3 + \Lambda t)} \right\}.$$

Let us come back to the general expression of $\mathbb{E}(N(t))$ and to the sequence (d_n). After some algebra, we can write

$$\begin{aligned} d_n &= \sum_{1 \leq j \leq n} j \varrho^j c_{n,j} \\ &= \sum_{1 \leq j \leq n} j \varrho^j \sum_{k=0}^{\lfloor \frac{n-j}{2} \rfloor} \varphi_{n,k} \\ &= \sum_{k=0}^{\lfloor \frac{n-1}{2} \rfloor} \varphi_{n,k} \sum_{j=1}^{n-2k} j \varrho^j \\ &= \frac{\varrho}{(1-\varrho)^2} \sum_{k=0}^{\lfloor \frac{n-1}{2} \rfloor} \varphi_{n,k} \left\{ 1 + \varrho^{n-2k} \left[(n-2k)\varrho - n + 2k - 1 \right] \right\}. \end{aligned}$$

So, we finally obtain the expression

$$\mathbb{E}(N(t)) = \frac{\varrho}{(1-\varrho)^2} \sum_{n\geq 1} \mathrm{e}^{-\Lambda t} \frac{(\Lambda t)^n}{n!} \times \sum_{k=0}^{\lfloor \frac{n-1}{2} \rfloor} \varphi_{n,k} \Big\{ 1 + \varrho^{n-2k} \big[(n-2k)\varrho - n + 2k - 1 \big] \Big\}. \quad [8.21]$$

8.5. An uniformization-based path using duality

First of all, recall that based on uniformization, using the expression

$$P_{i,j}(t) = \sum_{n\geq 0} \mathrm{e}^{-(\lambda+\mu)t} \frac{\big((\lambda+\mu)t\big)^n}{n!} \big(U^n\big)_{i,j},$$

all we need to do is to compute $\big(U^n\big)_{i,j}$. This is a typical combinatorial problem.

REMARK 8.3.– The point with computing $\big(U^n\big)_{i,j}$ using combinatorial techniques is the loop at state 0, which makes that different paths from i to j with the same length n may have different probabilities. This is the bottleneck of the direct combinatorial approach. In section 8.4, we follow a generating function approach (the most important tool in combinatorics), working with infinite matrices, to find those probabilities. In this specific birth–death topology, the problem is greatly simplified by means of the duality concept (next section).

Concerning the combinatorial side of the discrete-time analysis mentioned here and in the sequel of the chapter, there is also an important literature. Related to our topic here, see for instance Böhm *et al.* (1997) and the references therein.

8.5.1. *Duality*

Let $\big(P_{i,j}(t)\big)_{i,j\in\mathbb{N}}$ be the transition function associated with some Markov process X on $\mathbb{N}$, that is, $P_{i,j}(t) = \mathbb{P}(X(t) = j \mid X(0) = i)$, also denoted $\mathbb{P}_i(X(t) = j)$. The transition function $\big(P_{i,j}(t)\big)_{i,j\in\mathbb{N}}$ is *stochastically increasing* iff for all $i,j \in \mathbb{N}$, for all $t \in \mathbb{R}_{\geq 0}$ and for all $k \in \mathbb{N}$, the inequality $\mathbb{P}_i(X(t) \geq k) \leq \mathbb{P}_{i+1}(X(t) \geq k)$ holds.

In Siegmund (1976), we have the following result: if $\big(P_{i,j}(t)\big)_{i,j\in\mathbb{N}}$ is stochastically increasing, then there exists another transition function $\big(P^*_{i,j}(t)\big)$

associated with another process X^* with values also on $\mathbb{N}$, possibly defective (see Remark 8.5), such that for all $i, j \in \mathbb{N}$ and $t \in \mathbb{R}_{\geq 0}$,

$$\mathbb{P}_i\big(X^*(t) \leq j\big) = \mathbb{P}_j\big(X(t) \geq i\big)^4. \qquad [8.22]$$

Moreover, (i) P^* is also stochastically increasing and (ii) the reciprocal is true. We say that $\big(P^*_{i,j}(t)\big)$ (respectively, X^*) is the Siegmund-dual of $P(t)$ (respectively, of X). When this happens, we can represent the elements of P^* as linear functions of those of P, and vice versa, as follows:

– Given P,

$$P^*_{i,j}(t) = \sum_{k=0}^{i-1} \Big[P_{j-1,k}(t) - P_{j,k}(t)\Big], \qquad [8.23]$$

for $i, j = 0, 1, 2, \ldots$. This is (Anderson 1991, (4.2), page 251), with a slight change in notation. In the finite case where the state space of X has n states, say $\{0, 1, \ldots, n-1\}$, [8.23] holds for $j \leq n-1$; when $j = n$, we have $P^*_{i,n-1}(t) = 1 - \sum_{k=i}^{n-1} P_{n-1,k}(t)$. In this case, for the last row of P^* we have $P^*_{n,j}(t) = 0$ if $j < n$ and $P^*_{n,n}(t) = 1$.

– Given P^*,

$$P_{i,j}(t) = \sum_{k=0}^{i} \Big[P^*_{j,k}(t) - P^*_{j+1,k}(t)\Big]. \qquad [8.24]$$

REMARK 8.4.– This translates into the same relations but between the infinitesimal generators associated with the transition functions, say A and A^*, that is, replacing in previous expressions $P_{u,v}(t)$ by $A_{u,v}$ and $P^*_{u,v}(t)$ by $A^*_{u,v}$. The same also holds for discrete-time chains; in that case, we write $\big(P^n\big)_{u,v}$ in place of $P_{u,v}(t)$ and $\big(P^{*n}\big)_{u,v}$ in place of $P^*_{u,v}(t)$ (Anderson 1991).

Finally, observe that relations [8.23] and [8.24] can be also written as

$$\mathbb{P}_i(X^*(t) = j) = \mathbb{P}_j(X(t) \leq i) - \mathbb{P}_{j+1}(X(t) \leq i), \qquad [8.25]$$

$$\mathbb{P}_i(X(t) = j) = \mathbb{P}_{j-1}(X^*(t) \leq i-1) - \mathbb{P}_j(X^*(t) \leq i-1). \qquad [8.26]$$

In Anderson (1991), this concept of duality is developed and exploited, in particular for birth and death processes, for which the dual always exist, being moreover particularly simple. Graphically, the dual of the standard birth–death process with birth rates λ_i, $i > 0$ and μ_j, $j \geq 1$ is depicted in Figure 8.4. See that the Siegmund-dual of the original process has, at least, one absorbing state.

4. There are slight differences here with Anderson (1991) in the indexing.

REMARK 8.5.– There is no room here to provide technical details about the fact that the dual may be a defective process, the meaning of this, and about the nature of the dual of the dual. Readers can refer to Anderson (1991) for more information about these points. At this stage, and as far as we are concerned by $M/M/*$ queues, we can take the transformation given in Figure 8.4 as the definition of the dual of a general birth–death process, and relations [8.23] and [8.24] (or [8.25] and [8.26]) as the main properties of the transformation.

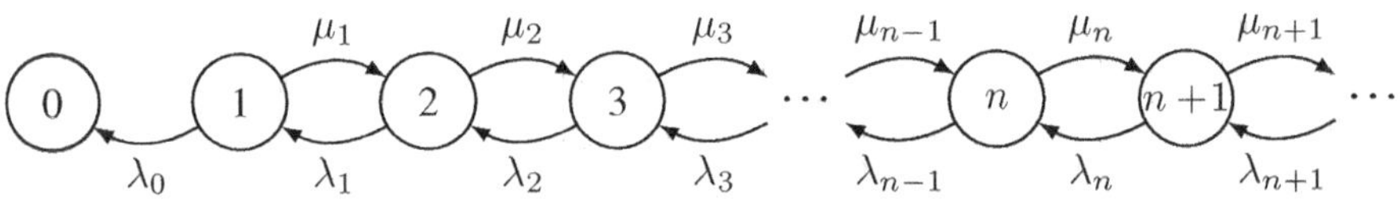

Figure 8.4. *The Siegmund-dual of the standard birth–death process with birth rates* λ_i, $i \geq 0$ *and* μ_j, $j \geq 1$

To complete the introduction of the duality concept, let us show an example using a small Markov process that is not in the birth–death family.

REMARK 8.6.– As it should be clear from the definition, the dual of a process does not exist in general, because this needs that specific monotonicity properties must be satisfied. This is illustrated in the following example (taken from Green *et al.* (2003)), where on the left side of Figure 8.5 we depict the general 3-states Markov process. Its dual exists under specific conditions on the transition rates given in the figure.

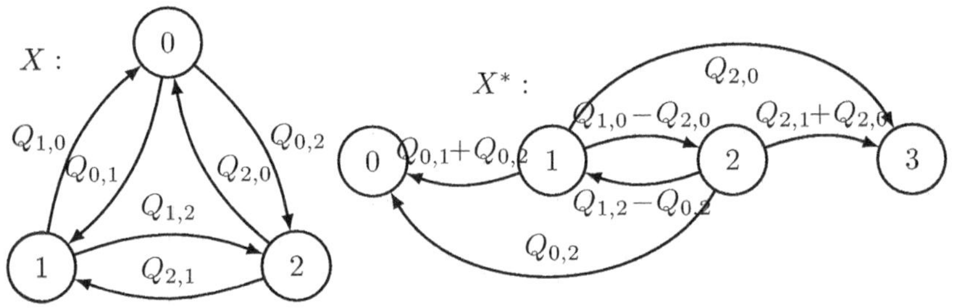

Figure 8.5. *This is a pair* (X, X^*) *where the dual exists under specific conditions. On the left, the general three-states process. On the right, its dual, that exists iff* $Q_{2,0} < Q_{1,0}$ *and* $Q_{0,2} < Q_{1,2}$. *Observe that the existence conditions depend on the numbering of the nodes.*

As a particular case of Figure 8.5, an example is shown in Figure 8.6 where the dual never exists.

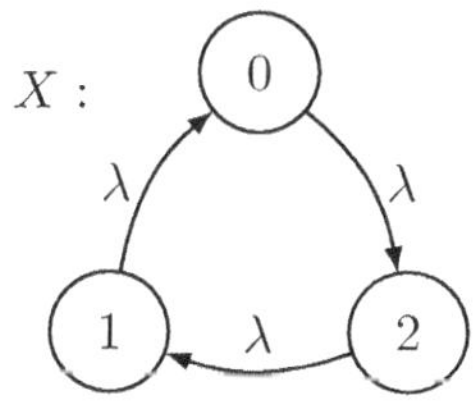

Figure 8.6. *Process X has no dual, whatever the value of λ (and, obviously, whatever the numbering of the states)*

8.5.2. *The path toward the transient state distributions using duality*

The idea is to start by the uniformization approach, that is, from relations [8.4] or [8.3]. We have model X, and we move to the uniformized discrete-time Markov chain Z, whose transition probability matrix is U. Then, we build its Siegmond-dual Z^*, with transition probability matrix U^*, we compute the probabilities $\left(U^{*n}\right)_{i,j}$ by any method, and we obtain $\left(U^n\right)_{i,j}$ using [8.24] (recall that the linear relationships between X and X^* are the same in continuous and in discrete-time, as developed in Remark 8.4), that is, using

$$\left(U^n\right)_{i,j} = \sum_{k=0}^{i} \left[\left(U^{*n}\right)_{j,k} - \left(U^{*n}\right)_{j+1,k}\right]. \qquad [8.27]$$

This path is useful if evaluating $\left(U^{*n}\right)_{i,j}$, for instance using combinatorial techniques, is easier than directly attacking the evaluation of $\left(U^n\right)_{i,j}$. A good example of this situation is the case of the $M/M/1$ model (see next section). The method has allowed to obtain formulas (closed-forms) in cases where this was considered a non-trivial task in the literature.

Before discussing some examples, a remark underlining the fact that we can uniformize first and then go to the dual, or first compute the dual and then uniformize. That is, we can follow the path $X \to Z \to Z^*$ or $X \to X^*$ first, and then, uniformize the latter. The result is again Z^*.

REMARK 8.7.– The operators "uniformization" and "dual" commute. This is immediate from the definitions:

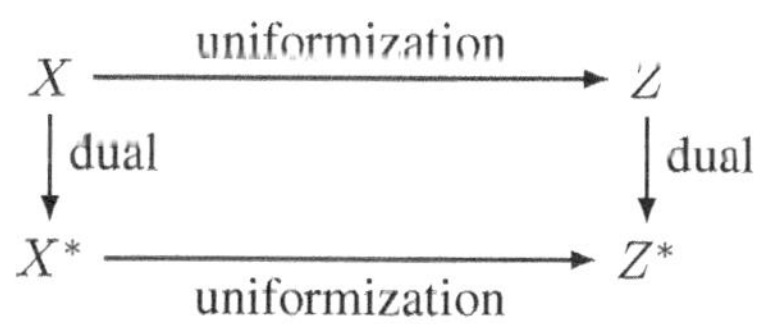

Let us now briefly describe this Siegmund dual-based approach in some fundamental cases. The goal is to provide details on the corresponding solving processes.

8.5.3. *Application to the* $M/M/1$ *queueing system*

We want to derive $p_j(t)$, the probability that there are j customers in an $M/M/1$ model at time t. As usual, to simplify the description, we assume that the $M/M/1$ is empty at time 0 ($X(0) = 0$). We then have

$$p_j(t) = \sum_{n=j}^{\infty} \mathrm{e}^{-(\lambda+\mu)t} \frac{(\lambda + \mu)^n t^n}{n!} (U^n)_{0,j},$$

where U is the transition probability matrix of the uniformization Z of X, whose graph is given in Figure 8.3.

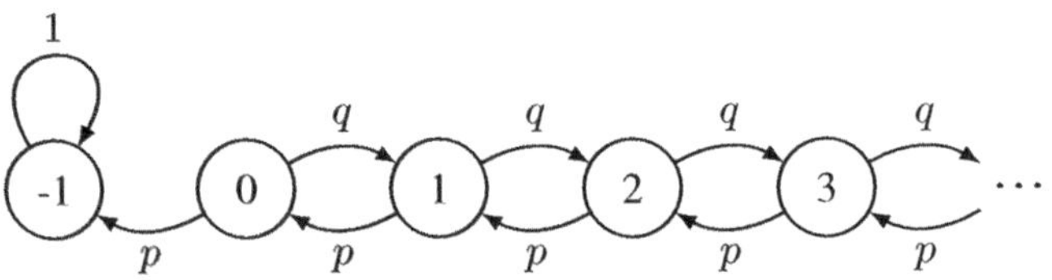

Figure 8.7. *The dual Z^* of the uniformization Z of the $M/M/1$ process, which is also the uniformization of the dual of the $M/M/1$ process with respect to the same uniformization rate $\lambda + \mu$*

Now, the dual of Z is Z^*, whose graph is shown in Figure 8.7. Given the homogeneity in the structure of this last model and the fact that the process is absorbing, with no loop in its transient class, the problem finally reduces to particularly simple path counting. Standard lattice path combinatorial techniques can then be used for this task. From the general formula connecting a process and its dual, and more precisely, using Remark 8.4 and relation [8.27] for $i = 0$, knowing the transient distribution of Z^* we recover that of Z through

$$\left(U^n\right)_{0,j} = \left(U^{*n}\right)_{j,0} - \left(U^{*n}\right)_{j+1,}, \qquad [8.28]$$

where U^* is the transition probability matrix of chain Z^*. Obtaining the probabilities $\left(U^{*n}\right)_{\ell,0}$ is an immediate application of the well-known *reflection principle* in the analysis of random walks (see, for instance, Feller (1968); see also Towsley (1987). This is the path we followed in Krinik *et al.* (2005); we refer readers to this paper for the details and for the references therein. The symbolic expressions we get are a bit more complex than those obtained in Leguesdron (1993), for instance than [8.15], but of course, they are equivalent, and it is a matter of simple combinatorial simplifications to verify it.

8.5.4. *Application to the $M/M/1/H$ queueing system*

The application of the same procedure to this finite case is also performed in Krinik *et al.* (2005), and it needs more involved developments, but the global path is the same. In the following sequence of pictures, we show the graphs of the $M/M/1/H$ model (Figure 8.8), its uniformization with respect to the rate $\lambda + \mu$ (Figure 8.9), its dual process (Figure 8.10) and the uniformization of the dual (Figure 8.11), which is also the dual of the uniformized chain.

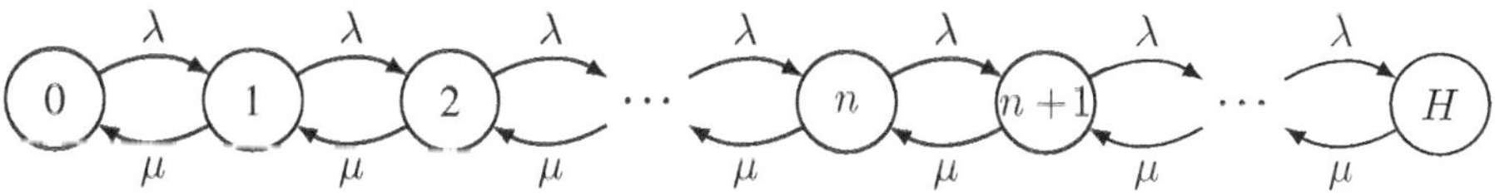

Figure 8.8. *The $M/M/1/H$ model, parameters λ and μ*

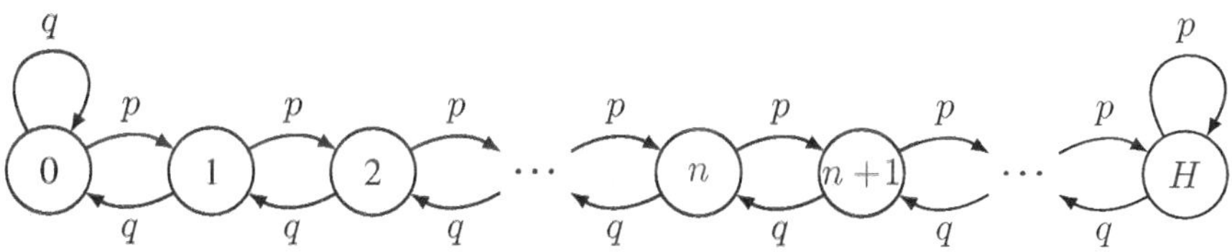

Figure 8.9. *The uniformized chain of the $M/M/1/H$ canonical process depicted in Figure 8.8, with uniformization rate $\Lambda = \lambda + \mu$, $p = \lambda/\Lambda$ and $q = \mu/\Lambda = 1 - p$*

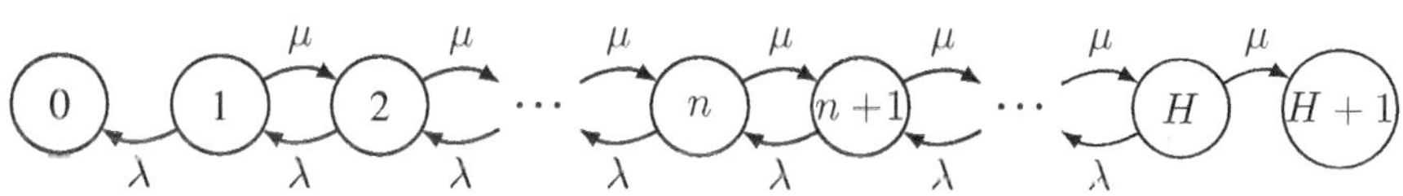

Figure 8.10. *The dual process of the $M/M/1/H$ given in Figure 8.8*

As a check, readers are invited to verify the duality connexions given in relations [8.23] and [8.24] on the simplest case of $H = 1$, that is, on the smallest and non-trivial irreducible process with two states. In Figure 8.12, we depict the process and its corresponding dual, with a single transient state "1" and two absorbing states "0" and "2". On process X, we have

$$P_{0,0}(t) = \frac{\mu}{\lambda + \mu} + \frac{\lambda}{\lambda + \mu} e^{-(\lambda + \mu)t}.$$

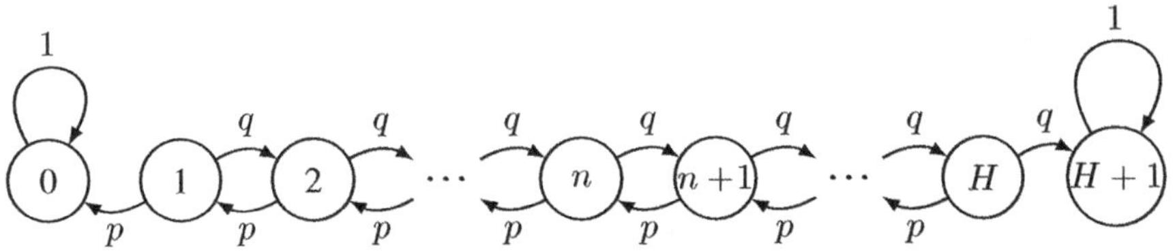

Figure 8.11. *The dual of the uniformized chain shown in Figure 8.9, which is also the uniformization of the process shown in Figure 8.10 (with respect to the same uniformization rate $\lambda + \mu$), as explained in remark 8.7*

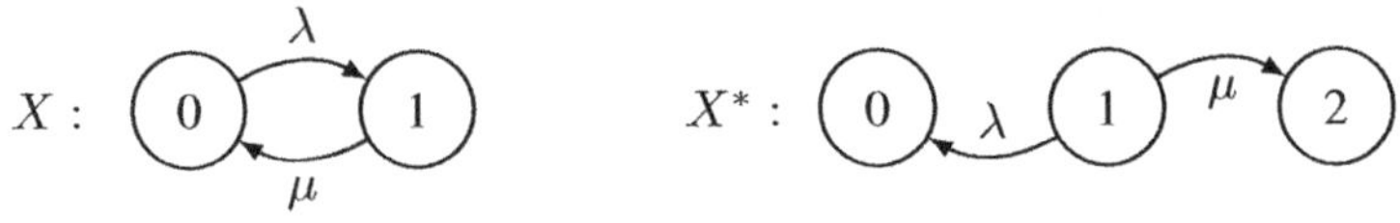

Figure 8.12. *The smallest non-trivial irreducible Markov process with two states (left), and its dual (right); $X(0) = 0$ and $X^*(0) = 1$*

On X^*,

$$P^*_{1,0}(t) = \frac{\lambda}{\lambda + \mu}\left(1 - e^{-(\lambda + \mu)t}\right).$$

The connection between X and X^* translates here into the relation $P_{0,0}(t) + P^*_{1,0}(t) = 1$, immediate to check.

For the analysis of the $M/M/1/H$ using duality, we need a generalization of the reflection principle, because in this case, we have two boundaries instead of a single one in the $M/M/1$ situation. The tool required to simplify the combinatorial expressions that appear in this model is a technical trick given as Lemma 1 in Krinik *et al.* (2005). It says that if we define

$$\binom{u}{v}_+ = \begin{cases} \binom{u}{v} & \text{if } u \geq v, \\ 0 & \text{if } v < 0 \text{ or } v > u, \end{cases}$$

then,

$$\sum_{g:\ g=a \bmod m} \binom{\ell}{g} = \frac{1}{m} \sum_{u=1}^{m} \omega_m^{-ua}(1+\omega_m^u)^\ell,$$

where positive integers a, m, ℓ satisfy $a < m$ and where $\omega_m = e^{2\pi i/m}$ is the mth root of unity.

The obtained form of the transient distribution of this model is trigonometric, formally similar to the one mentioned before, coming back to Takács (1962) or Morse (1958). Using basic trigonometric identities and some algebra, we can go from one to the other (see Krinik *et al.* (2005) and the references therein).

8.5.5. *Application to an $M/M/1/H$ model with catastrophes*

As a last example, consider the $M/M/1/H$ model where the server is subject to breakdowns that happen with rate γ. When a breakdown occurs, the system is emptied, that is, the Markov process jumps to state 0. This is another model where the use of duality allows us to derive the transient state distribution (Krinik *et al.* 2005). The model is depicted in Figure 8.13.

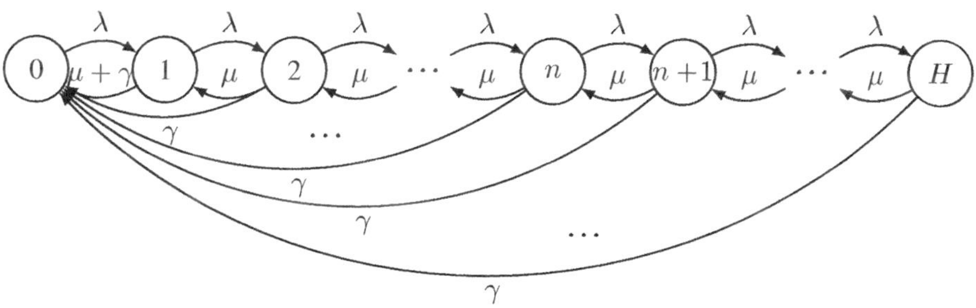

Figure 8.13. *The $M/M/1/H$ model with catastrophes; parameters: arrival rate λ, service rate μ and catastrophe rate γ*

The dual of this process is given in Figure 8.14. As for the $M/M/1/H$ model, the dual has two absorbing states.

The uniformization of the dual is depicted in Figure 8.15. The derivation process follows similar lines as for the $M/M/1/H$ model. Starting with an empty system, the transient distribution of the model is given by the following expression:

$$p_j(t) = \pi_j - \frac{2\mu\varrho^{j/2}}{H+1} \sum_{1\le h\le H} \left\{ \frac{e^{-(\lambda+\mu+\gamma-2\sqrt{\lambda\mu}\cos(\frac{h\pi}{H+1}))t}}{\lambda+\mu+\gamma-2\sqrt{\lambda\mu}\cos(\frac{h\pi}{H+1})} \times \right. \quad [8.29]$$

$$\left. \times\sqrt{\varrho}\sin(\frac{h\pi}{H+1})\left[\sin(\frac{jh\pi}{H+1}) - \sqrt{\varrho}\sin(\frac{(j+1)h\pi}{H+1})\right]\right\}, \quad [8.30]$$

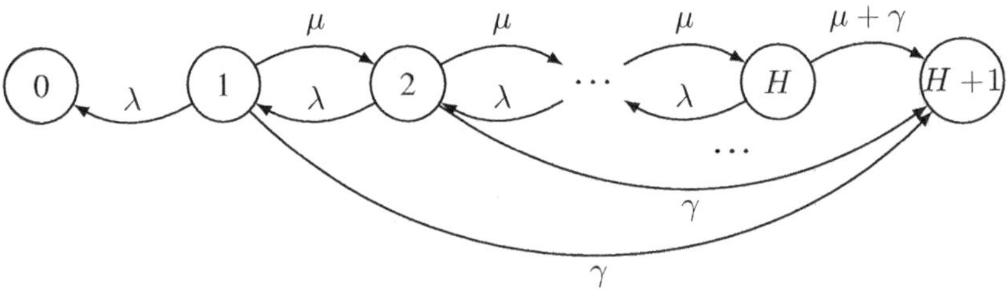

Figure 8.14. *The dual of the $M/M/1/H$ model with catastrophes is depicted in Figure 8.13*

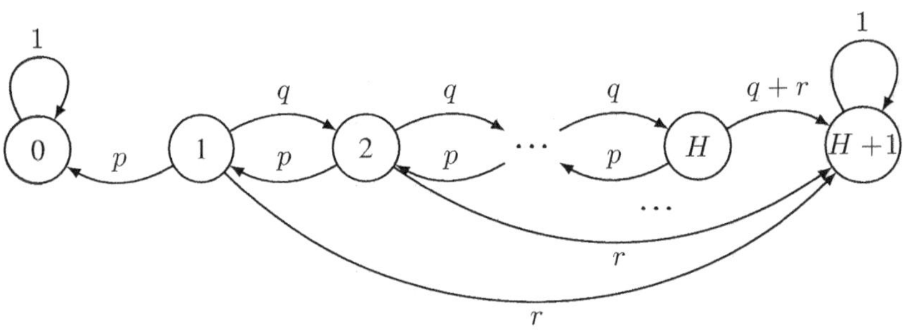

Figure 8.15. *The uniformization of the dual of the $M/M/1/H$ model with catastrophes. See Figure 8.13 for the model and Figure 8.14 for its dual. The uniformization rate is $\Lambda = \lambda + \mu + \gamma$ and the notation is $p = \lambda/\Lambda$, $q = \mu/\Lambda$ and $r = \gamma/\Lambda$. Recall that this model is also the dual of the uniformization of the initial model (using obviously the same uniformization rate).*

Compare this expression with the result for the $M/M/1/H$ given in [8.12]. See the details in the paper, as well as the general $P_{i,j}(t)$ expression, the steady-state one $(\pi_j)_{j\ge 0}$, etc.

8.6. Other transient results

In this section, we briefly refer to other related developments.

8.6.1. *Busy period of the* $M/M/1$

The distribution of the busy period of a queue is another typical transient metric, even if the busy period can be extremely long depending on the values of λ and μ. First of all, it is well known from the analysis of the Markov process $(N(t))$ that the busy period is a.s. finite iff $\lambda \leq \mu$. If $\lambda > \mu$, then the probability that the busy period is finite is μ/λ. Assuming $\lambda < \mu$, the case of a positive recurrent Markov process, and by denoting BP the busy period, we have, from Leguesdron (1993), the following closed-form of the cdf of BP:

$$\mathbb{P}(BP \leq t) = \sum_{n \geq 1} \mathrm{e}^{-(\lambda+\mu)t} \frac{(\lambda+\mu)^n t^n}{n!} \sum_{k=0}^{\lfloor \frac{n-1}{2} \rfloor} c_k p^k q^{k+1}, \qquad [8.31]$$

with c_k the kth Catalan number $c_k = \binom{2k}{k}/(k+1)$ and, as usual, $p = \lambda/(\lambda+\mu)$ and $q = \mu/(\lambda+\mu) = 1-p$.

Observe that the busy period is closely related to the absorption time of the dual of the $M/M/1$ depicted in Figure 8.4, replacing λ_i by λ and μ_j by μ. This provides another way of deriving [8.31] (see also section 3 in (Krinik *et al.* 2004)). The busy period of the $M/M/1$ has obviously the same distribution as the *congestion period* over level $\ell \geq 1$, the time spent by the process above $\ell \geq 1$ (that is, on the set of states $\{j \in \mathbb{N} : \ j \geq \ell\}$, another interesting transient metric). Of course, this distribution is different and dependent on ℓ if we consider other processes such as other Markovian queues, including the bounded $M/M/1/H$ one. We omit from developing this point further in order to keep the chapter a moderate size.

8.6.2. *Max backlog of the* $M/M/1$ *over a finite time interval*

A fine tuning transient metric for a queueing system is given by the random variable defined at a fixed time t

$$M(t) = \max\{\, N(s),\ s \leq t\,\},$$

that is, the maximum level reached by the number of customers in the queue on the finite interval $[0, t]$. This process was analyzed for the $M/M/1$ model in Rubino (2002) using uniformization. The result is a numerical scheme allowing us to evaluate the distribution of $M(t)$. The idea is to work on the auxiliary process $Y = Y(t)$ where $Y(t) = \big(N(t), M(t)\big)$, depicted in Figure 8.16. To do so, we derive an evaluation scheme using the uniformization of this process (Figure 8.17) and then, we exploit the regularities that it exhibits. Again, uniformization is the key for the analysis. See a numerical example in Figure 8.18 for a heavy-loaded $M/M/1$ with load $\varrho = 0.95$.

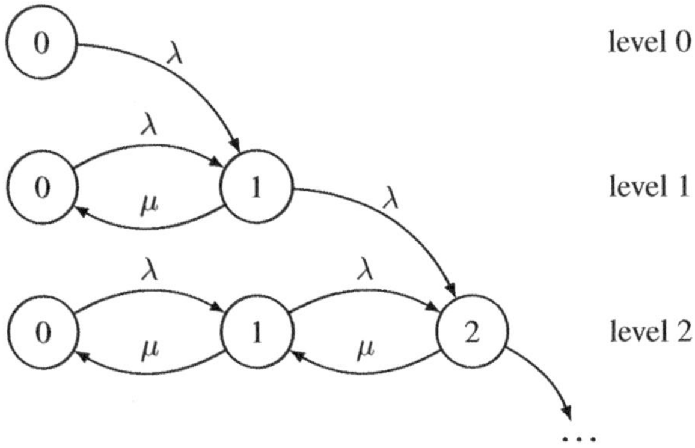

Figure 8.16. *Auxiliary 2-dimensional process* Y

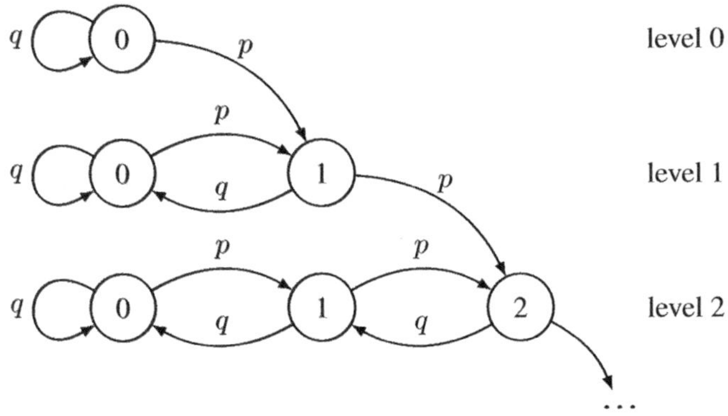

Figure 8.17. *The uniformization of process* Y *depicted in Figure 8.16 with respect to* $\lambda + \mu$, *with* $p = \lambda/(\lambda + \mu)$ *and* $q = 1 - p = \mu/(\lambda + \mu)$

8.6.3. $M/E/1$

To give an example of transient analysis of queueing systems leading to closed-forms and going beyond $M/M/$ structures, let us consider the case of

Erlangian services, that is, of the model $M/E/1$. The arrival rate is, as usual, λ, and the service distribution has K phases each with rate $K\mu$. The approach followed in Griffiths *et al.* (2006b) is based on generating functions and an extension to the modified Bessel functions of the second kind (Griffiths *et al.* 2006a), defined by

$$\tilde{I}_n^{u,v} = \left(\frac{z}{2}\right)^{n+u-v} \sum_{r=0}^{\infty} \frac{(z/2)^{r(u+1)}}{(u(r+1)-v)!\,\Gamma(n+r+1)}, \qquad [8.32]$$

where $z \in \mathbb{C}$, $n \in \mathbb{Z}$, $u \in \mathbb{N}_{>0}$, $v \in \{1, 2, \ldots, u\}$, and $\Gamma()$ is the classic Gamma function. In the paper they use another extension to modified Bessel functions, of the first kind now, proposed in Luchak (1956), defined by

$$I_n^u = \left(\frac{z}{2}\right)^n \sum_{r=0}^{\infty} \frac{(z/2)^{r(u+1)}}{r!\,\Gamma(n+ru+1)}, \qquad [8.33]$$

where $z \in \mathbb{C}$, $n \in \mathbb{N}$ and $u \in \mathbb{N}_{>0}$.

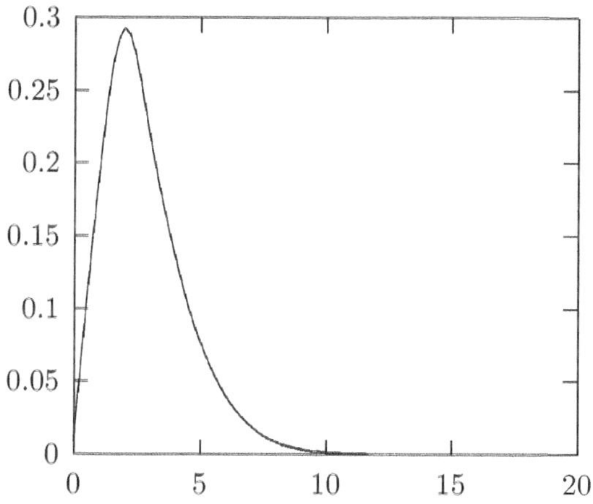

Figure 8.18. *A numerical example where $\lambda = 0.95$ and $\mu = 1$; we plot $\mathbb{P}(M(5) = \ell)$, for level ℓ from 0 to 20 (the curve appears as "continuous" just for a better presentation)*

The analysis is on the queue initially empty, and leads to closed-forms for the function $p_{j,s}(t)$ where j is the number of customers in the system and s is the server's phase, at time t, for $j \geq 1$, plus function $p_0(t)$ for the empty queue at time t (Griffiths *et al.* 2006b) for the details. The general case (system not necessarily empty at $t = 0$) is a non-trivial extension of the case described here; it has already been discussed in Griffiths *et al.* (2008).

8.7. Conclusion

This chapter discussed the approaches carried out to find closed-form representations of the transient state distributions of some basic Markovian queueing models. Many different paths have been followed since the first results in this direction obtained in the 1950s: combinatorial (including the use of transforms), based on spectral analysis, etc. The focus of the chapter is on the uniformization method that translates the continuous-time problem into a discrete-time problem, a fruitful idea offering several nice properties and dealing with new expressions, that we believe have advantages over traditional ones. In some cases where closed-forms have not been obtained yet, this path leads to efficient numerical procedures. In this chapter, we emphasized on another tool – *duality* or *Sigmund duality*; it changes the topology of the Markovian target, and in some cases this greatly simplifies the analysis.

For a topic like this there are certainly interesting works that have not been cited here that seem more relevant for the analyzed problems than those referenced in the text. However, remember here we are biased by a few central ideas making these works less applicable to our discussion.

8.8. References

Abate, J., Kijima, M., Whitt, W. (1991). Decompositions of the $M/M/1$ transition function. *Queueing Systems*, 9, 323–336.

Abate, J., Whitt, W. (1989). Calculating time dependent performance measures for the $M/M/1$ queue. *IEEE Transactions on Communications*, 37(10), 1102–1104.

Anderson, W.J. (1991). *Continuous-Time Markov Chains: An Applications-Oriented Approach*. Springer, New York.

Bailey, N. (1964). *The Elements of Stochastic Processes*. John Wiley & Sons, Inc, New York.

Bailey, N.T.J. (1954). A continuous time treatment of a single queue using generating functions. *Journal of the Royal Statistical Society: Series B*, 16(12), 288–291.

Bhattacharya, R.N., Waymire, E.C. (2009). *Stochastic Processes with Applications*. Society for Industrial and Applied Mathematics, Philadelphia.

Böhm, W., Krinik, A., Mohanty, S.G. (1997). The combinatorics of birth-death processes and applications to queues. *Queueing Systems*, 26(3-4), 255–267.

Boucherie, R.J. (1993). A note on the transient behavior of the Engset loss model. *Communications in Statistics. Stochastic Models*, 9(1), 145–156.

Boucherie, R.J., Taylor, P.G. (1993). Transient product from distributions in queueing networks. *Discrete Event Dynamic Systems*, 3(4), 375–396.

Champernowne, D.G. (1956). An elementary method of solution of the queueing problem with a single server and constant parameters. *Journal of the Royal Statistical Society. Series B*, 3(3), 125–128.

Clarke, A.B. (1953). The time dependent waiting line problem. Ann. Arbor Report M720–1R39, University of Michigan.

Cohen, J.W. (1982). *The Single Server Queue*. North-Holland, Amsterdam.

Conolly, B.W., Langaris, C. (1993). On a new formula for the transient state probabilities for $M/M/1$ queues and computational implications. *Management Science*, 30(1), 237–246.

de Souza e Silva, E., Gail, R. (2001). The uniformization method in performability analysis. In *Performability Modelling: Techniques and Tools*, Haverkort, B., Marie, R., Rubino, G., Trivedi, K. (eds). John Wiley & Sons, New York, 31–58.

Economou, A., Fakinos, D. (2008). Alternative approaches for the transient analysis of Markov chains with catastrophes. *Journal of Statistical Theory and Practice*, 2(2), 183–197.

Feller, W. (1968). *An Introduction of Probability Theory and its Applications*. John Wiley & Sons, New York.

Green, M.L., Krinik, A., Mortensen, C., Rubino, G., Swift, R.J. (2003). Transient probability functions: A sample path approach. *Discrete Mathematics and Theoretical Computer Science*, 127–136.

Griffiths, J.D., Leonenko, G.M., Williams, J.E. (2006a). New generalization of the modified Bessel function and its generating function. *Fractional Calculus and Applied Analysis*, 8(3), 267–276.

Griffiths, J.D., Leonenko, G.M., Williams, J.E. (2006b). The transient solution to $M/E_k/1$ queue. *Operations Research Letters*, 34, 349–354.

Griffiths, J.D., Leonenko, G.M., Williams, J.E. (2008). Time-dependent analysis of non-empty $M/E_k/1$ queue. *Quality Technology & Quantitative Management*, 5(3), 309–320.

Gross, D., Harris, C.M., Shortle, J.F., Thompson, J.M. (2018). *Fundamentals of Queueing Theory*, 5th ed. John Wiley & Sons, New York.

Jensen, A. (1953). Markoff chains as an aid in the study of Markoff processes. *Scandinavian Actuarial Journal*, 1953(suo1), 87–91.

Karlin, S., Taylor, H.M. (1981). *A Second Course in Stochastic Processes*. Academic Press, New York.

Kijima, M. (1997) *Markov Processes for Stochastic Modeling*. Chapman & Hall, New York.

Krinik, A., Mortensen, C., Rubino, G. (2004). Connections between birth-death processes. In *Stochastic Processes and Functional Analysis*, Krinik, A.C., Swift, R.J. (eds). Marcel Dekker, New York, 219–240.

Krinik, A., Rubino, G., Marcus, D., Swift, R. J., Kasfy, H., Lam, H. (2005). Dual processes to solve single server systems. *Journal of Statistical Planning and Inference*, 135(1), 702–713.

Kulkarni, V.G. (2017). *Modeling and Analysis of Stochastic Systems*, 3rd ed. Chapman and Hall/CRC, Boca Raton.

Lebah, M., Pellaumail, J. (1993). Transient behavior for some jackson networks. *Performance Evaluation*, 17(2), 115–122.

Ledermann, W., Reuter, G.E. (1954). Spectral theory for the differential equations of simple birth and death process. *Philosophical Transactions of the Royal Society of London Series A*, 246(914), 321–369.

Leguesdron, P., Pellaumail, J., Rubino, G., Sericola, B. (1993). Transient analysis of the $M/M/1$ queue. *Advances in Applied Probability*, 25(3), 702–713.

Luchak, G. (1956). The solution of the single-channel queuing equations characterized by a time-dependent poisson-distributed arrival rate and a general class of holding times. *Operations Research*, 4(6), 711–732.

Morse, P. (1955). Stochastic properties of waiting lines. *Journal of the Operations Research Society of America*, 3(3), 255–261.

Morse, P. (1958). *Queues, Inventories and Maintenance*. John Wiley & Sons, New York.

Parthasarathy, P.R. (1987). A transient solution to an $M/M/1$ queue: A simple approach. *Adv. Appl. Prob*, 19, 997–998.

Pegden, C.D., Rosenshine, M. (1982). Some new results for the $M/M/1$ queue. *Management Science*, 28, 821–828.

Prabhu, N.U. (1965). *Queues and Inventories*. John Wiley & Sons, New York.

Robert, P. (2003). *Stochastic Networks and Queues. Applications of Mathematics*. Springer, Berlin.

Rubino, G. (2002). Evaluation of the maximum level reached by a queue over a finite period. *International Conference on Dependable Systems and Networks (DSN 2002)*, Bethesda, MD.

Saaty, T.L. (1982). *Elements of Queueing Theory with Applications*. Dover Publications Inc., New York.

Sharma, O.P. (1990). *Markovian Queues. Mathematics and Its Applications*. Ellis Horwood, Richmond, Texas.

Siegmund, D. (1976). The equivalence of absorbing and reflecting barrier problems for stochastically monotone Markov processes. *Ann. Prob.*, 6, 914–924.

Swift, R.J. (2001). Transient probabilities for a simple birth-death-immigration process under the influence of total catastrophes. *International Journal of Mathematics and Mathematical Sciences*, 25(10), 689–692.

Takács, L. (1962). *Introduction to the Theory of Queues.* Oxford University Press, New York.

Towsley, D. (1987). An application of the reflection principle to the transient analysis of the $M/M/1$ queue. *Naval Research Logistics*, 34, 451–456.

van de Coevering, M.C.T. (1995) Computing transient performance measures for the $M/M/1$ queue. *OR Spektrum*, 17, 19–22.

Whittle, P. (1986). *Systems in stochastic equilibrium. Probability and Mathematical Statistics*. John Wiley & Sons, Chichester.

List of Authors

Larisa AFANASEVA
Department of Probability, Faculty of Mathematics and Mechanics
Lomonosov Moscow State University
Russia

Vladimir ANISIMOV
Centre for Design & Analysis
Amgen Inc.
London
United Kingdom

Srinivas R. CHAKRAVARTHY
Departments of Industrial and Manufacturing Engineering & Mathematics
Kettering University
Flint, MI
USA

Boyan DIMITROV
Department of Mathematics
Kettering University
Flint, MI
USA

Antonis ECONOMOU
Department of Mathematics
National and Kapodistrian University of Athens
Greece

Demetres D. KOUVATSOS
NetPEn - Networks and Performance Engineering Research Group
School of Electrical Engineering and Computer Science
University of Bradford
United Kingdom

Achyutha KRISHNAMOORTHY
Department of Mathematics
CMS College
Kottayam
India

Nikolaos LIMNIOS
L.M.A.C. (Laboratoire de Mathématiques Appliquées de Compiègne)
University of Technology of Compiègne
France

Ismail A. MAGEED
NetP.En - Networks and Performance Engineering Research Group
School of Electrical Engineering and Computer Science
University of Bradford
United Kingdom

Evsey MOROZOV
Institute of Applied Mathematical Research
Russian Academy of Sciences
Petrozavodsk
and
Petrozavodsk State University
Russia

Viswanath C. NARAYANAN
Department of Mathematics
Government Engineering College
Thrissur
India

Gerardo RUBINO
Inria
Rennes
France

Dhanya SHAJIN
Department of Mathematics
S. N. College
Chempazhanthy
Thiruvananthapuram
India

Bart STEYAERT
Department of TELIN
Ghent University
Belgium

Index

A, B

C, D

E, F

G, H, I

L, M

N, P

Q

R

S

T, U